Inverse Design Methods for the Built Environment

The inverse design approach is new to the built environment research and design community, though it has been used in other industries including automobile and airplane design. This book, from some of the pioneers of inverse design applications in the built environment, introduces the basic principles of inverse design and the specific techniques that can be applied to built environment systems. The authors' inverse design concept uses the desired enclosed environment as the design objective and inversely determines the systems required to achieve the objective.

The book discusses a number of backward and forward methods for inverse design. Backward methods, such as the quasi-reversibility method, the pseudo-reversibility method, and the regularized inverse matrix method, can be used to identify contaminant sources in an enclosed environment. However, these methods cannot be used to inversely design a desired indoor environment. Forward methods, such as the computational-fluid-dynamics (CFD)-based genetic algorithm (GA) method, the CFD-based adjoint method, the CFD-based artificial neural network (ANN) method, and the CFD-based proper orthogonal decomposition (POD) method, show the promise in the inverse design of airflow and heat transfer in an enclosed environment.

The book describes the fundamentals of the methods for beginners, provides exciting design examples for the reader to duplicate, discusses the pros and cons of each design method, and points out the knowledge gaps for further development.

Qingyan Chen is the Vincent P. Reilly Professor of Mechanical Engineering at Purdue University, USA. He also serves as the Editor-in-Chief of the international journal *Building and Environment*.

Zhiqiang Zhai is a Professor of Architectural Engineering at the University of Colorado at Boulder, USA.

Xueyi You is a Professor of Environmental Engineering at Tianjin University, China.

Tengfei Zhang is a Professor of HVAC at Dalian University of Technology, China.

Inverse Design Methods for the Built Environment

Qingyan Chen, Zhiqiang Zhai, Xueyi You and Tengfei Zhang

Routledge
Taylor & Francis Group
LONDON AND NEW YORK

First published in paperback 2024

First published 2018
by Routledge
4 Park Square, Milton Park, Abingdon, Oxon OX14 4RN

and by Routledge
605 Third Avenue, New York, NY 10017

Routledge is an imprint of the Taylor & Francis Group, an informa business

British Library Cataloguing-in-Publication Data
A catalogue record for this book is available from the British Library

Library of Congress Cataloging in Publication Data
A catalog record for this book has been requested

ISBN: 978-1-138-20498-0 (hbk)
ISBN: 978-1-03-292459-5 (pbk)
ISBN: 978-1-315-46801-3 (ebk)

DOI: 10.4324/9781315468013

Typeset in Times New Roman
by Swales & Willis Ltd, Exeter, Devon, UK

Contents

Preface

In the design of the built environment, the traditional approach is to assume a system and determine the environment created by the system. One can then compare the designed environment with the desired one. If the difference between the designed and desired environments is significant, the system parameters can be adjusted to reduce this difference. The system parameters may be adjusted several times in order to make the difference acceptable. This is the traditional trial-and-error method. The method relies heavily on the designer's experience and is very time consuming; furthermore, the final design may not be optimal.

In our teaching practice, we found that it is quite challenging for new designers to use the trial-and-error method, especially for the design of air distributions in the built environment. An ideal approach is to start with the design objective and use a method that can automatically search for a suitable system. Such a system will be able to produce an air distribution in the built environment that satisfies the design objective. This approach is an inverse design process, in contrast with the traditional trial-and-error method. We tested the idea about 10 years ago and found that an accurate inverse design approach was not possible. This was because the Navier-Stokes equation cannot be inversely solved; the process leads to entropy reduction, which is a violation of the second law of thermodynamics. Although we can modify the Navier-Stokes equation so that an inverse solution is possible, the approach has very limited applications, such as species transport, and the results thus obtained have great uncertainties.

Therefore, we began to seek a forward solution of the Navier-Stokes equation for the inverse design that would not lead to entropy reduction. Our goal was to identify the optimal design automatically. From 2012 to 2016, the authors and several of their PhD students tested different inverse or optimal design methods, such as the genetic algorithm method, adjoint method, artificial neural network method, and proper orthogonal decomposition method. These methods all use computational fluid dynamics to solve the Navier-Stokes equation. Fortunately, they all seem to work well for the inverse design of air distributions in the built environment. However, each method has advantages and disadvantages. To eliminate the disadvantages, we have investigated combinations of two or more methods and proposed new numerical algorithms. Our systematic approach could provide designers with a new tool. The authors' primary motivation in writing this book was to share with readers the theory that we have developed. The book also uses a number of examples to demonstrate the application of different inverse methods to the design of air distributions in the built environment.

This book could not be published without the efforts of several former and current PhD students of the authors. Dr. Wei Liu of Tianjin University and Purdue University was the coordinator of the book. He is the main author of Chapters 1 and 3 and the primary contributor to Chapter 6. Dr. Yu Xue of Tianjin University is the main author of Chapter 2 and a major contributor to Chapter 6. Part of his work was accomplished when he was a Visiting Student at the University of Colorado, Boulder. Dr. Tianhu Zhang of Tianjin University is the primary author of Chapter 4 and also contributed significantly to Chapter 6. Ms. Yun Wei of Dalian University of Technology is the main contributor of Chapter 5. The authors would like to acknowledge the contributions made by these current and former students.

The work presented in this book was a part of the results of a five-year project entitled, "Key Problems in Air Environment Control for Large Airliner Cabins." The project was supported financially by the National Basic Research Program of China (the 973 Program) through Grant No. 2012CB720100 from the Ministry of Science and Technology of China.

Qingyan Chen, Purdue University and Tianjin University
Zhiqiang Zhai, University of Colorado at Boulder
Xueyi You, Tianjin University
Tengfei Zhang, Dalian University of Technology

Nomenclature

a	element of the coefficient vector **a**
a	coefficient vector of dummy variables or eigenvector
aa	index of the input of the neuron
a_r	coefficient of a snapshot field Φ_r
A	cause-effect governing matrix
AA	number of input of the neuron
b	index of the spatial mode
b	vector used for illustrating the mathematical properties
bb	index of the input of the ANN
bi	bias
B	number of required spatial modes
BB	number of input of the ANN
c	coefficient of the POD mode
c_b	amplitude or coefficient of the spatial mode $\boldsymbol{\varphi}_b$
cc	index of the output of the ANN
C	concentration response vector
CC	number of output of the ANN
C_f	symbol of the constrained function
d	derivative of the piecewise cubic function
d	vector used for illustrating the mathematical properties
dd	index of sample
D	rate of strain tensor
DA	desired output vector of ANN
DD	number of sample
e	number of Lagrange linear shape functions per element
e	symbol of a vector used for illustrating the mathematical properties
ee	index of the design variable
E	element
E_s	sum square error
E_w	regularization term
EE	number of the design variable
f_{cl}	clothing area factor
f	dummy vector for the velocity, temperature, etc., at all grid points that constitute a vector
F	body force
F_i	*ith* component of the body forces

g	lower limit of the variable x in illustration of the coefficient interpolation method
$\mathbf{g}$	gravity vector
g_i	*ith* component of the gravity vector
$gE(k)$	global node number of node k of element E
gg	index of the design objective
gs	gradient of the sum square error
GG	number of the design objective
h	air supply inlet size
h_s	spacing between adjacent knots
$h_{s,m}$	maximum spacing between adjacent knots
hh	index of the individual
H	Hessian matrix
H_c	cubic Hermite basis function
H_p	piecewise cubic Hermite interpolation
HH	number of the individual
Hy	hypervolume
ii	iteration index
I	identity matrix
I_{cl}	thermal resistance of clothing
I_n	cell index
I_s	sampling interval
I_t	subinterval of the variable x in interpolation
J	Jacobian matrix
k	node number
l	upper limit of the variable x in illustration of the coefficient interpolation method
L	augmented objective function
Le	Lebesgue measure
$\mathbf{L}$	regularized matrix
m	metabolic heat production
M	iteration number
M_m	molar mass
n	number of eigenvalues for which the total energy can be desired
$\vec{n}$	unit vector in the normal direction
$\mathbf{N}$	Navier-Stokes equations, vector form
N_1	continuity equation
N_2, N_3, N_4	momentum equations
N_5	energy equation
N_6	scalar transport equation
N_e	number of eigenvalues for which the total energy can be desired
N_f	code for the first sample
N_g	number of grid points
N_p	population size for sampling
N_r	random number
N_s	sample size
N_v	number of vectors in a data ensemble
O	objective function

p	pressure
p^{***}, p^{****}	intermediate pressure
P	number of patterns
q	quadrature point
$\mathbf{q}$	release rate vector
q_p	a given set of data values in interpolation
Q	airflow rate
Q_s	number of strata
r	index of snapshots
R	number of snapshots
R_g	ideal gas constant
R	real number
$\mathbf{R}^m$	m-dimension real number space
R_l	learning rate
R_m	momentum rate
s	elements in the averaged autocorrelation matrix
$\mathbf{s}$	averaged autocorrelation matrix of the fields
S	shape function
S_T	energy source
t	forward time
$\vec{t}$	unit vector in the in-plane direction
tr	training error
T	air temperature
T^*	intermediate temperature
T_{cl}	temperature of clothing
T_r	mean radiation temperature
u	velocity in x direction
$\mathbf{U}$	air velocity, vector
U^*,U^{**},U^{***},U^{****}	intermediate air velocity
v	velocity in y direction
w	velocity in z direction
W	external work accomplished
x, y, z	index of coordinates
$\mathbf{x}$	position vector
x_p	partition points in interpolation
X_a	arrival point
X_d	departure point
$\mathrm{y}_{\mathrm{pareto}}$	Pareto front
$\mathrm{y}_{\mathrm{ref}}$	reference point
zi, zj	weight indices from 1 to Z
zz	index of weight
Z	number of weight

Greek symbols

α, β	positive constant
γ	thermal expansion coefficient
Γ	flow boundary

δ	variation
δ_E	an arbitrary small number in real Euclidean space
Δ	difference quotient
ε	stabilization coefficient
ε_p	errors in approximating the pressure p
ε_U	errors in approximating the velocity U
ζ_1, ζ_2	regularization coefficients
κ	effective conductivity
λ	step size, positive constant
λ_L	Lagrangian multiplier or eigenvalue of **s**
Λ	regularized coefficient
μ	combination coefficient
ν	effective viscosity
ξ	design variables, vector
ρ	air density
σ	quadrature weight
τ	backward time
Υ	diffusion coefficient
ϕ_1, ϕ_2	a defined function in interpolation
$\boldsymbol{\varphi}$, $\boldsymbol{\Xi}$	orthogonal basis or spatial mode or POD mode
$\boldsymbol{\varphi}_b$	spatial mode
Φ	scalar/vector field
Φ_c	constructed field
Φ_τ	age of air
χ	state variables
Ψ_i^n	the amplitude of error at node *i* and time step *n*
Ψ	interior flow domain
ω	computational domain
ϖ	weighting factor
Ω	design domain

Subscripts

0	desired value
1, 2, 3, 4, 5	component of a vector
a	adjoint parameter
t	in-plane component
initial	initialized value
n	normal component
old	value at previous design cycle
new	value updated in current design cycle
op	operating value
x, y, z	components in the x, y, and z directions, respectively

Superscripts

i, j	index of a data sample
k	index of a spatial mode

Abbreviations

0-eqn	zero equation model
2D	two-dimensional
ACH	air exchange rate
ADGA	internal integrated method of genetic algorithm and adjoint method
Age	age of air
ANN	artificial neural network
ARE	average relative error
ASHRAE	American Society of Heating, Refrigerating, and Air-Conditioning Engineers
CD	crowding degree
CFD	computational fluid dynamics
CFL	Courant–Friedrichs–Lewy number
CPU	central processing unit
DE	design error
FDM	finite difference method
FEM	finite element method
FFD	fast computational fluid dynamics
FLC	fuzzy logic controller
GA	genetic algorithm
HVAC	heating, ventilation, and air conditioning
MOGA	multi-objective genetic algorithm
MSE	mean squared error
NS	Navier-Stokes
NIPC	non-incremental pressure-correction scheme
NIPC-SL	non-incremental pressure-correction scheme with semi-Lagrangian scheme
OpenFOAM	Open source Field Operation And Manipulation
PD	percentage of dissatisfied people due to draft
PISO	pressure implicit with splitting of operator
PMV	predicted mean vote of thermal comfort
PMVc	modified predicted mean vote for aircraft cabins
POD	proper orthogonal decomposition
PR	pseudo-reversibility
PRESTO	pressure staggering option
QR	quasi-reversibility
RANS	Reynolds-averaged Navier-Stokes
RE	relative error

RIPC	rotational incremental pressure-correction scheme
RNG	re-normalization group
SIMPLE	semi-implicit method for pressure-linked equations
SIPC	standard incremental pressure-correction scheme
SL	semi-Lagrangian scheme
SLPISO	semi-Lagrangian scheme based PISO
SOGA	single-objective genetic algorithm
SR	success rate
TI	turbulence intensity

1 Methods for inverse/optimal design of an enclosed environment[1]

The conventional design of enclosed environments uses a trial-and-error approach that is time consuming and may not meet the design objective. Inverse design or the optimal design concept uses the desired enclosed environment as the design objective and inversely determines the systems required to achieve the objective. This chapter discusses a number of backward and forward methods for inverse or optimal design. Backward methods, such as the quasi-reversibility method, pseudo-reversibility method, and regularized inverse matrix method, can be used to identify contaminant sources in an enclosed environment. However, these methods cannot be used to inversely design a desired indoor environment. Forward methods, such as the CFD-based (computational fluid dynamics) adjoint method, CFD-based genetic algorithm method, CFD-based artificial neural network method, and proper orthogonal decomposition method, show the promise in the inverse design of airflow and heat transfer in an enclosed environment. The CFD-based adjoint method is accurate and can handle many design parameters without increasing computing costs, but the method may find a locally optimal design that could meet the design objective with constrains. The CFD-based genetic algorithm method, on the other hand, can provide the global optimal design that can meet the design objective without constraints, but the computing cost can increase dramatically with the number of design parameters. The CFD-based artificial neural network method can be a surrogate model of CFD to reduce the computing time, but the accuracy of design result is compromised. The proper orthogonal decomposition method is a reduced-order method that can significantly lower computing costs, but at the expense of reduced accuracy. This chapter also discusses the possibility of reducing the computing costs of CFD-based design methods.

1.1 Introduction

People spend roughly 90% of their daily lives in enclosed environments such as buildings, transportation vehicles, etc. (Klepeis et al., 2001). A thermally comfortable, healthy, and productive enclosed environment is very important for occupants. Such an environment is typically achieved by controlling the air temperature, relative humidity, air speed, and chemical species concentrations in enclosed spaces with heating, ventilating, and air-conditioning (HVAC) systems. The energy used by HVAC systems accounts for over 40% of the total energy consumption in buildings in the United States (DOE, 2011). In transportation vehicles such as medium- and large-sized commercial aircrafts, the operating power of HVAC systems in a 400-passenger airliner is as high as 350 kW, which is about 75% of all non-propulsive energy consumption on board (Majeed and Eng, 2010).

Even with such high energy consumption and associated carbon dioxide emissions, the resulting enclosed environments are not satisfactory. Inside buildings, according to survey data from the International Facility Management Association (2003), the predominant complaint of office occupants is that "it's too hot and too cold, simultaneously." Not only is the thermal environment poor, but the studies have also found that people may suffer from sick building syndrome in 20% of our buildings (Hedge et al., 1996). This syndrome is frequently attributed to flaws in HVAC systems. Other possible causes related to indoor air quality are volatile organic compounds from building material outgassing, molds, and light industrial chemicals used in factories (Pejtersen et al., 1999; Wargocki et al., 2000). During overseas flights on commercial airliners, up to half of the passengers typically experience health problems, with 5% of them requiring medical attention (Leder and Newman, 2005). Respiratory diseases in particular can be transmitted in an airliner cabin (Mangili and Gendreau, 2005). For example, among the 120 passengers on a flight from Hong Kong to Beijing in 2003, 22 were infected with the severe acute respiratory syndrome (SARS) (Olsen et al., 2003), and in 2009 the H1N1-A virus spread to 76 countries on six continents within a month and a half (World Health Organization, 2009). Indoor environments, especially in buildings with high occupant density and transport vehicles, are in need of further improvements.

At the same time, requirements for indoor environments are changing with improvements in our standard of living. Our current design goal for indoor environments is to ensure that 80% of building occupants are comfortable and satisfied. This standard is far too low. In order to approach a satisfaction level of 100%, the design should provide a personalized environment that is thermally comfortable, healthy, productive, and safe for all occupants, with low energy consumption and minimal adverse environmental impact (Spengler and Chen, 2000).

A trial-and-error process has traditionally been used to design HVAC systems for enclosed environments. A designer may have to assume an HVAC system for the environment according to his/her experience. The system's ability to produce the desired thermal comfort and indoor air quality level will then be evaluated through the distribution characteristics of air temperature, relative humidity, air velocity, and contaminant concentrations in the enclosed environment by using analytical and empirical models, computer simulations, or by means of experimental measurements (Chen, 2009). By adjusting the HVAC system parameters iteratively on the basis of the evaluation results, the designer can achieve a desired enclosed environment. However, this trial-and-error process may involve a large number of iterations between the initial and final designs, highly depending on the designer's knowledge and experience. If a sophisticated performance evaluation tool is used, such as computational fluid dynamics (CFD), the design process can take days or weeks, but the resulting enclosed environment may still not meet the design objective.

To design an HVAC system for a desirable enclosed environment, the best approach is to start with the design objective (e.g., the desired thermal comfort and indoor air quality level). The inverse or optimal design method can achieve the desirable enclosed environment in a single series of evaluation that determines the necessary HVAC system. This chapter provides a review of the state of the art of inverse/optimal design methods, shows some examples of their applications, and indicates possible directions for future exploration.

1.2 Inverse design for an enclosed environment

A typical inverse problem study seeks to identify unknown causes of known consequences (Friedr, 1978). Fig. 1.1 shows a system and its input and output. The inverse problem study is either for problem reconstruction or identification. The goal of a reconstruction problem is to find the input with given system parameters and output. The goal of an identification problem is to determine system parameters with given input and output.

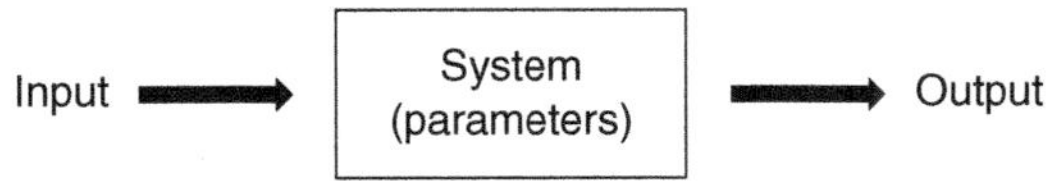

Figure 1.1 Configuration of a scientific problem

The inverse design of an enclosed environment is a reconstruction problem. The input could be the inlet location and size and the supply air parameters from an HVAC system, envelope temperature, the location and strength of a contaminant source, etc. The output could be the desired thermal comfort, air quality, energy efficiency, and so on. The system that connects the input and output is airflow and heat and mass transfer in the enclosed space governed by the Navier-Stokes (NS) equation. The airflow and heat and mass transfer can be obtained by means of analytical and empirical models, computer simulations, or experimental measurements. Because of the limited space in this chapter, we have used only the NS-equation-based methods.

The inverse design can be further divided into backward methods and forward methods. Backward methods include the quasi-reversibility (QR) method (Lattès and Lions, 1969), pseudo-reversibility (PR) method (Kato et al., 2001), and regularized inverse matrix method (Tikhonov and Arsenin, 1977). Forward methods include the adjoint method (Bellman, 1956), genetic algorithm (GA) (Holland, 1975), artificial neural network (ANN) (Mehra and Wah, 1992), and proper orthogonal decomposition (POD) analysis (Lumley, 1967). In the next section, we will introduce these inverse design methods together with several examples.

1.2.1 Backward methods

Backward methods attempt to solve the inverse NS equation by a direct inversion such as with a negative time step in the time-marching problems. However, the simple direct solution is impossible because the actual world is not reversible. A viable solution imposes a reduction in entropy that is against the second-law of thermodynamics. Thus, the governing equation has to be twisted to avoid solving directly the inversed governing equation. The modified method can be the quasi-reversibility and pseudo-reversibility method.

1.2.1.1 Quasi-reversibility method

The quasi-reversibility method solves the irreversible governing equation using a stabilization technique. This method was first used to solve ill-posed partial differential equations for heat conduction problems (Lattès and Lions, 1969). It was also used to solve the inverse heat conduction in a quasi-boundary-value-problem variant of the original formulation (Clark and Oppenheimer, 1994). Later, the method was applied to identify groundwater contaminant sources (Skaggs and Kabala, 1995). To identify a contaminant source by using the contaminant information detected by a sensor, one could inversely solve the governing species transport equation (Eq. (1.1). The inversion to Eq. (1.1) is ill-posed because the time-forward process is dispersive and an inversion to a dispersive process is unstable. The quasi-reversibility method replaces the most unstable diffusion term with a fourth-order term, as shown in Eq. (1.2) (Zhang and Chen, 2007a). Then the equation can be solved reversibly.

$$\frac{\partial \Phi}{\partial t} + \frac{\partial}{\partial x_i}(U_i \Phi) = \frac{\partial}{\partial x_i}(\Upsilon_\Phi \frac{\partial \Phi}{\partial x_i}) \tag{1.1}$$

$$\frac{\partial \Phi}{\partial \tau}+\frac{\partial}{\partial x_i}(U_i\Phi)=\varepsilon\frac{\partial^4 \Phi}{\partial x_i^4} \tag{1.2}$$

where Φ is the species concentration, t is forward time, i.e., $\Delta t > 0$, x_i is the coordinate, U_i is a velocity component, Υ_Φ is the diffusion coefficient, τ is backward time, i.e., $\Delta\tau < 0$, and ε is the stabilization coefficient, which determines the strength of stabilized operation.

The above quasi-reversibility method was demonstrated to identify a gaseous pollutant source in an aircraft cabin released at floor level from t = 0–0.04 s according to the scenario shown in Fig. 1.2(a). The contaminant concentration distribution at t = 6 s provided by a forward CFD simulation, as shown in Fig. 1.2(b), was used as the initial condition to conduct the inverse simulation. Fig. 1.2(c) shows the possible source location illustrated by the highest level of concentration contour based on the inverse solution. Although the contaminant strength calculated by the inverse method becomes dispersive, the location of the contaminant source can be identified. The quasi-reversibility method was also extended to inversely locate an airborne particle source (Zhang et al., 2012). The quasi-reversibility method requires a known flow field as the priori, because Eq. (1.2) is only for the species concentration. However, if a design of an enclosed environment also requires the inverse prediction of the flow field, this method is not applicable.

1.2.1.2 Pseudo-reversibility method

Instead of reversing time as in the quasi-reversibility method, the pseudo-reversibility method solves inverse contaminant transport with reversed flows. The pseudo-reversibility method also

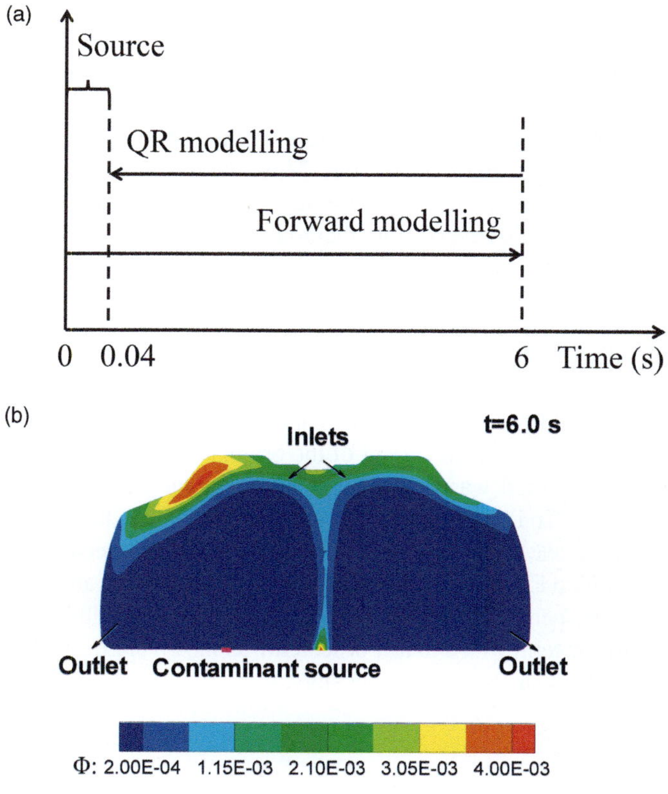

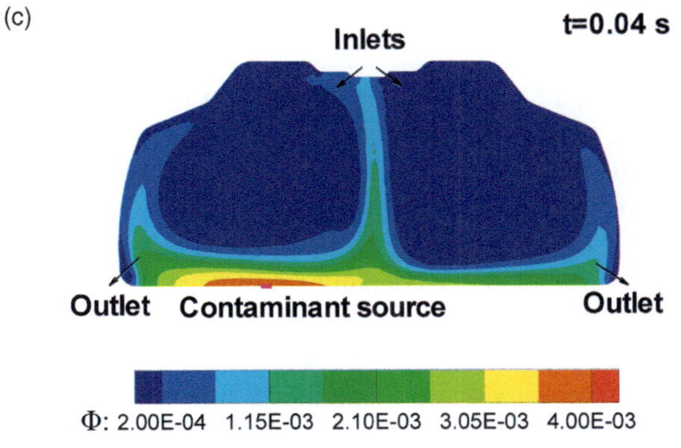

Figure 1.2 Demonstration of the quasi-reversibility method in identifying a contaminant source (Skaggs and Kabala, 1995): (a) simulation schematics, (b) initial contaminant concentration field, and (c) inversely computed contaminant concentration at t = 0.04 s

requires a known flow field as the priori. Kato et al. (2001; 2002; 2004) used the method to assess local pollution from upwind regions with backward trajectory analysis of the flows in the atmospheric environment. The method has also been used in groundwater contaminant transport (Bagtzoglou et al., 1992; Wilson and Liu, 1994), where both convection and diffusion transport of contaminants is solved with flow reversion. Zhang and Chen (2007b) compared the pseudo-reversibility method with the quasi-reversibility method in the inverse identification of a contaminant source in an aircraft cabin. They found that both methods were dispersive, and accuracy was poor if there was significant diffusion in contaminant transport.

1.2.1.3 Regularized inverse matrix method

The regularized inverse matrix method performs an inversion to the cause-effect governing matrix **A** and improves the stability of the inverse operation by regularization. Suppose the release rate vector **q** and the concentration response vector **C** can be described by Eq. (1.3). Since the matrix **A** is ill-posed, a regularized matrix **L** is added, and the strength of the regularized operation is controlled by a regularized coefficient Λ (Tikhonov, 1963; Tikhonov and Arsenin, 1977). With an appropriate Λ, the release rate **q** can be solved once a **C** is provided (Eq. (1.4)).

$$C = Aq \tag{1.3}$$

$$q = (A^T A + \Lambda^2 L^T L)^{-1} A^T C \tag{1.4}$$

where the superscript T in Eq. (1.4) represents the transpose operation. The cause-effect matrix **A** is a function of the flow, source, and sensor location. More details on calculation of the matrix **A** with the convolution theory can be found in (Ishida and Kato, 2008; Hiyama et al., 2008; Zhang et al., 2013). The regularized matrix **L** is usually taken as the second-order time derivative format (Zhang et al., 2013), which is well posed and reversible.

Zhang et al. (2013) applied this method to inversely quantify the temporal rate profile of a continuously released gaseous pollutant source on the basis of limited sensor information in an office. As shown in Fig. 1.3(a), a gaseous contaminant was released at the head level of an occupant; the corresponding temporal release rate is represented by the black line in Fig. 1.3(c). The gas concentration (Fig. 1.3(b)), as monitored by a sensor located at the outlet, was used as input for the inverse simulation. As shown in Fig. 1.3(c), the temporal release rate identified by the method (in the red line) agrees well with the actual rate.

In summary, backward methods are applicable for solving the reversed scalar transport equations and work well in inverse identification of a contaminant source. The primary limitation of the backward methods is that they only provide an approximate solution and require a known flow field. The inverse design of an enclosed environment usually requires inverse prediction of the flow field, in which case backward methods are not applicable or need further development. The computing costs for the backward methods are similar to typical CFD simulations plus a continued simulation of species dispersion. Therefore, the backward methods are affordable for most indoor environment studies.

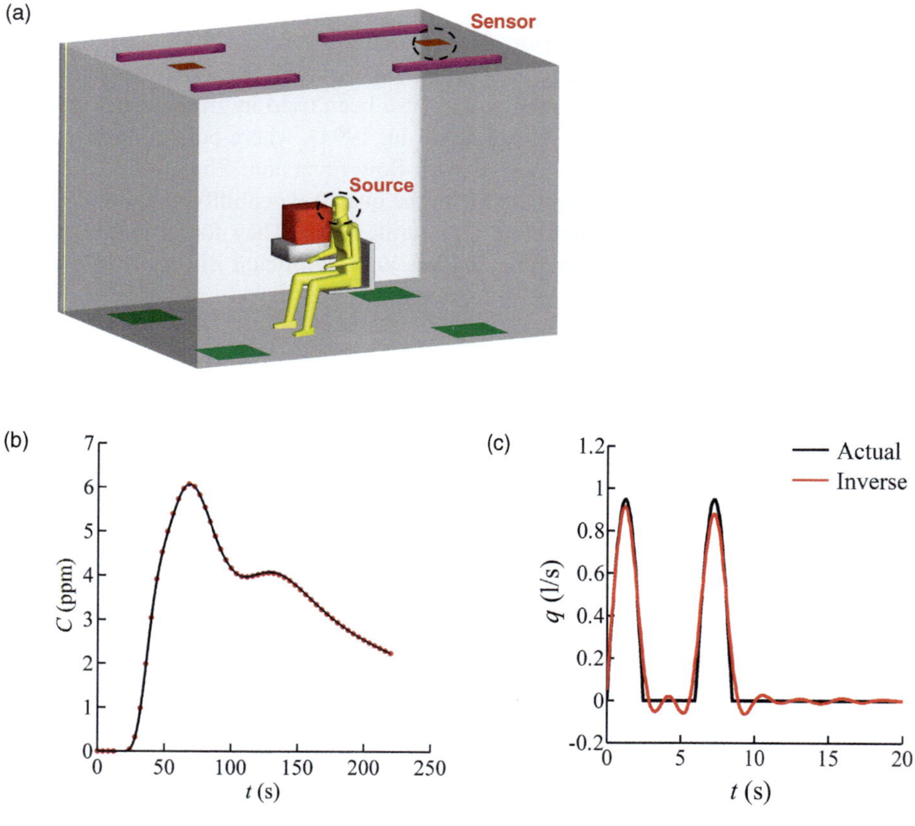

Figure 1.3 Demonstration of the regularized inversed matrix method in a release of a tracer gas in an office room to simulate a contaminant from human respiration: (a) three-dimensional layout of the room, (b) monitored tracer gas concentration at the outlet, and (c) inversely identified temporal release rate compared with the actual release rate (Zhang et al., 2007)

1.2.2 Forward methods

This section discusses the forward methods based on the NS equations with CFD simulation techniques for inverse/optimal design. These methods have more extensive applications than the backward methods. Forward methods can be further classified primarily as the CFD-based adjoint method, the CFD-based genetic algorithm method, the CFD-based artificial neural network, and the POD analysis method. The first two methods integrate CFD with an optimization algorithm. The last two methods use a surrogate model of CFD and a reduced order model to increase the computing efficiency, respectively. Forward methods convert an inverse design problem to a control problem by constructing an objective function. Various objective functions can be constructed for a design objective. For example, if the design objective is to reach a good thermal comfort in an occupied space, the objective function, *O*, can be constructed using the predicted mean vote (PMV) value that is function of air velocity, temperature, and relative humidity distributions in the occupied zone, such as:

$$O(\xi) = \int_{\Omega} (\text{PMV})^2 d\Omega \tag{1.5}$$

where ξ is a vector that represents the design variables (e.g., supply air velocity and temperature) and Ω is the occupied zone volume. Inverse design identifies the optimal ξ that minimizes $O(\xi)$ (i.e., neutral thermal condition). The design approach varies with the used forward method.

1.2.2.1 CFD-based adjoint method

The adjoint method computes the derivative of the objective function (Eq. (1.5)) over the design variables, ξ, so that the method can search the direction for gradually minimizing the objective function. Because of the strong nonlinearity in the NS equations, the derivative $dO/d\xi$ cannot be straightforwardly computed. The adjoint method introduces a Lagrange multiplier with p_a as adjoint pressure, $\mathbf{U}_\mathbf{a}$ as adjoint velocity, T_a as adjoint temperature, and Φ_a as adjoint scalar. The Lagrange multiplier can transform the constrained control problem into an unconstrained control problem. The augmented objective function can be reformulated as:

$$L(\xi) = O(\xi) + \int_{\omega} (p_a, \mathbf{U}_\mathbf{a}, T_a, \Phi_a) \cdot \mathbf{N} d\omega \tag{1.6}$$

where ω represents the flow domain and $\mathbf{N} = (N_1, N_2, N_3, N_4, N_5, N_6)^T$ represents the NS equations in vector form (N_1 the continuity equation; N_2, N_3, and N_4 the momentum equations; N_5 the energy equation; and N_6 the scalar transport equation). One can refer to Othmer et al. (2007) for the final form of the adjoint equations. To minimize the objective function on the design variables requires solving the following equation by gradually adjusting design variable ξ:

$$\frac{dO}{d\xi} \approx \frac{dL}{d\xi} = \frac{\partial O}{\partial \xi} + \int_{\omega} (p_a, \mathbf{U}_\mathbf{a}, T_a, \Phi_a) \frac{\partial \mathbf{N}}{\partial \xi} d\omega \tag{1.7}$$

The adjoint method can be very efficient in finding an optimal solution, although the method may identify the local optima.

The adjoint method has been used in various fields since it was first introduced by Lions (1971) for systems governed by partial differential equations. Pironneau (1974) was the first to use the adjoint method for design in fluid dynamics. Jameson (1988; 1995) applied the adjoint method in aerodynamic design with the aim to find the geometry that would minimize drag. Jameson (1999) and Reuther et al. (1999a; 1999b) developed the adjoint approach for the airfoil optimization, wing design, and complete aircraft configurations. A number of other research groups have developed various adjoint CFD codes for design optimization (Anderson and Venkatakrishnan, 1999; Othmer, 2008). The method is successfully applied in heat transfer problems, shape optimization, and pollution source identification.

For the inverse reconstruction and design problem in an enclosed environment, Liu and Zhai (2008) and Zhai et al. (2012) developed the adjoint method to locate an instantaneous or dynamical indoor contaminant source. Liu and Chen (2015) established a CFD-based adjoint method for the optimal design of indoor airflow with thermo-fluid boundary conditions as the design variables and flow and/or temperature fields as the design objective. This study found that different initial inlet air conditions led to different optimal inlet air conditions, which implies the existence of multiple solutions. Later, Liu et al. (2016) included the air supply location and size as design variables. In a two-dimensional non-isothermal ventilated cavity case (Blay et al., 1992) as shown in Fig. 1.4(a), matching the simulation results with measured air velocity and temperature on the red dashed lines was used as the design objective. Figs. 1.4(b) and 1.4(c) show, respectively, the computed flow field with the initially guessed and the finally identified air supply location, size, and parameters. The design procedure required only 16 adjustments of the design variables, automatically performed during a single computation. The inlet location was inversely identified with reasonable accuracy for the targeted air velocity and temperature profiles at the two sections. The study also found that the computing costs did not vary with the number of design variables.

1.2.2.2 CFD-based genetic algorithm method

As another optimization approach, the genetic algorithm (GA) was first introduced by Holland (1975) in the 1970s. GA is a gradient-free optimization method that simulates natural evolution in the search of optimal solutions. In order to perform evolution operations on design variables, GA must encode design variables into a string, which is called an individual. A collection of individuals constitutes a population. GA optimizes the solution with evolution (or iteration) of the population. A fitness value (the value of the objective function) quantifies the "goodness" of an individual. According to the number of objective functions, GA can be classified as a single-objective genetic algorithm (SOGA) or a multi-objective genetic algorithm (MOGA) (Schaffer, 1985).

A number of studies have improved the genetic algorithm method. For instance, Grefenstette (1986) succeeded in increasing the efficiency of a genetic algorithm by adjusting its parameters with another genetic algorithm. Horn et al. (1994) developed the tournament selection technique for the selection process, which increased the distinguishability of the algorithm. Srinivas and Patnaik (1994) developed the adaptive genetic algorithm, in which the probabilities of crossover and mutation can be varied depending on the fitness values of the solutions. This new method can converge to the global optimum in far fewer generations with lesser likelihood of falling into local optima. Genetic algorithms inherently tend to produce a set of proper solutions, and Karr et al. (1991) recognized this potential in their development of regression correlations and applied genetic algorithms using least-squares curve-fitting techniques. Researchers have also attempted to increase the convergence speed without reducing the capability of a genetic algorithm in searching optimal global multi-solutions for multiple objectives. On the basis of the vector-evaluated genetic algorithm proposed by Schaffer (1985), other groups have made the genetic algorithm more capable of multi-objective optimization (Fonseca and Fleming, 1993; Zitzler and Thiele, 1999; Lu and Yen, 2003).

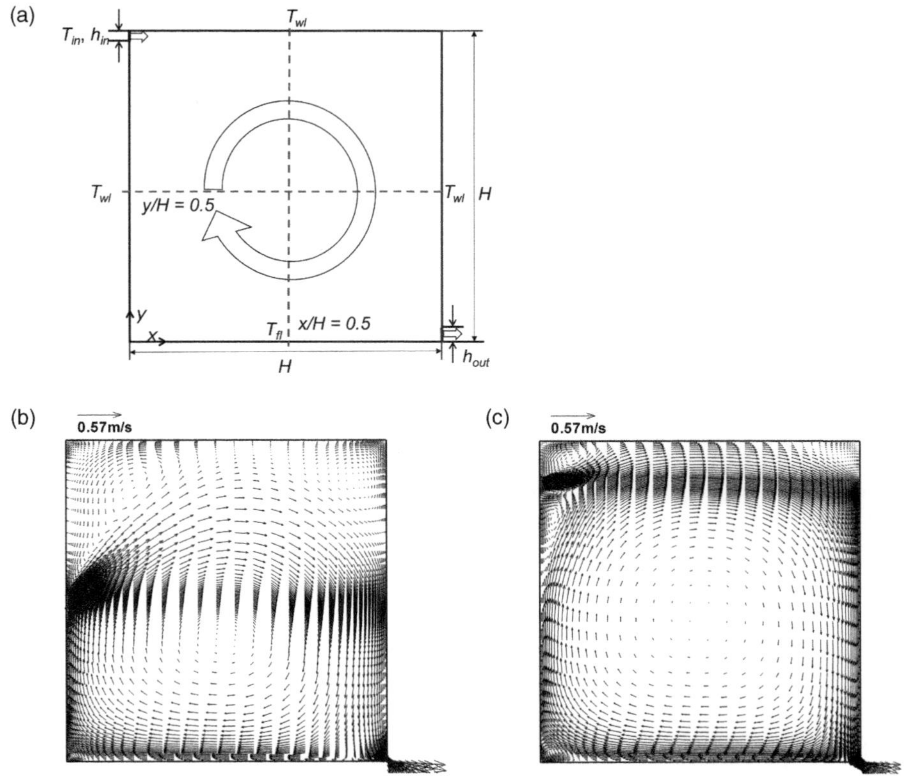

Figure 1.4 Demonstration of the CFD-based adjoint method in inverse reconstruction of a two-dimensional non-isothermal cavity case: (a) Sketch of the cavity (Blay et al., 1992) and (b) flow with the initial supply location and (c) flow with the final supply location (Liu et al., 2016)

Genetic algorithms have been used in various industrial and engineering applications, such as machine learning (Goldberg, 1989), urban planning (Schaumann et al., 1998), chemical kinetics reaction mechanisms (Polifke et al., 1998; Elliott et al., 2002; Elliott et al., 2003), and pollutant source identification in groundwater (Cieniawski et al., 1995; Mahinthakumar and Sayeed, 2005). With reference to buildings, Huang and Lam (1997) used the genetic algorithm for automatic tuning of controllers in HVAC systems, and Wright et al. (2002) identified pay-off characteristics between the building energy cost and the occupant thermal discomfort. Nassif et al. (2005) minimized energy use and maximized comfort in multiple zones with high loads. Fong et al. (2006) set suitable operating parameters for an HVAC system without sacrificing thermal comfort, and Ma and Wang (2011) developed a supervisor and optimal control strategy for energy efficiency.

The application of genetic algorithms in heat transfer and fluid flow started in the 1990s, such as the work by Queipo et al. (1994). This is probably due to the relatively long computational time in these applications. Other applications include heat exchange performance (West and Sherif, 2001a, 2001b; Ozkol and Komurgoz, 2005), airfoil design (Belegundu et al., 1994; Obayashi and Takanashi, 1995), car shape design (Campos et al., 2007), airflow field optimization of a melt blowing slot die (Sun and Wang, 2010), and geometric optimization of a gas turbine transition piece (Muñoz et al., 2011).

Malkawi et al. (2003) were probably the first to combine CFD with a genetic algorithm for indoor environmental design. Kato and Lee (2004) applied a similar approach to optimize a hybrid air-conditioning system with natural ventilation. Xue et al. (2013) implemented the inverse prediction and optimization of flow control conditions for confined spaces. Further, Zhai et al. (2014) optimized flow control conditions for a confined space (Fig. 1.5(a)) by satisfying multiple design indices, such as PMV, percentage dissatisfied with the draft (PD), and mean age of air. If the design objective is to have PMV = 0, PD < 15%, and mean age of air less than 125 s on the design domain represented by the circles in Fig. 1.5(a), then all points on the curve of PMV = 0 shown in Fig. 1.5(b) satisfy the first condition. The PMV = 0 and PD 15% curves intersecting at point Q2 meet the second condition with a supply air velocity of 1.75 m/s and air temperature of 24.2°. The curve of the mean age of air at 125 s intersecting with the curve of PMV = 0 at Q1 satisfies the third condition with a supply air velocity of 1.11 m/s and air temperature of 22.69°. As a result, any design along the red line between Q1 and Q2 would meet the design objectives. Since the indoor environment design is a multi-variable, multi-solution, multi-objective problem, the genetic algorithm can identify the global optima. Nevertheless, the CFD-based genetic algorithm method requires many CFD simulations during the population evolution, and the computational effort increases by orders of magnitude with an increase in the number of the design variables. An optimal design of an indoor environment using GA can take a week or more to reach the final solutions, depending on the case sophistication and accuracy requirements.

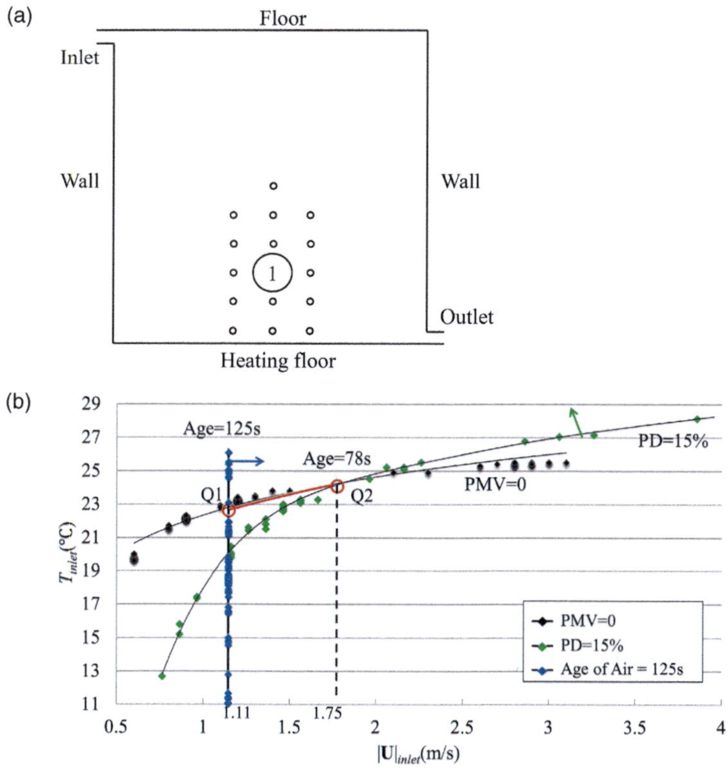

Figure 1.5 Demonstration of the CFD-based genetic algorithm method in inverse design of a two-dimensional ventilated cavity: (a) Sketch of the cavity and (b) optimal design parameters identified by the genetic algorithm method (the red line) (Zhai et al., 2014)

1.2.2.3 CFD-based artificial neural network method

The Artificial neural network (ANN) is a highly sophisticated paradigm that borrows features from human and animal brains to enable recognition of patterns within data. It learns to solve problems by developing a memory capable of associating a large number of input patterns with a resulting set of outputs or effects (Singh et al., 2004). More than 40 types of neural networks have been developed, including Perceptron, Hopfield, and the Feedback neural network. ANN can establish the mapping relationship between the input and output by training and adapting itself to different input–output pairs, and it can be trained to solve problems that are difficult for human beings or conventional computers (Khashman, 2010; Yi and Malkawi, 2011). The complicated linear or nonlinear relationship between the input and output can be mimicked by the neurons in ANN, which can be presented by the equations of weights, biases, and transfer functions.

The ANN method has been applied widely in a number of areas, including finance, medicine, and environmental science (Maier and Dandy, 2000; Liang and Du, 2008; Mazhar et al., 2013). Rege and Tock (1996) used ANN to estimate the point source emission rates of hazardous gases in West Texas. Based on the results of this study, the use of ANN provides an attractive and highly effective tool to model atmospheric dispersion, in which a large number of variables interact in a nonlinear manner. Deng and Hwang (2006) used ANN to identify the unknown boundary conditions in the inverse heat conduction problems. The weak generalization capacity of ANN is improved by employing the Bayesian regularization algorithm. The results show that the proposed neural network analysis method is capable of predicting the unknown parameters in inverse problems with an acceptable error. Caliskan (2007) used ANN to identify the icing parameters in an A340 aircraft to keep the air-conditioning performance close to the performance prior to icing. The simulation results show the successful ANN identification of the icing parameters. Singh et al. (2004) demonstrated that the ANN was capable of identifying unknown groundwater pollution sources at multiple locations using erroneous measurement data. Mjalli et al. (2007) used ANN to capture the operation characteristics of a real wastewater plant. The ANN model provided accurate predictions of the effluent stream, in terms of biological oxygen demand, chemical oxygen demand, and total suspended solids.

For enclosed environments, the relationship between the boundary conditions and environment parameters can be established by CFD, but the computing time is usually expensive. ANN can be used as a surrogate model of the CFD to reduce the computing time. It has been used to predict indoor velocity, temperature, thermal comfort, energy cost, and contaminant source (Srivastava, 2003). Liang and Du (2005) used ANN to maintain the indoor comfort level within the desired range under both heating/cooling modes. Qin et al. (2012) used ANN to describe the annual dynamic process with detailed parameter data and a reasonable amount of calculation time. It can reasonably predict reasonable dynamic energy consumption and thermal environment parameters. Bastani et al. (2012) used ANN as a statistical analysis integrated with a multi-zone airborne contaminant transport model to predict the pollutant source. Ayata et al. (2007) explored the potential use of ANN in natural ventilation. The results indicated that the ANN approach can be utilized as an efficient tool for predicting indoor air velocity distributions for natural ventilation. Ozbalta et al. (2012) used ANN to predict indoor temperature and relative humidity, and the results showed the R^2 values between predicted and actual values of indoor temperature and indoor relative humidity were computed as 0.94 and 0.96, respectively. Thomas and Soleimani-Mohseni (2007) compared the prediction accuracy of ANN and other models. The results showed that ANN models gave more accurate temperature predictions. Neto et al. (2008) compared ANN model and a detailed building HVAC design and simulation software as a forecasting tool for the energy demand. The results of ANN showed a fair agreement between energy consumption forecasts and actual values. Zhang and You (2014a) used ANN to identify the indoor boundary conditions by the data of velocity and temperature measured at observation locations. The Bayesian Regularization training algorithm was used to train ANN. Four sensors (A, B, C, and D), as shown in Fig. 1.6, were set to measure the velocity and temperature. Four different boundary conditions were studied. The comparison of the true boundary conditions and the boundary conditions obtained by ANN are shown in Table 1.1. It is shown that high prediction accuracy was realized.

1.2.2.4 Proper orthogonal decomposition method

An option for reducing the CFD simulation effort is to apply the proper orthogonal decomposition (POD) method (Lumley, 1967). The POD method provides a quick approach to describing forward mapping of a thermo-fluid distribution from various boundary setting conditions. POD establishes a basis for the modal decomposition of an ensemble of data, such as thermo-flow fields obtained from CFD simulations or measurements. A thermo-flow field can be expressed as a combination of the orthogonal spatial modes with their amplitudes or coefficients (Holmes et al., 1996), as follows:

$$\Phi_c = \sum_{b=1}^{B} c_b \varphi_b \tag{1.8}$$

where Φ_c is the constructed field, c_b is the amplitude or coefficient of the spatial mode $\boldsymbol{\varphi}_b$, b is the index of the spatial mode, and B is the number of required spatial modes.

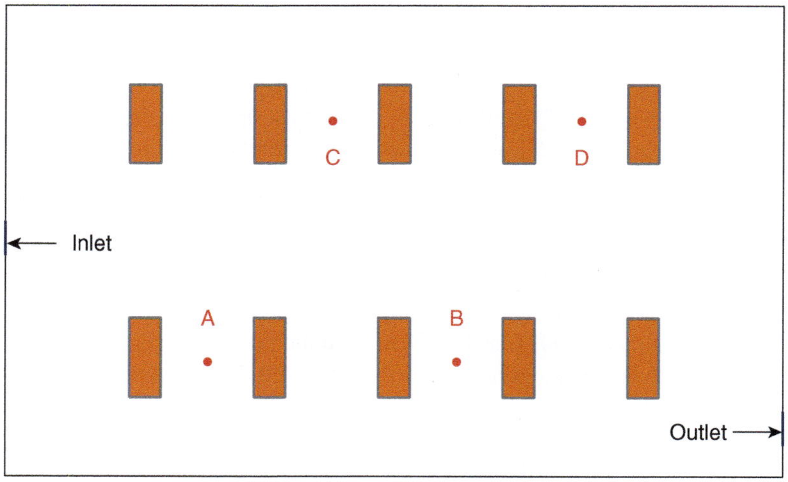

Figure 1.6 Schematic diagram of the indoor model used to identify boundary condition (Zhang and You, 2014a)

Table 1.1 Comparison of true boundary conditions and those obtained by ANN (Zhang and You, 2014)

	Case 1		*Case 2*		*Case 3*		*Case 4*	
	$\lvert\mathbf{U}\rvert$ *(m/s)*	*T (K)*	$\lvert\mathbf{U}\rvert$ *(m/s)*	*T (K)*	$\lvert\mathbf{U}\rvert$ *(m/s)*	*T (K)*	$\lvert\mathbf{U}\rvert$ *(m/s)*	*T (K)*
True value	0.18	294.5	0.25	298.6	0.36	296.3	0.44	302.3
ANN result	0.17	294.4	0.24	298.9	0.35	296.5	0.43	302.6

To reduce the effort required to solve for the spatial modes, Sirovich (1987) proposed the snapshot method for extracting the spatial modes as samples. Generally, the number of snapshots is far less than the number of spatial grid points, so considerable computing expense can be saved. The spatial mode is calculated as a combination of each snapshot field with its coefficient as:

$$\varphi = \sum_{r=1}^{R} a_r \Phi_r \tag{1.9}$$

where a_r is the coefficient of a snapshot field Φ_r, and R is the number of snapshots. Once the coefficients of the spatial modes are provided, a field can be constructed on the basis of Eq. (1.8). According to the orthogonality principles, the coefficient of a spatial mode corresponding to a snapshot field is:

$$c_b = (\Phi_r, \varphi_b) \tag{1.10}$$

The coefficients of the POD modes can correspond to a number of causal variables, for example, the air-supply parameters. By interpolating the coefficients for the rest air-supply parameters that are not used to extract the POD modes, the rest thermo-flow distribution can be instantly constructed.

The POD analysis was first proposed by Lumley (1967) nearly a half-century ago to analyze the structure of inhomogeneous turbulent flows. Later, this technique became popular in both experimental and numerical fluid mechanics. Citriniti and George (2000) utilized the POD method to analyze the measurement data provided by a hotwire anemometry in a turbulent shear layer. Lam (2013) used POD to construct the measured velocity and concentration distributions of a round jet. Early applications of the method for numerical fluid dynamics were for near-wall and jet-induced mixing flows (Bakewell and Lumley, 1967; Herzog, 1986; Glauser and George, 1987). Later, POD was applied to predict some basic fluid flows and heat transfer. For example, Cazemier et al. (1998) used POD to construct a low-dimensional model for cavity-driven flows, in which 80 spatial modes were extracted from 700 snapshots. Couplet et al. (2005) proposed a calibrated reduced-order POD-Galerkin model to predict the flows around a square cylinder and past a backward-facing step, respectively. Ding et al. (2008) proposed the POD-based data interpolation strategy to predict the thermo-flow fields for natural convection and forced convection flows. Liberge and Hamdouni (2010) investigated the dynamic flow induced by a rigid oscillating cylinder. The POD-based adaptive controller was also developed for real-time dynamic flow adjustment and prediction (Ravindran, 2000; Ravindran, 2002).

For enclosed environments, POD is mainly used for fast prediction of indoor thermo-flow and pollutant concentration, optimization of air-supply parameters, development of controllers for dynamic ventilation control, etc. Elhadidi and Khalifa (2005) applied the POD analysis to efficiently predict the velocity and temperature distributions inside an empty office. Sempey et al. (2009) performed a POD-based prediction of temperature distribution in air-conditioned rooms in a fixed-flow context. Allery et al. (2005) tracked particle motion in a two-dimensional ventilated cavity where the inside airflow was provided by a POD construction. Li et al. (2013a) integrated a genetic algorithm into the POD prediction of thermo-fluid flow to efficiently optimize the air-supply velocity and temperature. Ahuja et al. (2012) developed POD-based controllers to eliminate heat disturbance of an indoor environment by solar radiation, occupants, lights, etc. Tallet et al. (2012) used the POD analysis to predict fluid flow and dynamically control the opening of a window in a two-dimensional ventilated cavity. Li et al. (2013b) proposed a POD-based temperature prediction model for dynamic control of room temperature.

Wang et al. (2014) developed a POD-based inverse design method to determine appropriate air-supply parameters in an enclosed environment. With a quick cause-effect mapping of the indoor environmental performance from boundary settings, the air-supply parameters that meet the design targets are identified. Fig. 1.6(a) shows a three-dimensional half cabin section in which the air-supply velocity and temperature for one slot opening were inversely determined. Fig. 1.6(b) shows the results for the left air-supply opening S_1 that meet the design objective of |PMV|< 0.5 in the head (pink) region in Fig. 1.7(a). The supply air velocity and temperature in the green dot area of Fig. 1.7(b) can lead to satisfactory PMV in the head region. A full CFD simulation took approximately 8 hours to solve a single case when providing the snapshot fields, while the POD required less than 2 seconds to construct a new case. This study used 40 CFD simulations to construct 2,871 POD interpolations. The total computing time was 321 h, most of which was for the CFD simulations. If full CFD simulations were conducted, the computing time for 2,871 CFD simulations would be 23,000 h. Thus, the POD method is very efficient.

The computing costs of the forward simulations are generally much higher than those of the backward simulations. This is because tens or even hundreds of forwarded CFD

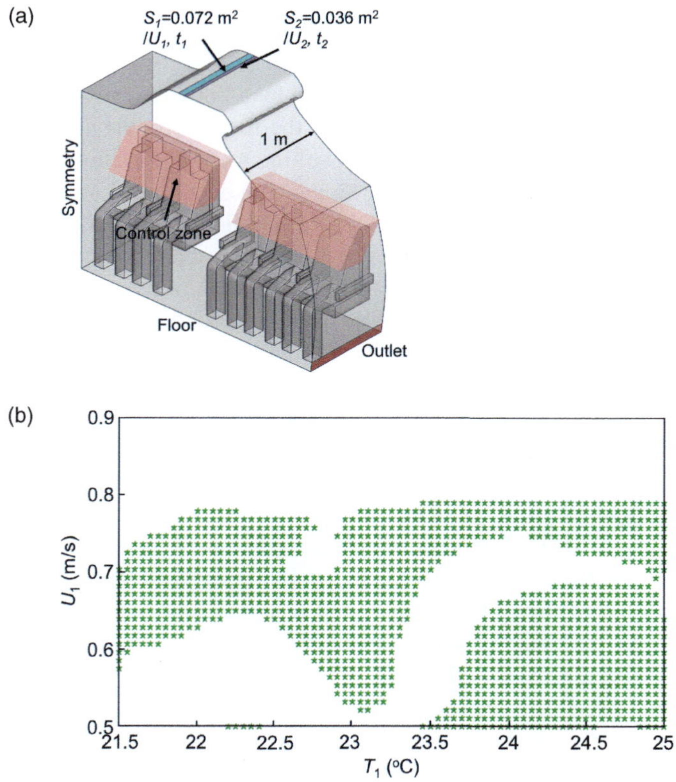

Figure 1.7 Demonstration of proper orthogonal decomposition method in inversely determining air-supply velocity and temperature in an aircraft cabin section (Wang et al., 2014): (a) geometry of the cabin section and (b) inversely solved air speed and temperature for S_1 (where a green dot represents an acceptable solution)

simulations are needed. The adjoint method uses the same computing time regardless of the number of the parameters solved, while the GA method is very sensitive to the number of the parameters. The GA method is thus more computationally demanding for a complex problem. The POD method is far less expensive because it uses interpolation. However, the accuracy is not as good as the other two methods discussed here.

1.3 Discussion

An inverse design can be carried out based on either the backward method or the forward method. Either method has its cons and pros. The backward method reverses the governing cause-effect relationship to find the causes and thus the reversion is straightforward. Since the reversed cause-effect relationship is ill posed, a regularization technique by adding extra terms to stabilize the inverse solution is required. However, the regularization destroys the original cause-effect relationship and consequently only an approximate solution can be provided. The accuracy of the backward method varies with the problem and the strength of the added regularized operation. There is no iterative computation required except in some circumstances when searching for an optimal regularized parameter is necessary. Hence, the backward method can provide results more efficiently than the forward method. However, the current backward method is limited to identification of a species source as shown in this chapter or thermal boundary conditions without involving flow.

For indoor environment studies, the backward methods are useful in many applications. For example, one could use the methods to track an infectious disease source in a hospital. Security authority can use the methods to locate the sensor locations for detecting chemical/biological warfare agents. The methods can also be used for comfort studies. For instance, the methods can identify where to place a radiant panel in a large workshop to keep a small work zone warm in winter.

The inverse design based on the forward method or optimal design method does not handle the inversion directly but turns into the forward cause-effect relationship to find the design variables that satisfy the objectives. Thus, this is an optimization approach. The solved governing equations are forward and well-posed, which has a solution and the solution is normally stable. Since a certain boundary condition would lead to one forward solution, an inverse design requires many trials with different boundary settings. An optimization technique like the adjoint method and genetic algorithm can be utilized to accelerate the search for a desired solution. The adjoint method is a gradient-based optimization, which may trap in a local optimum if no external interruption is imposed. The genetic algorithm searches for an optimal solution globally but at the expense of many solution trials, so it is computationally intensive based on CFD. The ANN and POD methods presented in this chapter do not utilize the optimization method. The ANN mimics the behavior of the CFD model with less computing time. But the accuracy of design is reduced slightly. The POD provides a very quick cause-effect mapping based on the extracted spatial modes and interpolation of their coefficients. However, if the data used does not contain the characteristics of the desirable solution, the method would not be able to provide desirable design. In addition, the interpolation is not free from errors.

The forward methods have great potential for indoor environment design. For example, in a completely new setting where no experience is available, the methods can provide an ideal solution. Should one wish to design a uniform flow around a human body with personalized ventilation, the existing design method would require hundreds of trials. The forward methods in theory have the ability to find the air supply locations and parameters to achieve the flow in one single computation. The beauty of the forward methods is stable and solvable. The major problem of the forward methods is computationally demanding so that the methods have yet widely applied to design complex systems. However, the methods have great potential in the near future. Therefore, this book focuses on the forward methods.

To increase efficiency, one option is to use fast computational fluid dynamics (FFD) (Zuo and Chen, 2009) as a substitute for CFD. FFD solves NS equation by using a three-step, time-advancement scheme (Ferziger and Perić, 2002) that splits the momentum equation into three discretized equations. FFD first solves the advection term by a semi-Lagrangian approach (Staniforth and Côté, 1991) and then solves the source and diffusion term with an implicit scheme. Finally, FFD solves the pressure term and continuity equation together by using a pressure-projection method. By combining the solution of the three discretized equations, the method could get a final answer for the NS equation. The algorithm does not require iteration so the computing speed is fast. It was initiated for the computer game industry, which seeks speed rather than accuracy. After further development by our team, FFD is also rather accurate now (Zuo and Chen, 2010; Zuo et al., 2010; Zuo et al., 2012; Jin et al., 2012; Jin et al., 2013; Jin and Chen, 2014; Jin et al., 2015a, 2015b). According to Zuo and Chen (2009), FFD can be 50 times faster than CFD and can predict the flow field with acceptable accuracy. Another option is to replace the CFD simulation with an artificial neural network. Zhou and Haghighat (2009) combined a genetic algorithm and artificial neural network to optimize ventilation systems in office spaces. Boithias et al. (2012) also proposed a genetic-algorithm-based method to optimize the architecture, training parameters, and inputs of an artificial neural network for predicting energy use and indoor discomfort during building control optimization. However, the use of an artificial neural network instead of CFD would increase the uncertainty and the risk of accumulative error (Nguyen et al., 2014). To solve this problem, Zhang and You (2014b) used both an artificial neural network and CFD to obtain the design objectives of new individuals. This integrated method could ensure accuracy in calculating the design objectives and reduce computational costs by 65%.

Another approach to reducing the computational effort of the CFD-based genetic algorithm method is to take advantage of the combined genetic algorithm and adjoint method. Giannakoglou et al. (2000) developed the combined genetic method for the optimal design of aerodynamic shapes. The genetic algorithm was used for the initial design, and the adjoint method was applied for fine tuning, leading to faster convergence. The integrated CFD-based GA and adjoint method can be much faster than the pure CFD-based GA. In addition, the integrated method can find the global optima, thus overcoming the shortcomings of the pure CFD-based adjoint method. Xue and Zhai (2014) combined the genetic algorithm with the adjoint method and revealed that this new integrated method is at great advantage when the number of variables for optimization is large.

A combination of the backward methods with the forward methods can be a further direction for future research. The backward methods can provide an approximate inverse solution efficiently. Hence, the inverse solution from the backward methods may be input into the forward methods as the initial guess of the final solution. Then the forward methods are employed to further improve the solution accuracy. The above strategy may shorten the time spent in circumscribing the rough ranges of design variables that satisfy the design targets. Such a scenario is viable at least for inverse identification of species sources. For example, Zhang et al. (2015) backwardly determined the pollutant source release rates at all possible source locations first and then identified the correct source in a forward matching process.

This book set only global objectives over the domain as the design target. However, the trends of modern enclosed environments are towards to personalized customization, especially in high-density public building or transportation vehicle compartments, where internal flow and thermal boundary condition design should be paid great attention. In principle, the methods presented in this chapter can be expanded to account for each individual's demanding objectives. The traditional general ventilation and boundary condition control may not fulfill the task. The designers may have to turn to personal environmental control. An inverse design that can take into account each individual's demanding objectives and its combination with a personal environmental control is a good topic for future study. The inverse design can demonstrate evidently its advantage in such a case.

This book contains six chapters that detail the forward/optimal design methods reviewed above for enclosed environment design. Chapter 2 discusses the CFD-based genetic algorithm method, Chapter 3 the CFD-based adjoint method, Chapter 4 the CFD-based artificial neural network method, Chapter 5 the CFD-based proper orthogonal decomposition method, and Chapter 6 the enhancement of the inverse design methods.

1.4 Conclusions

This chapter critically reviewed the state-of-the-art methods for inverse design of an enclosed environment and demonstrated the application of these methods through several examples. Both the backward and forward methods were criticized. The advantages and disadvantages of each method were analyzed in terms of applicability, computing efficiency, and accuracy. The study has led to the following primary conclusions:

Backward methods are useful for identifying contaminant sources and inverse design of thermal boundary conditions, but not for entire flow field design. The current backward methods can only handle an inverse design in known flow context.

Forward methods are promising for inverse design of enclosed environments. The CFD-based adjoint method may only identify the local optima of the design objective, but the computing effort remains the same regardless of the number of the design variables. The CFD-based genetic algorithm method can find the global optima of the design objective; however, the computing effort is significantly large. The ANN and POD methods can reduce the computing effort, although the resulting accuracy can be poor.

Several recommendations are provided to increase the modeling efficiency for CFD-based methods, such as using fast fluid dynamics techniques. A combination of two or more methods may also reduce the computing effort without compromising prediction accuracy.

Note

1 This chapter is partially reproduced from the paper entitled State-of-the-art methods for inverse design of an enclosed environment, *Building and Environment,* 91: 91–100. The reprint is with permission from Elsevier.

References

Ahuja S, Cliff E, Narayanan S. Reduced-order modeling for control of indoor building airflows. *Proceedings of the 5th National Conference of IBPSA-USA (SIMBUILD 2012)*. 2012;1–8.

Allery C, Beghein C, Hamdouni A. Applying proper orthogonal decomposition to the computation of particle dispersion in a two-dimensional ventilated cavity. *Communications in Nonlinear Science and Numerical Simulation*. 2005;10(8):907–920.

Anderson WK, Venkatakrishnan V. Aerodynamic design optimization on unstructured grids with a continuous adjoint formulation. *Computers and Fluids*. 1999;28(4):443–480.

Ayata T, Arcaklioglu E, Yildiz O. Application of ANN to explore the potential use of natural ventilation in buildings in Turkey. *Applied Thermal Engineering*. 2007;27(1):12–20.

Bagtzoglou AC, Dougherty DE, Tompson AF. Application of particle methods to reliable identification of groundwater pollution sources. *Water Resources Management*. 1992;6(1):15–23.

Bakewell HP, Lumley JL. Viscous sublayer and adjacent wall region in turbulent pipe flow. *Physics of Fluids*. 1967;10(9):1880–1889.

Bastani A, Haghighat F, Kozinski JA. Contaminant source identification within a building: toward design of immune buildings. *Building and Environment*. 2012;51(2012):320– 329.

Belegundu AD, Murthy DV, Salagame RR, Contans EW. Multi-objective optimization of laminated ceramic composites using genetic algorithms. *Proceedings of the 5th AIAA/USAF/NASA/ISSMO Symposium on Multidisciplinary Analysis and Optimization*. 1994;1015–1022.

Bellman R. Dynamic programming and Lagrange multipliers. *Proceedings of the National Academy of Sciences of the United States of America*. 1956;42(10):767.

Blay D, Mergui S, Niculae C. Confined turbulent mixed convection in the presence of a horizontal buoyant wall jet. *Fundamentals of Mixed Convection, ASME HTD*. 1992;213:65–72.

Boithias F, Mankibi ME, Michel P. Genetic algorithms based optimization of artificial neural network architecture for buildings' indoor discomfort and energy consumption prediction. *Building Simulation: An International Journal*. 2012;5:95–106.

Caliskan F. Neural network based icing identification and fault tolerant control of a 340 aircraft. *International Journal of Electrical, Computer, and Systems Engineering*. 2007;1(2):98–103.

Campos F, Geremia P, Skaperdas E. Automatic optimization of automotive designs using mesh morphing and CFD. *Proceedings of the 2nd ANSA mETA International Congress*. 2007;14–15.

Cazemier W, Verstappen RWCP, Veldman AEP. Proper orthogonal decomposition and low-dimensional models for driven cavity flows. *Physics of Fluids*. 1998;10(7):1685–1699.

Chen Q. Ventilation performance prediction for buildings: A method overview and recent applications. *Building and Environment*. 2009;44(4):848–858.

Cieniawski SE, Eheart JW, Ranjithan S. Using genetic algorithms to solve a multiobjective groundwater monitoring problem. *Water Resources Research*. 1995;31(2):399–409.

Citriniti JH, George WK. Reconstruction of the global velocity field in the axisymmetric mixing layer utilizing the proper orthogonal decomposition. *Journal of Fluid Mechanics*. 2000;418(9):137–166.

Clark G, Oppenheimer S. Quasireversibility methods for non-well-posed problems. *Electronic Journal of Differential Equations*. 1994;8:1–9.

Couplet M, Basdevant C, Sagaut P. Calibrated reduced-order POD-Galerkin system for fluid flow modeling. *Journal of Computational Physics*. 2005;207(1):192–220.

Deng S, Hwang Y. Applying neural networks to the solution of forward and inverse heat conduction problems. *International Journal of Heat and Mass Transfer*. 2006;49(25–26): 4732–4750.

Ding P, Wu X, He Y, Tao W. A fast and efficient method for predicting fluid flow and heat transfer problems. *Journal of Heat Transfer*. 2008;130(3):1–17.

DOE. Building Energy Data Book. *U.S. Department of Energy*. 2011.

Elhadidi B, Khalifa HE. Application of proper orthogonal decomposition to indoor airflows. *ASHRAE Transactions*. 2005;625–634.

Elliott L, Ingham DB, Kyne AG, Mera NS, Pourkashanian M, Wilson CW. A real coded genetic algorithm for the optimization of reaction rate parameters or chemical kinetic modeling in a perfectly stirred reactor. *Genetic and Evolutionary Computation Conference*. 2002;138–144.

Elliott L, Ingham DB, Kyne AG, Mera NS, Pourkashanian M, Wilson CW. Incorporation of physical bounds on rate parameters for reaction mechanism optimization using genetic algorithms. *Combust Science Technology*. 2003;175(4):619–648.

Ferziger JH, Perić M. *Computational methods for fluid dynamics* (Vol. 3). Berlin: Springer. 2002.

Fong KF, Hanby VI, Chow TT. HVAC system optimization for energy management by evolutionary programming. *Energy and Buildings*. 2006;38(3):220–231.

Fonseca CM, Fleming PJ. Multiobjective genetic algorithms. *Proceedings of the IEE colloquium on Genetic Algorithms for Control Systems Engineering*. 1993.

Friedr J. *Stable solution – inverse problems*. Braunschweig: Vieweg & Sohn, Verlagsgesellschaft mbH. 1978.

Giannakoglou K, Pappou T, Giotis A, Koubogiannis D. A parallel inverse-design algorithm in aeronautics based on genetic algorithms and the adjoint method. *Proceedings of the European Congress on Computational Methods in Applied Sciences and Engineering (ECCOMAS 2000)*. 2000;1–14. Barcelona, Spain.

Glauser MN, George WK. Orthogonal decomposition of the axisymmetric jet mixing layer including azimuthal dependence. *Advances in Turbulence*. 1987;357–366.

Goldberg DE. *Genetic algorithms in search, optimization, and machine learning*. Boston, MA: Addison-Wesley Longman Publishing Co. 1989.

Grefenstette JJ. Optimization of control parameters for genetic algorithms. *IEEE Transactions on Systems, Man, and Cybernetics*. 1986;122–128.

Hedge A, Erickson WA, Rubin G. Predicting sick building syndrome at the individual and aggregate levels. *Environment International*. 1996;22(1):3–19.

Herzog S. The large scale structure in the near-wall region of turbulent pipe flow. PhD dissertation. Department of Mechanical Engineering, Cornell University. 1986.

Hiyama K, Ishida Y, Kato, S. Coupling 3D transient pollutant transport in a room into a flow network model with concentration response factor method. *ASHRAE Transactions*. 2008;114(2):119–129.

Holland JH. *Adaptation in natural and artificial systems: An introductory analysis with applications to biology, control, and artificial intelligence*. Ann Arbor: U Michigan Press. 1975.

Holmes P, Lumley JL, Berkooz G. *Turbulence, coherent structures, dynamical systems and symmetry*. Cambridge, UK: Cambridge University Press. 1996.

Horn J, Nafpliotis N, Goldberg DE. A niched Pareto genetic algorithm for multiobjective optimization. *Proceedings of the First IEEE Conference on Evolutionary Computation*. 1994;82–87.

Huang W, Lam HN. Using genetic algorithms to optimize controller parameters for HVAC systems. *Energy and Buildings*. 1997;26(3):277–282.

International Facility Management Association (IFMA). IFMA survey ranks top 10 office complaints. 2003. Retrieved July 1, 2012, from http://www.buildings.com/tabid/3334/ArticleID/1689/Default.aspx.

Ishida Y, Kato S. Method for coupling three-dimensional transient pollutant transport into one-dimensional transport simulation based on concentration response factor. *ASHRAE Transactions*. 2008;114(1): 259–272.

Jameson A. Aerodynamic design via control theory. *Journal of Scientific Computing*. 1988;3(3): 233–260.

Jameson A. Optimum aerodynamic design using CFD and control theory. *AIAA paper*. 1995;1729:124–131.

Jameson A. Re-engineering the design process through computation. *Journal of Aircraft*. 1999;36(1): 36–50.

Jin M, Chen Q. Accelerating fast fluid dynamics with a coarse-grid projection scheme. *HVAC&R Research*. 2014;20(8):932–943.

Jin M, Liu W, Chen Q. Improvement of fast fluid dynamics with a conservative semi-Lagrangian scheme. *International Journal of Numerical Methods for Heat and Fluid Flow*. 2015a;25(1):2–18.

Jin M, Liu W, Chen Q. Simulating buoyancy-driven airflow in buildings by coarse-grid fast fluid dynamics. *Building and Environment*. 2015b;85:144–152.

Jin M, Zuo W, Chen Q. Improvements of fast fluid dynamics for simulating airflow in buildings. *Numerical Heat Transfer, Part B: Fundamentals*. 2012;62(6):419–438.

Jin M, Zuo W, Chen Q. Simulating natural ventilation in and around buildings by fast fluid dynamics. *Numerical Heat Transfer, Part A: Applications*. 2013;64(4):273–289.

Karr CL, Stanley DA, Scheiner BJ. *Genetic algorithm applied to least squares curve fitting*. U.S. Department of the Interior, Bureau of Mines. 1991.

Kato S, Kajii Y, Itokazu R, Hirokawa J, Koda S, Kinjo Y. Transport of atmospheric carbon monoxide, ozone, and hydrocarbons from Chinese coast to Okinawa island in the Western Pacific during winter. *Atmospheric Environment*. 2004;38:2975–2981.

Kato S, Lee JH. Optimization of hybrid air-conditioning system with natural ventilation by GA and CFD. *Proceedings of the 25th AIVC Conference, Ventilation and Retrofitting*. 2004;15–17.

Kato S, Pochanart P, Kajii Y. Measurements of ozone and nonmethane hydrocarbons at Chichi-jima Island, a remote island in the western Pacific: long-range transport of polluted air from the Pacific Rim region. *Atmospheric Environment*. 2001;35(34):6021–6029.

Kato S, Pochanart P, Hirokawa J, Kajii Y, Akimoto H, Ozaki Y, Obi K, Katsuno T, Streets D, Minko NP. The influence of Siberian forest fires on carbon monoxide concentrations at Happo, Japan. *Atmospheric Environment*. 2002;36(2):385–390.

Khashman A. Neural networks for credit risk evaluation: investigation of different neural models and learning schemes. *Expert Systems with Applications*. 2010;37 (9):6233–6239.

Klepeis NE, Nelson WC, Ott WR, Robinson JP, Tsang AM, Switzer P, Engelmann WH. The National Human Activity Pattern Survey (NHAPS): A resource for assessing exposure to environmental pollutants. *Journal of Exposure Analysis and Environmental Epidemiology*. 2001;11(3):231–252.

Lam KM. Application of POD analysis to concentration field of a jet flow. *Journal of Hydro-environment Research*. 2013;7(3):174–181.

Lattès R, Lions JL. The method of quasi-reversibility: Applications to partial differential equations. In R. Bellman (Ed.) *Modern analytic and computational methods in science and mathematics*. New York: American Elsevier Publishing Company. 1969.

Leder K, Newman D. Respiratory infections during air travel. *Internal Medicine Journal*. 2005;35(1):50–55.

Li K, Su H, Chu J, Xu C. A fast-POD model for simulation and control of indoor thermal environment of buildings. *Building and Environment*. 2013b;60(2):150–157.

Li K, Xue W, Xu C, Su H. Optimization of ventilation system operation in office environment using POD model reduction and genetic algorithm. *Energy and Buildings*. 2013a;67(12):34–43.

Liang J, Du R. Thermal comfort control based on neural network for HVAC application. *Proceedings of the 2005 IEEE Conference on Control Applications*. Toronto, Canada: IEEE, 2005, 819–824.

Liang J, Du R. Design of intelligent comfort control system with human learning and minimum power control strategies. *Energy Conversion and Management*. 2008;49(4):517–528.

Liberge E, Hamdouni A. Reduced order modeling method via proper orthogonal decomposition (POD) for flow around an oscillating cylinder. *Journal of Fluids and Structures*. 2010;26(2):292–311.

Lions JL. *Optimal control of systems governed by partial differential equations problèmes aux limites*. Berlin: Springer. 1971.

Liu W, Chen Q. Optimal air distribution design in enclosed spaces using an adjoint method. *Inverse Problems in Science and Engineering*. 2015;23(5):760–779.

Liu W, Jin M, Chen C, Chen Q. Optimization of air supply location, size, and parameters in enclosed environments using a CFD-based adjoint method. *Journal of Building Performance Simulation*. 2016;9(2):149–161.

Liu X, Zhai Z. Location identification for indoor instantaneous point contaminant source by probability-based inverse computational fluid dynamics modeling. *Indoor Air*. 2008;18(1):2–11.

Lu H, Yen GG. Rank-density-based multiobjective genetic algorithm and benchmark test function study. *IEEE Transactions on Evolutionary Computation*. 2003;7(4):325–343.

Lumley JL. The structure of inhomogeneous turbulent flows. *Atmospheric Turbulence and Radio Wave Propagation*. 1967;166–178.

Ma Z, Wang S. Supervisory and optimal control of central chiller plants using simplified adaptive models and genetic algorithm. *Applied Energy*. 2011;88:198–211.

Mahinthakumar GK, Sayeed M. Hybrid genetic algorithm – Local search methods for solving groundwater source identification inverse problems. *Water Resources Planning and Management*. 2005;131(1):45–57.

Maier HR, Dandy GC. Neural networks for the prediction and forecasting of water resources variables: A review of modeling issues and applications. *Environmental Modelling & Software*. 2000;15(1):101–124.

Majeed O, Eng P. Aircraft environmental control systems. *Carleton University Aero Lecture*. 2010;4003.

Malkawi AM, Srinivasan RS, Yi YK, Choudhary R. Performance-based design evolution: the use of genetic algorithms and CFD. *Proceedings of the Eighth International IBPSA Conference.* 2003;793–798.

Mangili A, Gendreau MA. Transmission of infectious diseases during commercial air travel. *The Lancet.* 2005;365(9463):989–996.

Mazhar F, Khan AM, Chaudhry IA, Ahsan M. On using neural networks in UAV structural design for CFD data fitting and classification. *Aerospace Science and Technology.* 2013; 30(1):210–225.

Mehra P, Wah BW. Artificial neural networks: Concepts and theory. *IEEE Computer Society Press.* 1992.

Mjalli FS, Al-Asheh S, Alfadala HE. Use of artificial neural network black-box modeling for the prediction of wastewater treatment plants performance. *Journal of Environmental Management.* 2007;83(3):329–338.

Muñoz AG, Ayala-Ramirez V, Alfaro-Ayala JA, Acosta BMT. Optimization of the transition piece applying genetic algorithms. *Applied Thermal Engineering.* 2011;31(16):3214–3225.

Nassif N, Kajl S, Sabourin R. Optimization of HVAC control system strategy using two-objective genetic algorithm. *HVAC&R Research.* 2005;11(3):459–486.

Neto AH, Fiorelli FAS. Comparison between detailed model simulation and artificial neural network for forecasting building energy consumption. *Energy and Buildings.* 2008;40(12): 2169–2176.

Nguyen AT, Reiter S, Rigo P. A review on simulation-based optimization methods applied to building performance analysis. *Applied Energy.* 2014;113:1043–1058.

Obayashi S, Takanashi S. Genetic algorithm for aerodynamic inverse optimization problems. *Proceedings of the 1st IEE/IEEE International Conference on Genetic Algorithm in Engineering Systems: Innovations and Applications.* 1995;7–12.

Olsen SJ, Chang HL, Cheung TYY, Tang AFY, Fisk TL, Ooi SPL, Dowell SF. Transmission of the severe acute respiratory syndrome on aircraft. *New England Journal of Medicine.* 2003;349(25):2416–2422.

Othmer C. A continuous adjoint formulation for the computation of topological and surface sensitivities of ducted flows. *International Journal for Numerical Methods in Fluids.* 2008;58(8):861–877.

Othmer C, de Villiers E, Weller HG. Implementation of a continuous adjoint for topology optimization of ducted flows. *Proceedings of the 18th AIAA Computational Fluid Dynamics Conference.* 2007.

Ozbalta TG, Sezer A, Yildiz Y. Models for prediction of daily mean indoor temperature and relative humidity: education building in Izmir, Turkey. *Indoor and Built Environment.* 2012;21(6):772–781.

Ozkol I, Komurgoz G. Determination of the optimum geometry of the heat exchanger body via a genetic algorithm. *Numerical Heat Transfer Part A – Applications.* 2005;48(3):283–296.

Pejtersen J, Brohus H, Hyldgaard CE. The effect of renovating an office building on occupants' comfort and health. *Proceedings of the 8th International Conference on Indoor Air Quality and Climate (Indoor Air 1999).* 1999;2:160–165.

Pironneau O. On optimum design in fluid mechanics. *Journal of Fluid Mechanics.* 1974;64(1):97–110.

Polifke W, Geng W, Dobbeling K. Optimization of rate coefficients for simplified reaction mechanisms with genetic algorithms. *Combust Flame.* 1998;113(1):119–135.

Qin R, Yan D, Zhou X, Jiang Y. Research on a dynamic simulation method of atrium thermal environment based on neural network. *Building and Environment.* 2012;50(2012): 214–220.

Queipo N, Devarakonda R, Humphrey JAC. Genetic algorithms for thermosciences research – application to the optimized cooling of electronic components. *International Journal of Heat and Mass Transfer.* 1994;37(6):893–908.

Ravindran SS. Reduced-order adaptive controllers for fluid flows using POD. *Journal of Scientific Computing.* 2000;15(4):457–478.

Ravindran SS. Adaptive reduced-order controllers for a thermal flow system using proper orthogonal decomposition. *SIAM Journal on Scientific Computing.* 2002;23(6):1924–1942.

Rege MA, Tock RW. A simple neural network for estimating emission rates of hydrogen sulfide and ammonia from single point sources. *Journal of the Air & Waste Management Association.* 1996;46(10):953–962.

Reuther J, Jameson A, Alonso JJ, Rimlinger MJ, Saunders D. Constrained multipoint aerodynamic shape optimization using an adjoint formulation and parallel computers, part 1. *Journal of Aircraft*. 1999a;36(1):51–60.

Reuther J, Jameson A, Alonso JJ, Rimlinger MJ, Saunders D. Constrained multipoint aerodynamic shape optimization using an adjoint formulation and parallel computers, part 2. *Journal of Aircraft*. 1999b;36(1):61–74.

Schaffer J. Multiple objective optimization with vector evaluated genetic algorithms. *Proceedings of the International Conference on Genetic Algorithm and Their Applications*. 1985.

Schaumann EJ, Balling RJ, Day K. Genetic algorithms with multiple objectives. *Proceedings of the Seventh AIAA/USAF/NASA/ISSMO Symposium on Multidisciplinary Analysis and Optimization*. 1998;2114–2123.

Sempey A, Inard C, Ghiaus C, Allery C. Fast simulation of temperature distribution in air conditioned rooms by using proper orthogonal decomposition. *Building and Environment*. 2009;44(2):280–289.

Singh RM, Datta B, Jain A. Identification of unknown groundwater pollution sources using artificial neural networks. *Journal of Water Resources Planning and Management-ASCE*. 2004;130(6): 506–514.

Sirovich, L. Turbulence and the dynamics of coherent structures. I-Coherent structures. II-Symmetries and transformations. III-Dynamics and scaling. *Quarterly of Applied Mathematics*. 1987;45: 561–571.

Skaggs TH, Kabala ZJ. Recovering the history of a groundwater contaminant plume: Method of quasi-reversibility. *Water Resource Research*. 1995;31(11):2669–2673.

Spengler JD, Chen Q. Indoor air quality factors in designing a healthy building. *Annual Review of Energy and the Environment*. 2000;25:567–600.

Srinivas M, Patnaik LM. Adaptive probabilities of crossover and mutation in genetic algorithms. *IEEE Transactions on Systems, Man, and Cybernetics*. 1994;24(4):656–667.

Srivastava AK. Detection of volatile organic compounds (VOCs) using SnO2 gas-sensor array and artificial neural network. *Sensors and Actuators B: Chemical*. 2003;96 (1):24–37.

Staniforth A, Côté J. Semi-Lagrangian integration schemes for atmospheric models – a review. *Monthly Weather Review*. 1991;119(9):2206–2223.

Sun Y, Wang X. Optimization of air flow field of the melt blowing slot die via numerical simulation and genetic algorithm. *Journal of Applied Polymer Science*. 2010;115(3):1540–1545.

Tallet A, Allard F, Allery C. Numerical simulation of real-time air flow control by POD/ROM applied to anisothermal ventilated cavity. *Proceedings of the 10th International Conference on Healthy Buildings (Healthy Building 2012)*. 2012;1960–1965.

Thomas B, Soleimani-Mohseni M. Artificial neural network models for indoor temperature prediction: investigations in two buildings. *Neural Computing & Applications*. 2007;16(1): 81–89.

Tikhonov AN. Solution of incorrectly formulated problems and the regularization method. *In Soviet Math. Dokl*. 1963;5:1035–1038.

Tikhonov AN, Arsenin VY. *Solutions of ill-posed problems*. Washington DC: Halsted Press. 1977.

Wang J, Zhou H, Zhang T, Wang S. Inverse design of aircraft cabin environment based on proper decomposition of thermo-flow fields. *Proceedings of the 13th International Conference on Indoor Air Quality and Climate (Indoor Air 2014)*. 2014. Hong Kong.

Wargocki P, Wyon DP, Sundell J, Clausen G, Fanger P. The effects of outdoor air supply rate in an office on perceived air quality, sick building syndrome (SBS) symptoms and productivity. *Indoor Air*. 2000;10(4):222–236.

West AC, Sherif SA. Optimization of multistage vapour compression systems using genetic algorithms. Part 1: Vapour compression system model. *International Journal of Energy Research*. 2001a;25(9):803–812.

West AC, Sherif SA. Optimization of multistage vapour compression systems using genetic algorithms. Part 2: Application of genetic algorithm and results. *International Journal of Energy Research*. 2001b;25(9):813–824.

Wilson JL, Liu J. Backward tracking to find the source of pollution. *Water Manage. Risk Remed.* 1994;1:181–199.

World Health Organization (WHO). *Chronology of Influenza A (H1N1)*. 2009.

Wright JA, Loosemore HA, Farmani R. Optimization of building thermal design and control by multi-criterion genetic algorithm. *Energy and Buildings*. 2002;34(9):959–972.

Xue Y, Zhai Z, Chen Q. Inverse prediction and optimization of flow control conditions for confined spaces using a CFD-based genetic algorithm. *Building and Environment*. 2013;64:77–84.

Xue Y, Zhai Z. Comparison and integration of genetic algorithm and adjoint algorithm for optimizing indoor environments. *Proceedings of the 13th International Conference on Indoor Air Quality and Climate (Indoor Air 2014)*. 2014. Hong Kong.

Yi YK, Malkawi AM. Integrating neural network models with computational fluid dynamics (CFD) for site-specific wind condition. *Building Simulation*. 2011;4(3):245–254.

Zhai Z, Liu X, Wang H, Li Y, Liu J. Experimental verification of tracking algorithm for dynamically-releasing single indoor contaminant. *Building Simulation*. 2012;5(1):5–14.

Zhai Z, Xue Y, Chen Q. Inverse design methods for indoor ventilation systems using CFD-based multi-objective genetic algorithm. *Building Simulation*. 2014;7(6):661–669.

Zhang T, Chen Q. Identification of contaminant sources in enclosed environments by inverse CFD modeling. *Indoor Air*. 2007a;17(3):167–177.

Zhang T, Chen Q. Identification of contaminant sources in enclosed spaces by a single sensor. *Indoor Air*. 2007b;17(6):439–449.

Zhang T, Li H, Wang, S. Inversely tracking indoor airborne particles to locate their release sources. *Atmospheric Environment*. 2012;55:328–338.

Zhang T, Yin S, Wang S. An inverse method based on CFD to quantify the temporal release rate of a continuously released pollutant source. *Atmospheric Environment*. 2013;77:62–77.

Zhang T, You X. Applying neural networks to solve the inverse problem of indoor environment. *Indoor and Built Environment*. 2014a;23(8):1187–1195.

Zhang T, You X. A simulation-based inverse design of preset aircraft cabin environment. *Building and Environment*. 2014b;82:20–26.

Zhang T, Zhou H, Wang S. Inverse identification of the release location, temporal rates and sensor alarming time of an airborne pollutant source. *Indoor Air*. 2015;25(4):415–427.

Zhou L, Haghighat F. Optimization of ventilation systems in office environment, Part II: Results and discussions. *Building and Environment*. 2009;44(4):657–665.

Zitzler E, Thiele L. Multiobjective evolutionary algorithms: A comparative case study and the strength Pareto approach. *IEEE Transactions on Evolutionary Computation*. 1999;3(4):257–271.

Zuo W, Chen Q. Real time or faster-than-real-time simulation of airflow in buildings. *Indoor Air*. 2009;19(1):33–44.

Zuo W, Chen Q. Fast and informative flow simulations in a building by using fast fluid dynamics model on graphics processing unit. *Building and Environment*. 2010;45(3):747–757.

Zuo W, Hu J, Chen Q. Improvements on FFD modeling by using different numerical schemes. *Numerical Heat Transfer, Part B: Fundamentals*. 2010;58(1):1–16.

Zuo W, Jin M, Chen Q. Reduction of numerical diffusion in the FFD model. *Engineering Applications of Computational Fluid Mechanics*. 2012;6(2):234–247.

2 CFD-based genetic algorithm method[1,2]

Optimizing indoor flow distributions with specific design goals requires systematic evaluation and prediction of the influences of critical flow control conditions such as flow inlet temperature and velocity. In order to identify optimal flow control conditions, the conventional approach simulates a large number of flow scenarios with different boundary conditions for comparison. This chapter presents a method that combines the genetic algorithm (GA) with computational fluid dynamics (CFD) technique, which can efficiently predict and optimize the flow boundary conditions with various objective functions. A mixing convection case in a confined space was used to evaluate the performance of the developed method. The study showed that the method could predict accurately the inlet boundary conditions, with targeted variable values in the space, upon a few CFD cases. The results revealed that the accuracy of the inverse prediction was influenced by the error of CFD simulation that needed be controlled within 20%. The study further used the predicted mean vote (PMV), the percentage dissatisfied of draft (PD), and the age of air around occupants as the design goals to optimize the inlet boundary conditions (e.g., supply velocity, temperature, and angle) of an office and an aircraft cabin. Using the developed program, this study proposed two new design methods for indoor environments: the constraint method and the optimization method. Both of the methods succeed in providing a variety of designs that satisfy all of the design goals with much less computing cost than the traditional method. The optimization method provides more accurate results while the constraint method needs less computational efforts.

2.1 Introduction

The genetic algorithm method was first introduced by Holland in the 1970s (Holland, 1975). This gradient-free optimization method simulates natural evolution process in the search for optimal solutions. In order to perform evolution operations on a design variable, a genetic algorithm must encode the variable into a string, which is called an individual. A collection of individuals constitutes a population. The algorithm optimizes the solution with evolution (or iteration) of the population. A fitness value (the value of the objective function) quantifies the "goodness" of an individual. Genetic algorithms can be separated into single-objective and multi-objective algorithms (Schaffer, 1985).

Many researchers have introduced improvements to the genetic algorithm. For instance, Grefenstette (1986) increased the efficiency of a genetic algorithm by adjusting its parameters with the use of a second genetic algorithm. Horn et al. (1994) developed the tournament technique for the selection process, which increased the distinguishability of the algorithm. Srinivas and Patnaik (1994) developed the adaptive genetic algorithm, in which the probabilities of crossover and mutation can be varied in accordance with the fitness values of the solutions. The new method can converge to the global optimum in far fewer generations and is less likely to fall into local optima. Recognizing that genetic algorithms inherently produce a set of proper solutions, Karr et al. (1991) applied the algorithms using least squares curve fitting techniques in their development of regression correlations. Researchers have also attempted to increase the speed of convergence without decreasing an algorithm's capability to find optimal global multi-solutions for multiple objectives. On the basis of the vector-evaluated genetic algorithm proposed by Schaffer (1985), other groups have made the genetic algorithm more capable of multi-objective optimizations (Fonseca and Fleming, 1993; Zitzler and Thiele, 1999; Lu and Yen, 2003).

Genetic algorithms have been gradually adopted in various industrial and engineering applications, such as machine learning (Goldberg, 1989), urban planning (Schaumann et al., 1998), chemical kinetics reaction mechanisms (Polifke et al., 1998; Elliott et al., 2002, 2003), and pollutant source identification in ground water (Cieniawski et al., 1995; Mahinthakumar and Sayeed, 2005). Among the applications in buildings, Huang and Lam (1997) used a genetic algorithm for automatic tuning of controllers in HVAC systems, and Wright et al. (2002) identified the pay-off characteristic between building energy cost and occupants' thermal discomfort. Nassif et al. (2005) minimized energy use and maximized comfort in multiple zones with high loads. Fong et al. (2006) set suitable operating parameters for a HVAC system without sacrificing thermal comfort, and Ma and Wang (2011) developed a supervisor and optimal control strategy for energy efficiency.

However, the utilization of the genetic algorithm in heat transfer and fluid flow started to appear only in the 1990s, such as in the work by Queipo et al. (1994). This was probably due to the relatively demanding computational time in applications such as heat exchange performance (West and Sherif, 2001; Ozkol and Komurgoz, 2005), airfoil design (Belegundu et al., 1994; Obayashi and Takanashi, 1995), car shape design (Campos et al., 2007), airflow field optimization of a melt blowing slot die (Sun and Wang, 2010), and geometric optimization of a gas turbine transition piece (Muñoz et al., 2011).

Malkawi et al. (2003) were probably the first to combine CFD with a genetic algorithm in the design of indoor environments. Kato and Lee (2004) applied a similar approach to optimizing a hybrid air-conditioning system with natural ventilation. Xue et al. (2013) used a genetic algorithm in the inverse prediction and optimization of flow control conditions for confined spaces. Zhai et al. (2014) also optimized flow control conditions in a confined space, satisfying multiple design indices such as predicted mean vote (PMV), percentage of dissatisfied people due to draft (PD), and mean age of air.

The design of ventilation parameters is an inverse problem, which means that the "cause" is determined from the "result." There are two categories of methods for solving inverse problems: backward methods and forward methods. Backward methods, such as the quasi-reversibility method (Lattès and Lions, 1969) and pseudo-reversibility method (Kato et al., 2002), attempt to solve the Navier-Stokes (NS) equations inversely, which violates the second law of thermodynamics. Such solutions can be realized only for scalar functions when the flow field is constant. Thus, the backward methods can be used only in contaminant source identification. Forward methods include trial-and-error methods and optimization methods. Traditional ventilation parameter design methods are typical trial-and-error methods, which are inefficient. Researchers are increasingly employing optimization methods to solve inverse problems such as geometry optimization (Jameson, 1995) or energy management (Wright et al., 2002). These problems are similar to that of ventilation parameter design.

Optimization algorithms generally fall into two categories: gradient-based methods and gradient-free methods. Gradient-based methods cannot deal well with nonlinear problems

(Wetter and Wright, 2003) and can be easily trapped in local optimal values (Wang and Jin, 2000). Airflow and heat transfer problems in confined spaces are highly nonlinear and thus require the utilization of gradient-free methods. As a gradient-free method, a genetic algorithm is capable of solving nonlinear and multi-solution problems, requiring less computing time than other methods to find the global optimum (Sakamoto et al., 1999). It seems that a genetic algorithm is the most suitable method for the inverse design of ventilation parameters.

2.2 Principles of the CFD-based genetic algorithm

2.2.1 General principles of a genetic algorithm

The basic element of a genetic algorithm is the "gene," which is generally coded with a binary 0 or 1 (as shown in Fig. 2.1). Several genes constitute a "genome," which is called a "design variable" in the inverse design problem. Several "genomes" or "design variables" constitute an individual that represents a candidate solution to the inverse design problem. A population of candidate solutions (individuals) is evolved toward better solutions. A genetic algorithm uses a fitness value to measure the quality of the solutions. The fitness value is usually the value of a specified objective function. In the inverse design of an enclosed environment, the objective function can be energy consumption, thermal comfort index, air quality such as age of air, etc.

A genetic algorithm uses evolution operations to generate new populations with higher average fitness values (Koza 1992). The evolution starts with a population of randomly generated individuals, i.e., a group of candidate solutions. The population in each iteration is called a "generation." In each generation, the fitness of each individual in the population is evaluated. The fitter individuals (i.e., the better solutions) are selected (by means of a "selection" operator) from the current population, and each individual's genome is modified (with "crossover" or "mutation" operators) to form a new generation. The new generation of candidate solutions is then used in the next iteration of the algorithm. Commonly, the algorithm terminates when either a maximum number of generations has been produced, or a satisfactory fitness level has been reached for the population. A typical genetic algorithm procedure breaks down as follows (as shown in Fig. 2.2):

(1) Generate a random initial population of potential solutions, which contains several individuals;
(2) Evaluate the fitness value of each individual with a specified objective/cost function;
(3) Check whether or not the population meets the prescribed optimization criterion; if not, apply genetic operations such as selection, crossover, and mutation to the population to create a new generation of potential solutions;
(4) Repeat steps (2) and (3) until the optimization criterion is met.

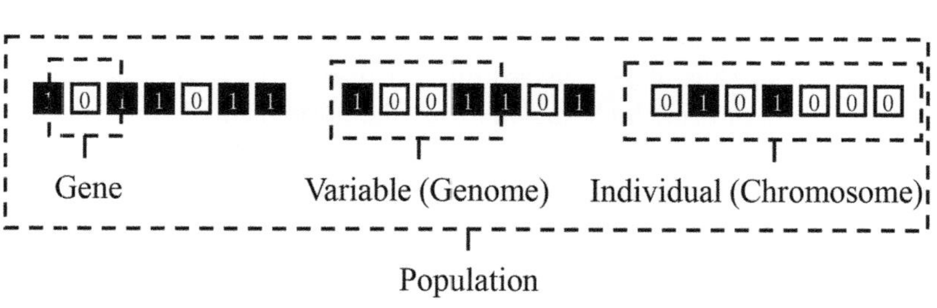

Figure 2.1 Basic constitution of a genetic algorithm

The function of each block can be explained as follows:

Initialization There will be several individuals in a population. The population size depends on the nature of the problem. Often, the initial population is generated randomly, allowing for the entire range of possible solutions (the search space). Occasionally, the solutions may be "seeded" in areas where optimal solutions are likely to be found.

Selection During each generation, a proportion of the existing population is selected to breed a new generation. Individual solutions are selected through a fitness-based process, where fitter solutions (as determined by fitness value) are typically more likely to be selected.

Crossover and mutation The next step is to generate the next-generation population of solutions from those selected through a combination of genetic operators: crossover (Fig. 2.3) and mutation (Fig. 2.4). For each new solution to be produced, a pair of "parent" solutions is selected for breeding from the pool selected previously. The "child" solution, produced by the above methods of crossover and mutation, typically shares the characteristics of its "parents." New parents are selected for each new child, and the process continues until a new population of solutions of appropriate size is generated. These processes ultimately result in a next-generation population of chromosomes that is different from the current generation. Generally, the average fitness will increase, since only the best solutions from the current generation are selected for breeding, along with a small proportion of less-fit solutions. These less-fit solutions ensure diversity within the genetic pool of the parents and therefore ensure the genetic diversity of the subsequent generation of children. It is worthwhile to fine-tune parameters such as mutation probability, crossover probability, and population size in order to find reasonable settings for the given problem class. A very small mutation rate may lead to genetic drift (which is non-ergodic in nature). A crossover rate that is too high may lead to premature convergence of the genetic algorithm. A mutation rate that is too high may lead to loss of good solutions.

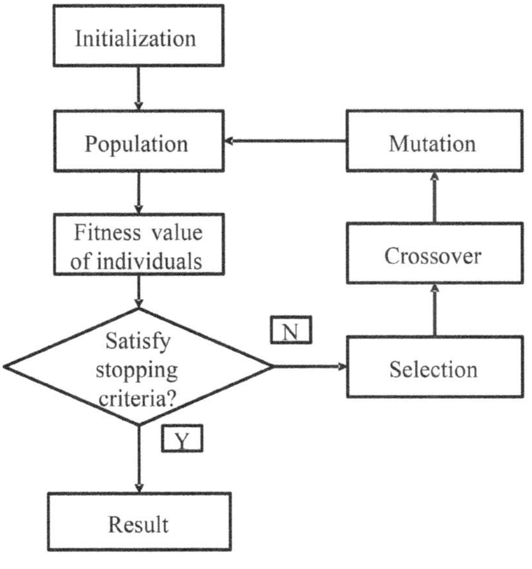

Figure 2.2 Flowchart of a genetic algorithm

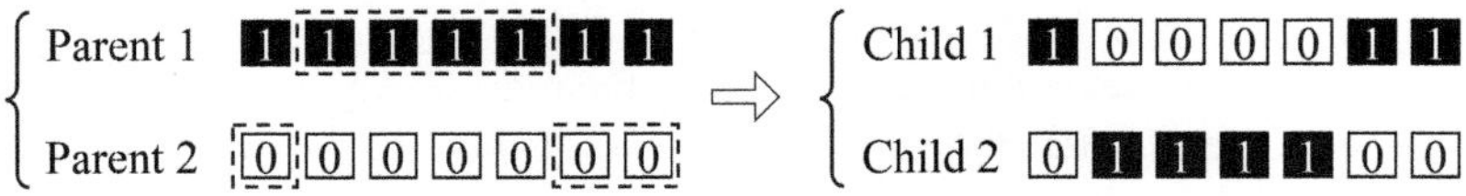

Figure 2.3 Crossover operation of a genetic algorithm

Figure 2.4 Mutation operation of a genetic algorithm

Termination The evolution operators are repeated until a termination condition is reached. Common terminating conditions are:

(1) A solution is found that satisfies minimum criteria: For inverse design cases, this terminate condition is suitable when researchers simply want to find a feasible solution for the design target.
(2) A fixed number of generations is reached: This terminate condition is suitable when computing resources are limited.
(3) An allocated budget limit (computation time/money) is reached: This terminate condition is suitable when computing resources are limited.
(4) The highest ranking solution's fitness reaches a condition such that iterations no longer produce better solutions: This terminate condition maximizes the possibility of obtaining the globally optimal result.

2.2.2 Integration of a genetic algorithm with CFD

Researchers cannot obtain fitness values for objectives such as the thermal comfort index or age of air directly from the values of design variables. They must use simulation tools, such as computational fluid dynamics (CFD) software, to obtain the velocity and temperature fields in the designed domain, and then calculate the fitness value according to the cost functions. Therefore, it is necessary to integrate a genetic algorithm (GA) with CFD to inversely design an indoor environment. This study developed a coupled computational platform that is based on GenOpt (Wetter, 2011) and FLUENT (2009) software. GenOpt is a generic optimization program that can be used to minimize an objective function computed by a simulation tool. The study coded a GA into GenOpt in order to perform the evolution procedures. FLUENT is a popular commercial CFD software package that is used to simulate the flow field in a designed domain.

Fig. 2.5 is a flowchart of the developed computational platform. For predicting and optimizing flow control conditions in confined spaces, the platform performs in the following main steps:

(1) GenOpt generates an initial population of potential solutions. Each individual of the population is composed of different flow inlet parameters, such as temperature, velocity, and direction, coded in binary.
(2) The CFD software simulates the flow field and temperature field of the designed domain with design variables of potential solutions.

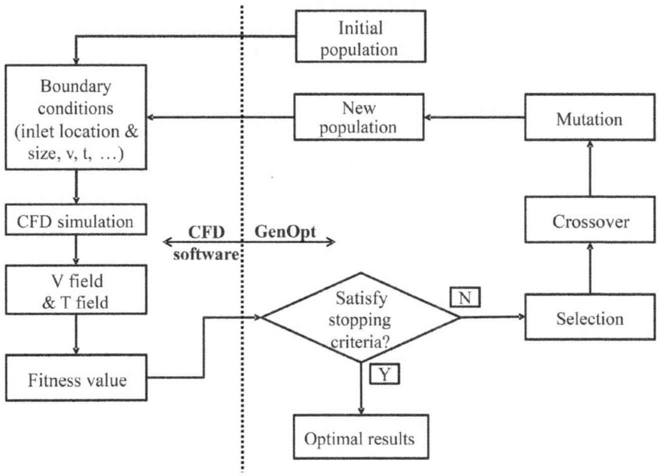

Figure 2.5 Flowchart of the GA-CFD optimization procedure

(3) Fitness values are calculated according to the prescribed cost/objective function.
(4) If the pre-set optimization criterion is met, GenOpt produces the optimal results. Otherwise, it applies genetic operations to the current population to create the next generation of the population. This generation will undergo the procedure in Steps (2) through (4) until the optimization goal is achieved.

In such an optimization, the overall computing time is determined (and thus measured) mainly by the number of CFD cases simulated. To ensure the viability of the method for real applications, the total number of CFD cases to be modeled must be reduced as much as possible. In addition to improving the genetic algorithm, during the GA-CFD simulation the calculated fitness values for individuals in populations are stored and reused if these individuals appear again in other or later populations.

2.2.3 Improvements to the integrated GA and CFD method

The use of a standard genetic algorithm for indoor environment study presents several challenges. In the coding procedure, a basic genetic algorithm usually encodes multiple variables into one long binary code. A multi-variable case requires large computing memory and other resources. When all of the variables are combined in a single long code, a standard GA crosses over the code for each individual at one or several points. This is a completely random procedure that is useful for variation but hinders the convergence from approaching the optimum (optima). A standard GA often uses the roulette-wheel selection method to choose suitable parents for creating the next population. In this method, the higher the fitness value of an individual, the greater the probability that it will be chosen as a parent to create the next generation of individuals. This process may be suitable for individuals with a large fitness-value difference, but it may also exhibit the character of random selection, lacking the benefits of evolution when the differences are marginal. Airflow and heat transfer in confined spaces involve multi-variables. Continuous changes in these variables may lead to small differences in fitness values among individuals. Indeed, many inverse prediction and optimization problems for indoor environments require subtle fine-tuning of critical flow control conditions. Hence, the standard GA needs to be improved to ensure better convergence.

This chapter modifies the coding, selection, and crossover procedure to accelerate the optimization convergence speed and reduce the computing effort, as illustrated in Fig. 2.6. In the coding step, various variables are encoded and managed independently, and thus it is easy to conduct GA operations on each variable. In the selection step, the roulette-wheel selection method has been replaced by the tournament selection method, which operates as follows:

(1) Select m individuals randomly from the current population (containing n individuals) to form a subset, where $2 \leqq m < n$;
(2) Identify the individual with the highest fitness value from the subset as one of the parents;
(3) Repeat Steps (1) and (2) until n individuals are obtained (allow repeated individuals in the selected n individuals).

The resulting n individuals form the parent population that will be used to generate a new population with crossover and mutation. In this way, individuals with a higher fitness value will be chosen even if the differences between any two individuals are small. In addition, at least m−1 individuals with the lowest fitness values will be rejected, which makes it possible to balance the optimization convergence speed and the global search capability of the algorithm.

For the crossover step, an "only son crossover method" was developed:

(1) Select an individual from the parent population randomly as the primary parent, and then choose another parent randomly;
(2) Cross over at one point randomly for each independent variable, so that the code of each variable is divided into two parts: a high-order part and a low-order part. The high-order part of the code exhibits the main features of the variable;
(3) Generate a "son" of the parents by using the high-order part of the primary parent and the low-order part of the other parent. Thus, the son inherits more features from the primary parent;
(4) Repeat Steps (1) through (3) until n son-individuals are obtained; then, after mutation, they form a new generation of the population.

The modifications above will substantially accelerate convergence because every step in the procedure contributes to the creation of a new generation with a higher average fitness value.

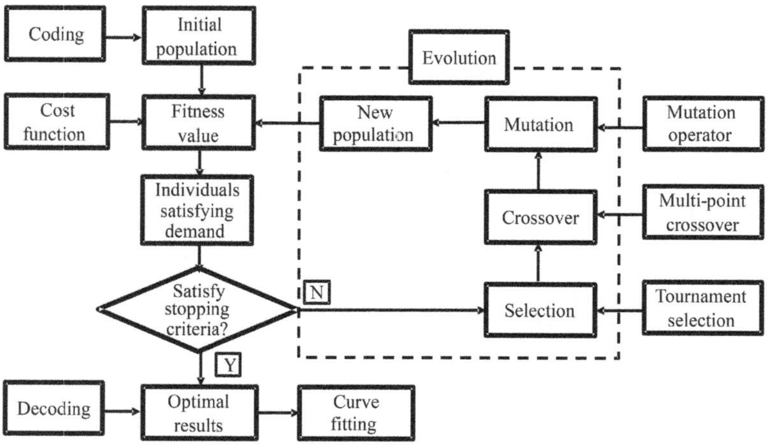

Figure 2.6 Flowchart of the modified genetic algorithm

In addition to exhibiting excellent capability in global optimization, genetic algorithms are suitable for developing regression correlations when applied to multi-solution problems (Karr et al., 1991; Gulsen et al., 1995). Regression with curve-fitting requires a sufficient number of optimal points to demonstrate the curve trend, which entails a large amount of computing time. If the expression of a curve is known in advance, the curve can be fitted with only a few points. This chapter describes an approach to effectively predicting the form of a correlation curve: an independent population is arranged to collect globally optimal individuals, and the optimization process stops when every individual of this independent population meets the requirement of the regression.

2.3 Applications of the CFD-based genetic algorithm

This section expounds on the application of the CFD-based genetic algorithm in the design of indoor environments. First, we tested and verified the ability of the integrated method to solve inverse problems for indoor environments and checked the errors. We then applied the CFD-based genetic algorithm to the optimal design of an enclosed environment.

2.3.1 Inverse identification of ventilation parameters in a built environment

2.3.1.1 Inverse identification example

The inlet conditions for a room ventilation system were predicted from simulated flow conditions in the room. We used a two-dimensional mixing convection case (Blay et al., 1992) to evaluate the performance of the developed simulation platform. This case is a widely used representative model of floor heating (natural convection) coupled with a mechanical ventilation system (forced convection) in a confined space. The case has a clear definition of geometries and boundary conditions, and thus it is ideal for assessing new methods and models for indoor environment study.

We obtained experimental results from the literature. The measurements were conducted in a laboratory chamber with dimensions of 1.04 m × 1.04 m × 0.7 m (x × y × z), equipped with a 18 mm wide inlet slot and a 24 mm wide outlet slot. There was a fairly good 2D flow at the central plate. The experiment measured the surface temperatures as $T_{roof} = T_{walls} = 15°C$, $T_{floor} = 35.5°C$, $T_{inlet} = 15°C$, and the supply air velocity as $|U|_{inlet} = 0.57$ m/s (normal to the inlet slot), as well as the temperature, x-velocity magnitude, and y-velocity magnitude at 10 points along the middle line on the central plate, as shown in Fig. 2.7. The turbulent kinetic energy and turbulent dissipation rate of the inlet were controlled at 0.00125 m^2s^{-2} and 0, respectively. We used our CFD-based genetic algorithm for the inverse prediction and optimization of the inlet conditions in this chamber.

For the inverse prediction, the wall temperatures were given from the experiment data, and the differences between the simulated and measured variables at the ten measurement points were used as the cost/objective function for the GA in order to identify the flow inlet conditions, including supply air temperature and velocity (magnitude and direction). The GA can find the optimal results only within the given range of variables, and thus the specified variable range must contain the actual inlet parameters. Reasonable ranges for the flow inlet parameters were specified, where the $|U|_{inlet}$ range was 0–1 m/s (the actual value was 0.57 m/s), the T_{inlet} range was 0–20°C (the actual value was 15°C), and the supply air angle was between 25° and −25° (0° is perpendicular to the inlet surface, counter-clockwise was positive, and the actual value was 0°). The cost function is defined as

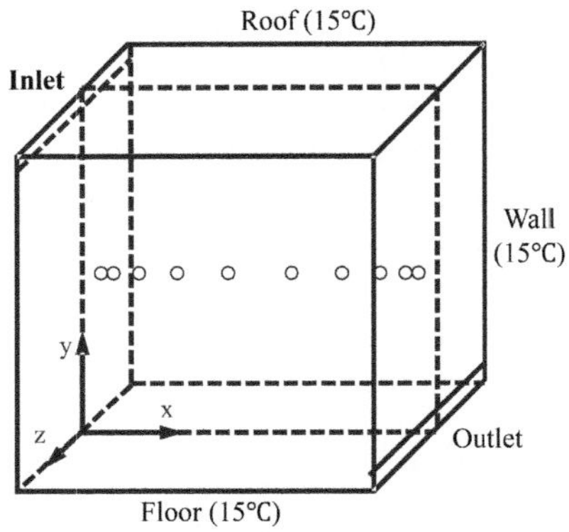

Figure 2.7 Illustration of the mixing convection case

Table 2.1 Inversely predicted results based on experimental data

Control variable	*Inversely predicted inlet variable*	*Inversely predicted result*	*Deviation*
U	$\mathbf{U}_{inlet}$	0.56 m/s	1.75%
T		0.30 m/s	47.37%
U	*T*	17.38°C	15.87%
T		16.25°C	8.31%
U	$\mathbf{U}_{inlet}$ & *T*	0.52 m/s; 5.35°C	8.77%; 64.3%
T		0.24 m/s; 15.5°C	57.89%; 3.33%
U & *T*		0.56 m/s; 16.32°C	1.75%; 8.8%

$$F = \frac{1}{3} \times \sqrt{\frac{\Sigma\left(u_i' - u_i\right)^2}{\Sigma u_i^2} + \frac{\Sigma\left(v_i' - v_i\right)^2}{\Sigma v_i^2} + \frac{\Sigma\left(T_i' - T_i\right)^2}{\Sigma T_i^2}} \left(i = 1,2,3\ldots,10\right) \tag{2.1}$$

where F is the cost function value, u_i' and v_i' are the simulated velocity at the ten points, T_i' is the simulated temperature at the ten points, and u_i, v_i, and T_i are the measured results. A smaller value of F indicates a higher fitness value, and the simulation goal was to find the solution with the minimum F.

Table 2.1 presents the inversely predicted results for seven cases, which were based on the differences between the simulated and measured results. The seven cases examined various combinations of the control variables (velocity and temperature values at the ten measurement points) and inversely predicted inlet variables (the velocity and temperature values at the inlet). As shown in Table 2.1, there were noticeable differences between the predicted and actual inlet conditions in the following cases: prediction of inlet velocity from the controlled temperature at the ten measurement points (47.37%); prediction of both inlet velocity and temperature from the controlled velocity at the 10 points (mainly for temperature, 64.3%); and prediction of both inlet velocity and temperature from the controlled temperature at the ten points (mainly for velocity, 57.89%). These differences can be understood by analyzing the governing equations (the momentum and energy equations) of the flow. In this mixing convection case, the forced convection mechanism is stronger than that of natural convection. Within the specified ranges of the inlet parameters, the indoor temperature distribution was less sensitive to supply air velocity than to supply air temperature, while the indoor velocity distribution was more sensitive to supply air velocity than to supply air temperature. Hence, predicting inlet velocity from indoor velocity was more accurate than predicting it from indoor temperature.

The differences between the simulated and measured control variables, which led to differences between the predicted and actual inlet conditions, were attributed to the errors inherent in both CFD simulations and experimental measurements. To verify the accuracy of the algorithm itself, we eliminated the errors by using the simulation results (rather than the measurements) of the 10 points with the actual inlet conditions as the reference for the inverse prediction. Table 2.2 shows the inversely predicted results under this ideal condition that eliminates the influence of CFD and experimental uncertainties. The CFD-based GA program was able to identify the actual inlet conditions precisely. It can be seen that the modified GA reduced the computing time by almost half in comparison to the standard GA.

2.3.1.2 Accuracy analysis

To further quantify the influence of CFD prediction accuracy on the inverse prediction, we used CFD results with different meshes to predict the inlet parameters, where the grid-independent CFD solution was treated as the reference and the target value. Fig. 2.8 shows a linear relationship between the CFD simulation uncertainty and the inverse prediction deviation. In general, the CFD uncertainty should be kept at 20% or less to ensure the inverse prediction deviation is below 15%.

2.3.2 Inverse design of an enclosed environment

Various optimization methods, including advanced tools and optimization algorithms, have been developed and applied to the design of optimal HVAC systems. For instance, Huang and Lam (1997) used a genetic algorithm to automatically tune proportional, integral, and derivative (PID) controllers in HVAC systems for optimal performance. Wright et al. (2002) investigated the application of a multi-objective genetic algorithm (MOGA) search method in identifying the optimum pay-off characteristic between building energy cost and occupant thermal discomfort. Fong et al. (2006) proposed a simulation-optimization approach in order to set suitable operating parameters for an HVAC system without sacrificing thermal comfort. To predict the non-uniformity of flow and heat (and thus the thermal comfort and air quality)

Table 2.2 Inversely predicted results based on simulation data

Control variable	*Inversely predicted inlet variable*	*Inversely predicted result*	*Deviation*
U	$\mathbf{U}_{inlet}$	0.57 m/s	0%
T		0.57 m/s	0%
U	*T*	15°C	0%
T		15°C	0%
U	$\mathbf{U}_{inlet}$ & *T*	0.57 m/s; 15°C	0%; 0%
T		0.57 m/s; 15°C	0%; 0%
U & *T*		0.57 m/s; 15°C	0%; 0%

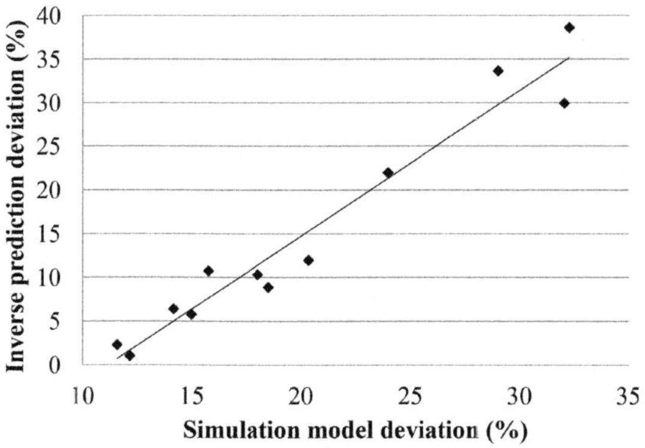

Figure 2.8 Influence of CFD uncertainty on inverse prediction error

in a space, the need for a CFD technique is inevitable. Zhou and Haghighat (2009) combined a GA, an artificial neural network, and CFD to optimize the ventilation system in an office space. Their design objective function was constructed to aggregate and weight several indices (for thermal comfort, indoor air quality, and ventilation energy usage). It is critical to consider all these key indices in a system design so as to achieve an overall comfortable and healthy environment. It is widely understood that a thermal balance calculation based on the assumption of perfect mixing cannot ensure comfort throughout the entire space when the ventilation rate is not sufficient to deliver fresh air to the breathing zones. Poor ventilation design can lead to thermal discomfort, the sensation of draft, and a lack of fresh air.

This section presents two new simulation-based air-supply system design methods for evaluating and optimizing thermal comfort (predicted mean vote – PMV), draft sensation (percentage of dissatisfied people due to draft – PD), and ventilation effectiveness (age of air) in a space. The newly developed and validated CFD-based MOGA tool was used for the system design (2013).

2.3.2.1 Inverse design of the indoor environment in a building

We again used the two-dimensional mixing convection case (Blay et al., 1992) to evaluate the proposed ventilation system design methods. The design goal was to optimize the inlet supply parameters in order to achieve good indoor environmental quality. The parameters to be optimized included inlet air temperature (T_{inlet}: 10–30°C) and inlet air velocity magnitude ($|\mathbf{U}|_{inlet}$: 0–5 m/s). Because the size of the inlet and the flow direction were fixed, a change in the velocity magnitude implied a change in ventilation rate. The optimization goals for Zone 1, as shown in Fig. 2.9, were $|PMV| < 0.1$, $PD < 15\%$, and age of air < 125 s. The mean value of the 16 points distributed throughout Zone 1 was used to represent the overall conditions in the zone.

2.3.2.1.1 DESIGN OBJECTIVE INDICES

Thermal comfort index: PMV The predicted mean vote is a widely used quantitative evaluation index for the assessment of indoor thermal comfort (Fanger, 1970). We used the PMV as the main design index to ensure thermal comfort (i.e., thermal balance) around an occupant. The PMV value is determined by:

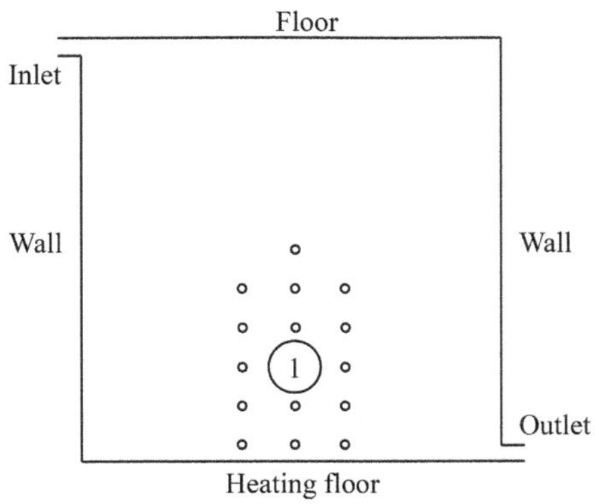

Figure 2.9 CFD model of the two-dimensional office case

$$\begin{aligned} \text{PMV} = \left(0.303e^{-0.036m} + 0.028\right) \times \{(m - W) - 0.00305 \times \left[5733 - 6.99(m - W) - p_w\right] \\ - 0.42 \times \left[(m - W) - 58.15\right] - 1.7 \times 10^{-5} \times m \times (5867 - p_a) - f_{cl} h_c (T_{cl} - T) \\ - 0.0014 \times m \times (34 - T) - 3.96 \times 10^{-8} f_{cl} \times \left[(T_{cl} + 273)^4 - (T_r + 273)^4\right]\} \end{aligned} \quad (2.2)$$

where

$$\begin{aligned} T_{cl} = 35.7 - 0.028(m - W) - I_{cl}\{3.96 \times 10^{-8} \times f_{cl} \times \left[(T_{cl} + 273)^4 - (T_r + 273)^4\right] \\ + f_{cl} h_c (T_{cl} - T)\} \end{aligned} \quad (2.3)$$

$$h_c = \begin{cases} 2.38 \times (T_{cl} - T)^{0.25} & 2.38(T_{cl} - T)^{0.25} > 12.10\sqrt{|\mathbf{U}|} \\ 12.10\sqrt{|\mathbf{U}|} & 2.38(T_{cl} - T)^{0.25} > 12.10\sqrt{|\mathbf{U}|} \end{cases} \quad (2.4)$$

$$f_{cl} = \begin{cases} 1.00 + 1.290 I_{cl} & I_{cl} \le 0.078 m^2\,°C/W \\ 1.00 + 1.290 I_{cl} & I_{cl} \le 0.078 m^2\,°C/W \end{cases} \quad (2.5)$$

where m is the metabolic heat production (Wm^{-2}); W is the external work accomplished (W); p_w is the water vapor pressure of ambient air (kPa); T stands for the temperature of ambient air (°C); T_{cl} is the temperature of clothing (°C); f_{cl} is the clothing area factor; T_r represents the mean radiation temperature; h_c is the convection heat transfer coefficient (Wm^{-2}K^{-1}); $\mathbf{U}$ is the air velocity (ms^{-1}); and I_{cl} is the thermal resistance of clothing (m^2·°C/W).

The average PMV is calculated from the absolute PMV values at several monitoring points around the body. The objective function should reach $|\text{PMV}| < 0.1$.

Draft sensitivity index: PD The percentage of people who are dissatisfied because of draft, PD, is also used as a design index for this case. Draft is an undesired local cooling of the human body caused by air movement. It has been identified as one of the most annoying factors during cooling. A person who senses draft in a space may demand a higher air temperature or even shut down the ventilation system. Fanger and Christensen (1986) attempted to establish the percentage of the population feeling draft when exposed to a given mean velocity. Furthermore, Fanger et al. (1989) investigated the effect of turbulence intensity on the sensation of draft. Turbulence intensity significantly affects draft sensation, as expressed by the following model:

$$\mathrm{PD} = (34 - T) \times (|\mathbf{U}| - 0.05)^{0.62} (0.37 \times |\mathbf{U}| \times TI + 3.14) \tag{2.6}$$

Where

$$TI = 100 \frac{|\mathbf{U}|_{sd}}{|\mathbf{U}|} \tag{2.7}$$

In the above equations, PD is the percentage of dissatisfied people due to draft, T is the temperature of ambient air (°C), $|\mathbf{U}|$ is the air velocity (ms^{-1}), and TI is the turbulence intensity (%). The variable $|\mathbf{U}|_{sd}$ is the standard deviation of the velocity measured with an omnidirectional anemometer having a 0.2 s time constant. In practice, for $|\mathbf{U}| < 0.05$ m/s, $|\mathbf{U}|$ is set at 0.05 in Eq. (2.5), and for PD > 100%, PD is set at 100%.

The design objective was to maintain an average PD of less than 15% at the monitoring points. Generally, the sensation of draft increases as inlet velocity increases, and thus the PD function limits the supply air velocity.

Air quality index: Age of air Indoor air quality is greatly influenced by ventilation effectiveness (Hedge et al., 1996). Among air quality indices such as the macro air exchange rate, air change efficiency, local air change efficiency, local ventilation index, and purging airflow rate, the age of air is mostly used to evaluate the actual "freshness" of air at certain locations in an enclosed space (Sandberg and Sjoberg, 1983). The age of air can be obtained by using a tracer gas technique or CFD simulation. CFD solves the following transport equation to calculate the age of air (Kato and Murakami, 1988):

$$\frac{\partial}{\partial x}(u\Phi_\tau) + \frac{\partial}{\partial x}(v\Phi_\tau) + \frac{\partial}{\partial x}(w\Phi_\tau) = \frac{\partial}{\partial x}\left(\Upsilon \frac{\partial \Phi_\tau}{\partial x}\right) + \frac{\partial}{\partial y}\left(\Upsilon \frac{\partial \Phi_\tau}{\partial y}\right) + \frac{\partial}{\partial y}\left(\Upsilon \frac{\partial \Phi_\tau}{\partial y}\right) + 1 \tag{2.8}$$

where u is the velocity in the x direction (m/s), v is the velocity in the y direction (m/s), w is the velocity in the z direction (m/s), Φ_τ is the age of air (s), and Υ represents the diffusion coefficient (m^2/s).

We predicted and monitored the age of air in each occupant's breathing zone. The ASHRAE Handbook (2011) requires that the ventilation rate for a person in an office be greater than 8 L/s. Assuming that a person occupies a 2 m^3 space in a dense area, such as a classroom, the greatest age of air would be about 250 s and the average age of air would be less than 125 s. We set the design objective for age of air at less than 125 s for better air quality. Generally, a decrease in inlet ventilation rate leads to an increase in age of air, and hence the use of age of air will assist in setting a lower limit for the ventilation rate. The energy use of an HVAC system is a function of supply air flow rate and temperature, which are reflected by PMV, PD, and age of air, directly or indirectly. Thus, the energy consumption of the system was not specified directly as an optimization objective, although it was considered in the selection of a final optimal solution.

2.3.2.1.2 THE CONSTRAINT METHOD

Our study employed a constraint method in order to focus on one main index while using the other indices to constrain the applicable ranges of parameters. The main index for Zone 1 was |PMV| < 0.1. Fig. 2.10 shows the solutions for different combinations of supply air temperature and velocity magnitude that were able to meet the PMV requirement in Zone 1. The supply air velocity magnitude and temperature exhibited the following logarithmic relationship:

$$T_{inlet} = 3.3202\ln(|\mathbf{U}|_{inlet}) + 22.344 \tag{2.9}$$

Analysis of Eq. (2.2) reveals a similar relationship between T and $|\mathbf{U}|$. Assuming that $m = 65\ \mathrm{Wm^{-2}}$ (the general value for low labor intensity), $W = 0$ W (for a sedentary person who is not performing mechanical work), $p_w = 10$ kPa, $I_{cl} = 0.1\ \mathrm{m^2 \cdot {}^\circ C/W}$, and T_r equals the air temperature near the walls, then, when PMV is set to zero, the following logarithmic relationship (as shown in Fig. 2.11) is observed from Eq. (2.2):

$$T = 4.5797 \times \ln(|\mathbf{U}|) + 28.446 \tag{2.10}$$

Eqs. (2.9) and (2.10) both have the same form:

$$y = A \times \ln(x) + B \tag{2.11}$$

This analysis shows that the inherent mathematical relationships among the parameters in the cost function can affect (and even determine) the relationships among the parameters at the boundary. If the correlation format of the input variables can be pre-determined (e.g., Eq. (2.11)), then, theoretically, only a few points (e.g., two points – two optimal cases – for this case) are needed to determine the coefficients (e.g., A and B for Eq. (2.11)) so that the entire pool of potential optimal cases (e.g., the curve in Fig. 2.10) can be determined. This can significantly reduce the optimization simulation effort in the CFD-GA method.

For all the points identified in Fig. 2.10, we calculated and recorded the PD and age of air values in Zone 1. The results indicate that both PD and age of air were strongly related to inlet flow velocity, as shown in Figs. 2.12 and 2.13, respectively, but were less dependent on supply air temperature.

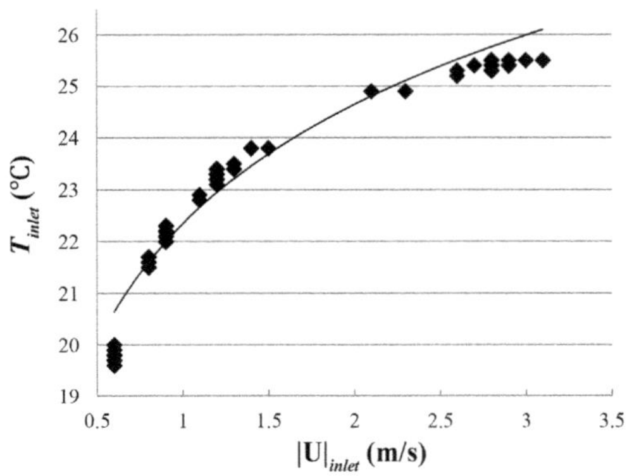

Figure 2.10 Logarithmic relationship between inlet velocity and temperature for |PMV| < 0.1 in Zone 1

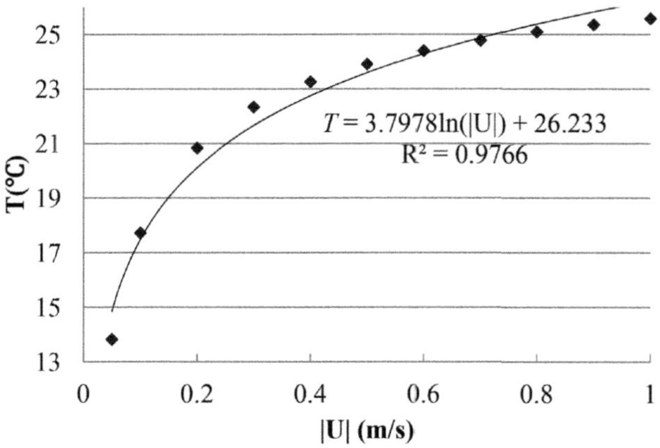

Figure 2.11 Logarithmic relationship between inlet velocity and temperature for PMV = 0 from the PMV equation (Eq. (2.2))

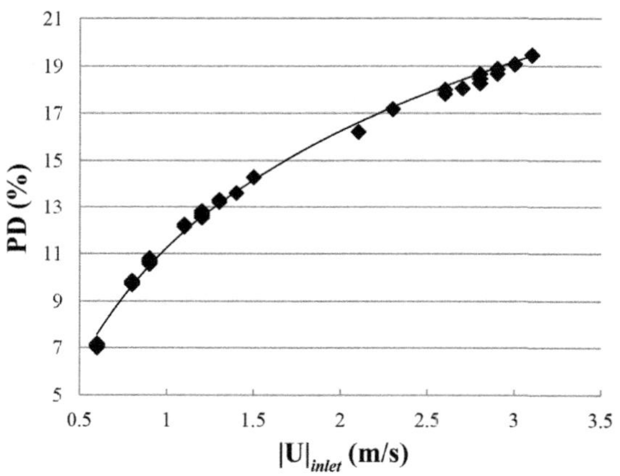

Figure 2.12 Relationship between inlet flow velocity and PD in Zone 1

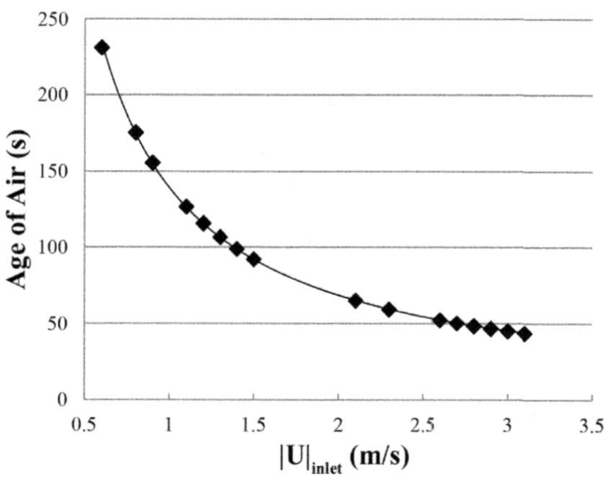

Figure 2.13 Relationship between inlet flow velocity and age of air in Zone 1

The regression equation of the curve in Fig. 2.12 is:

$$PD(\%) = 7.2169 \times \ln(|\mathbf{U}|_{inlet}) + 11.256 \tag{2.12}$$

And the regression equation of the curve in Fig. 2.13 is:

$$\text{Age of Air} = 139.09 \times |\mathbf{U}|_{inlet}^{-1.02} \tag{2.13}$$

For PD = 15%, the inlet flow velocity is 1.67 m/s according to Eq. 2.12, and the age of air is 82 s according to Eq. 2.13. This result provides the upper limit of the supply velocity. For age of air = 125 s, the inlet flow velocity is 1.11 m/s according to Eq. 2.13, and the PD is 12% according to Eq. 2.12. This result provides the lower limit of the supply velocity. Applying both constraints to the curve in Fig. 2.10 produces the range of input variables that can meet all three objective indices, as indicated by the red line between P1 (1.11 m/s, 22.69°C) and P2 (1.67 m/s, 24.05°C) in Fig. 2.14.

2.3.2.1.3 THE OPTIMIZATION METHOD

The multi-objective GA method, along with CFD, was applied to optimize the design using the three indices simultaneously. We essentially performed three CFD-GA simulations, using the three objective functions (PMV, PD, and age of air, respectively). We then plotted the applicable ranges of the supply parameters from each simulation on the same figure to determine the final optimal cases that met all the design criteria.

As with the PMV results in Fig. 2.10, the PD simulation identified a set of possible solutions with different combinations of supply air temperature and velocity magnitude that met the PD requirement in Zone 1 (i.e., that the average PD of Zone 1 was equal to 15%). In Fig. 2.15, every point on the upper left side of the curve satisfies the requirement that PD < 15% in Zone 1. This curve thus provides a boundary of acceptable supply air velocity and temperature values for PD consideration.

Following a similar approach to that of the earlier PMV simulation, we used a pre-defined expression between air velocity and temperature to reduce the number of cases required to produce the PD curve. This expression was obtained by analyzing the PD equation (Eq. (2.9)) as shown in Fig. 2.16, where Tu = 50 and PD = 15% (the specified upper limit). Here the trend is similar to that in Fig. 2.15.

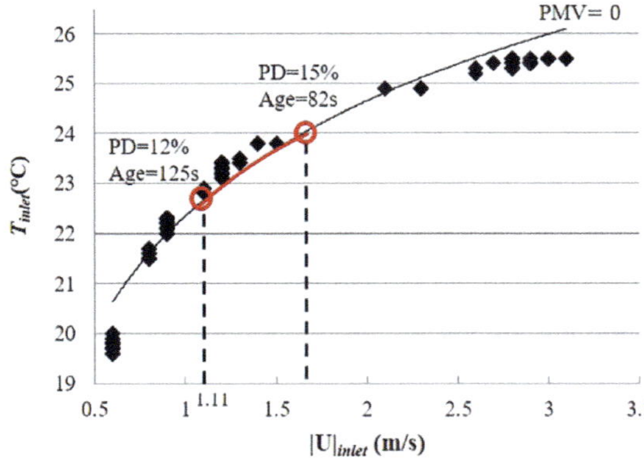

Figure 2.14 Optimal cases identified by the constraint method

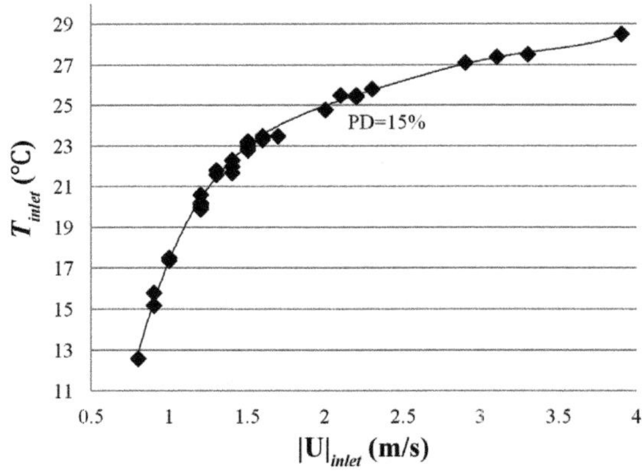

Figure 2.15 Correlation between supply air velocity and temperature for PD = 15% in Zone 1

A sixth-order polynomial can be used to fit the curve in Fig. 2.16:

$$T = -3368.9|\mathbf{U}|^6 + 10572|\mathbf{U}|^5 - 13495|\mathbf{U}|^4 + 8987|\mathbf{U}|^3 - 3317.5|\mathbf{U}|^2 + 658.94|\mathbf{U}| - 25.701 \quad (2.14)$$

Prediction of the curve in Fig. 2.15 using the pre-defined six-order polynomial expression, therefore, requires only six cases (points) to identify the six coefficients below:

$$T_{inlet} = -0.0733|\mathbf{U}|_{inlet}^{6} + 1.403|\mathbf{U}|_{inlet}^{5} - 10.993|\mathbf{U}|_{inlet}^{4} + 45.156|\mathbf{U}|_{inlet}^{3} - 103.05|\mathbf{U}|_{inlet}^{2} + 126.69|\mathbf{U}|_{inlet} - 41.753 \quad (2.15)$$

The simulation results do not reveal an explicit relationship between the age of air and supply air temperature. Rather, the age of air was influenced primarily by the supply air flow rate (i.e., velocity), as presented in Fig. 2.17. The supply flow velocity needed to be greater than 1.15 m/s to ensure an average age of air of less than 125 s in Zone 1.

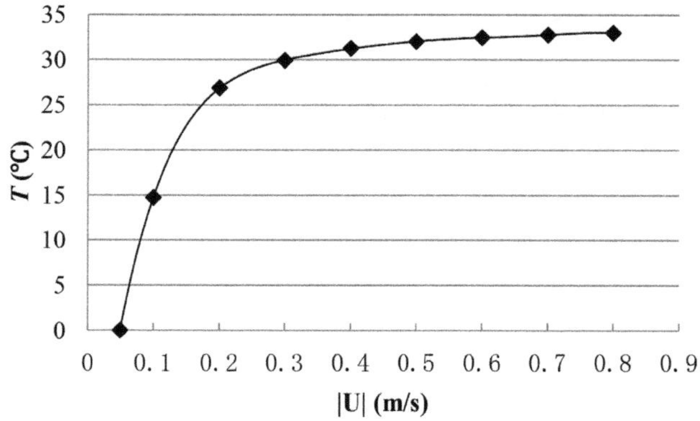

Figure 2.16 Correlation between air velocity and temperature for PD = 15% from the PD equation (Eq. (2.6))

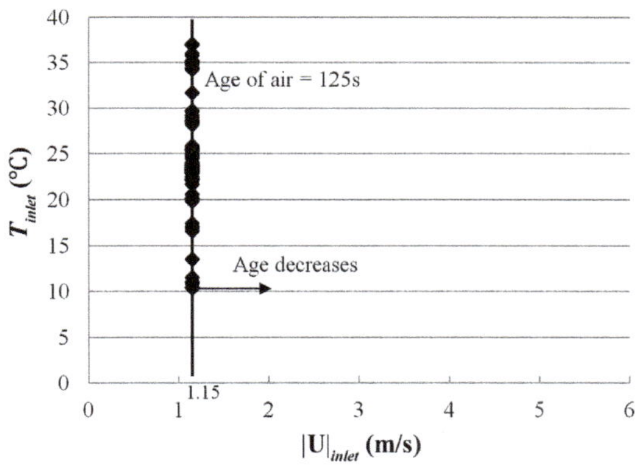

Figure 2.17 Correlation between supply air velocity and temperature for age of air = 125 s in Zone 1

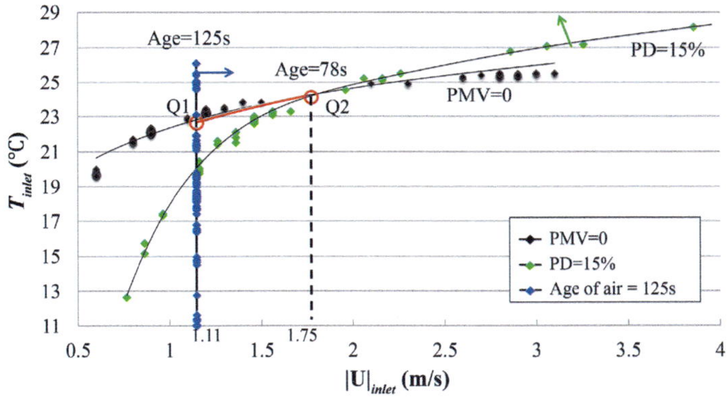

Figure 2.18 Optimal cases identified by the multi-objective optimization method

To identify the optimal design parameters for achieving multiple design objectives, the three optimal curves from Figs 2.15 through 2.17 were then combined into one figure. As shown in Fig. 2.18, all points on the curve of PMV = 0 ensure thermal comfort in Zone 1. The curve representing age of air = 125 s intersects with the curve of PMV = 0 at point P1 (i.e., 1.11 m/s, 22.69°C). This point represents the minimum supply air velocity; in other words, when the supply velocity is greater than 1.11 m/s, the average age of air in Zone 1 will be below 125 s. The points located on the left upper part of the PD = 15% curve can deliver a low draft sensation in Zone 1. Hence, the intersection point P2 (1.75 m/s, 24.2°C) of the PMV = 0 curve and the PD = 15% curve provides the upper limit of the acceptable supply air velocity. Only the points on the red line in Fig. 2.17 are able to meet all three design criteria, and they exhibit a very similar but slightly larger range than those in Fig. 2.14.

2.3.2.1.4 DISCUSSION

Both the constraint and optimization methods noticeably reduced the calculation effort. However, these methods cannot be applied to all conditions. The methods are on the basis of the understanding that "the inherent mathematical relationships among the parameters in the cost function can affect the relationships among the parameters at the boundary" (Xue et al., 2013). Some design variables, such as inlet flow direction, inlet size, and location, etc., may not have explicit corresponding parameters in the cost function. Furthermore, for some ventilation systems, such as displacement ventilation, the inlet flow velocity is so small that it may have little influence on the cost function. The curve-fitting method cannot be used to determine the intervals of the design variables in these situations. At the same time, it is not always possible to find solutions that satisfy all the design objectives, which means that neither the constraint method nor the optimization method would produce a result that satisfies multiple design objectives. In such circumstances, researchers must use an integrated method to search for optimal results. The case of an office with displacement ventilation is presented here as an example.

Fig. 2.19(a) shows the configuration of a small, two-person office that was used in an experiment by Yuan et al. (1999). A perforated displacement diffuser, 0.5 m wide and 1.1 m high, was placed in the middle of the right-side wall near the floor. The effective area ratio was 10%. The exhaust, 0.43 m × 0.43 m, was at the center of the ceiling. Occupants in the test room were simulated by two boxes, 0.4 m long, 0.35 m wide, and 1.1 m high, each heated by three, 25 W light bulbs. The measured surface temperature was between 28 and 30°C. Two personal computers were used to generate heat. One computer generated 108 W and the other 173 W. Six, 34 W fluorescent lamps provided overhead lighting. In addition, two tables and two file cabinets were placed in the room. The ventilation rate was 4 ACH, corresponding to a face velocity of 0.09 m/s at the diffuser. The supply air temperature was controlled at 17°C. The surface temperature of the window was in the range of 27.3–28.1°C, and the surface temperature of the movable wall ranged from 24.2 to 26.6°C. The surface temperatures of the other walls ranged from 23.3 to 26.0°C. The study simulated the flow field of the office and determined that the average thermal comfort value (the average PMV for the surrounding surface at a distance of 10 cm from the human body) of the occupants was 0.2, as shown in Fig. 2.19(b).

The parameters to be optimized included inlet air temperature (T_{inlet}: 15–25°C) and inlet flow direction (angle between x axis and flow direction: 0–45)°. The size of the inlet and the ventilation rate (4 ACH) were fixed. The design domains were the following: for thermal comfort, the surrounding surface at a distance of 10 cm from an occupant's body; for age of air, the surrounding surface at a distance of 5 cm from the head; and for draft sensitivity, the surrounding surface at a distance of 5 cm from the head and feet. The optimization goals were the average thermal comfort around occupants' bodies ($\overline{|PMV|} < 0.1$); average draft sensitivity around occupants' head and feet ($\overline{PD} < 20\%$); and average age of air around occupants' heads ($\overline{Age} < 125$ s).

The method did not produce any results that satisfied all the design objectives. The office utilized a displacement ventilation system, which resulted in the situation that the smallest age of air for the design domain was 227 s as shown in Fig. 2.20(a). However, every solution that was tried satisfied the design objective of $\overline{PD} < 20\%$. Fig. 2.20(b) shows the best result for draft sensitivity. These results indicate that researchers must set suitable design objectives for different cases. Otherwise, the method will not produce viable solutions.

When just considering the thermal comfort, the CFD-based genetic algorithm found several solutions that satisfied the design objective of $\overline{|PMV|} < 0.1$ after more than 200 solutions had been tested. The variable ranges were 19.5 to 26°C for temperature, and any direction for inlet flow. The best ventilation condition was 21.5°C for temperature and (1.0, 0, −0.8) for direction, and the corresponding $\overline{|PMV|}$ value was 0.098. The optimization procedure for thermal comfort is shown in Fig. 2.21. The $\overline{|PMV|}$ value was related only to inlet flow temperature,

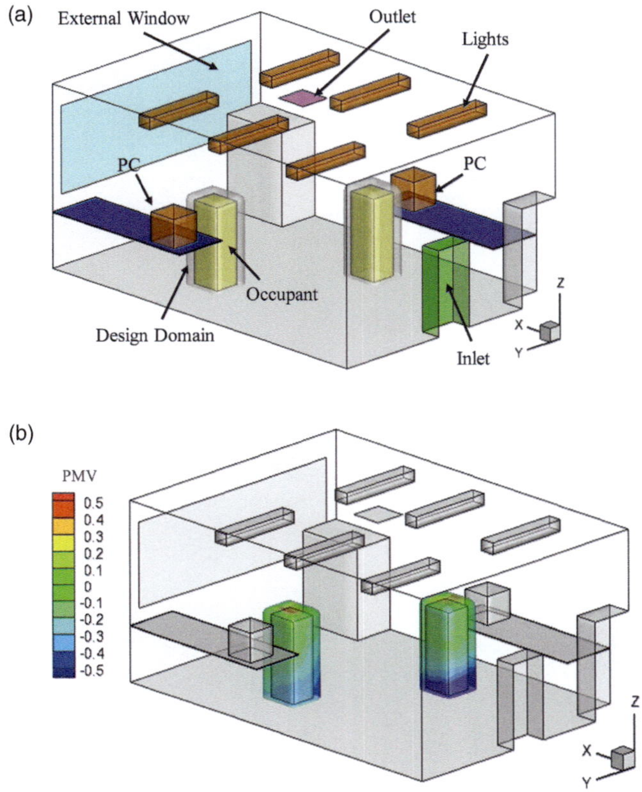

Figure 2.19 A small, two-person office used for inverse design of the air-supply parameters: (a) overview of the office model; (b) PMV value contour around the occupants

and there was no obvious relationship between inlet flow temperature and direction. In such a situation, researchers can only use a genetic algorithm to search for optimal solutions. The required calculation effort here would be much larger than when a GA is used for curve fitting.

2.3.2.2 Inverse design of an aircraft cabin environment

The above cases are relatively simple. To demonstrate the use of a genetic algorithm method to inversely design a complex enclosed environment, we consider a commercial aircraft cabin. An aircraft cabin has a typical public-transport environment: a confined space with passengers crowded together and a generally high ventilation rate. Poor ventilation of such an environment would lead to poor indoor air quality and a low level of thermal comfort. One of the problems most often cited in long-haul aircraft cabins is overheating of the head with simultaneous overcooling of the feet. This study used the integrated method to inversely design the ventilation system of an aircraft cabin in order to provide a comfortable and safe environment for passengers.

2.3.2.2.1 DESIGN OBJECTIVE INDICES

Thermal comfort index: PMVc The pressure in an aircraft cabin is about 20% lower than atmospheric pressure. The pressure difference reduces heat loss from the human body through

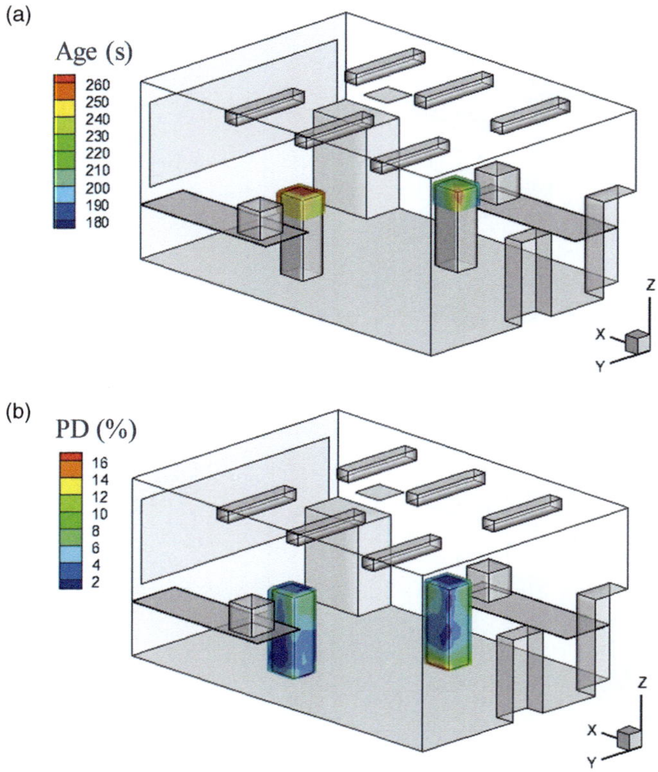

Figure 2.20 Best results for design indices: (a) age of air; (b) draft sensitivity

convection and increases heat loss through evaporation (Cui et al., 2014). Cui and Zhu (2015) developed a modified predicted mean vote for air cabins, PMVc. The value of PMVc can be determined from the PMV values for the summer and winter seasons, as follows:

$$\text{PMVc}(\text{summer}) = -0.0758\text{PMV}^2 - 0.6757\text{PMV} - 0.1262 \quad (2.16)$$

$$\text{PMVc}(\text{winter}) = -0.0696\text{PMV}^2 + 0.6906\text{PMV} - 0.1369 \quad (2.17)$$

Here, PMV is the predicted mean vote value calculated by Eq. 2.2.

Draft sensitivity index: PD or limited velocity Researchers can evaluate draft sensitivity in an aircraft cabin environment with the use of PD (Eq. (2.6)). Airworthiness standards (ASHRAE 2011) address draft sensitivity by limiting the velocity around the head (less than 0.2 m/s) and feet (less than 0.35 m/s). We used these two indices separately for the inverse design of two different aircraft cabin environments.

Air quality index: Age of air or ventilation rate Although the age of air (Eq. (2.8)) can be used as the index of total air quality, airworthiness standards (ASHRAE 2011) usually guarantee the air quality by setting the lowest ventilation rate at 9.4 L/(s·person) so that passengers receive enough oxygen.

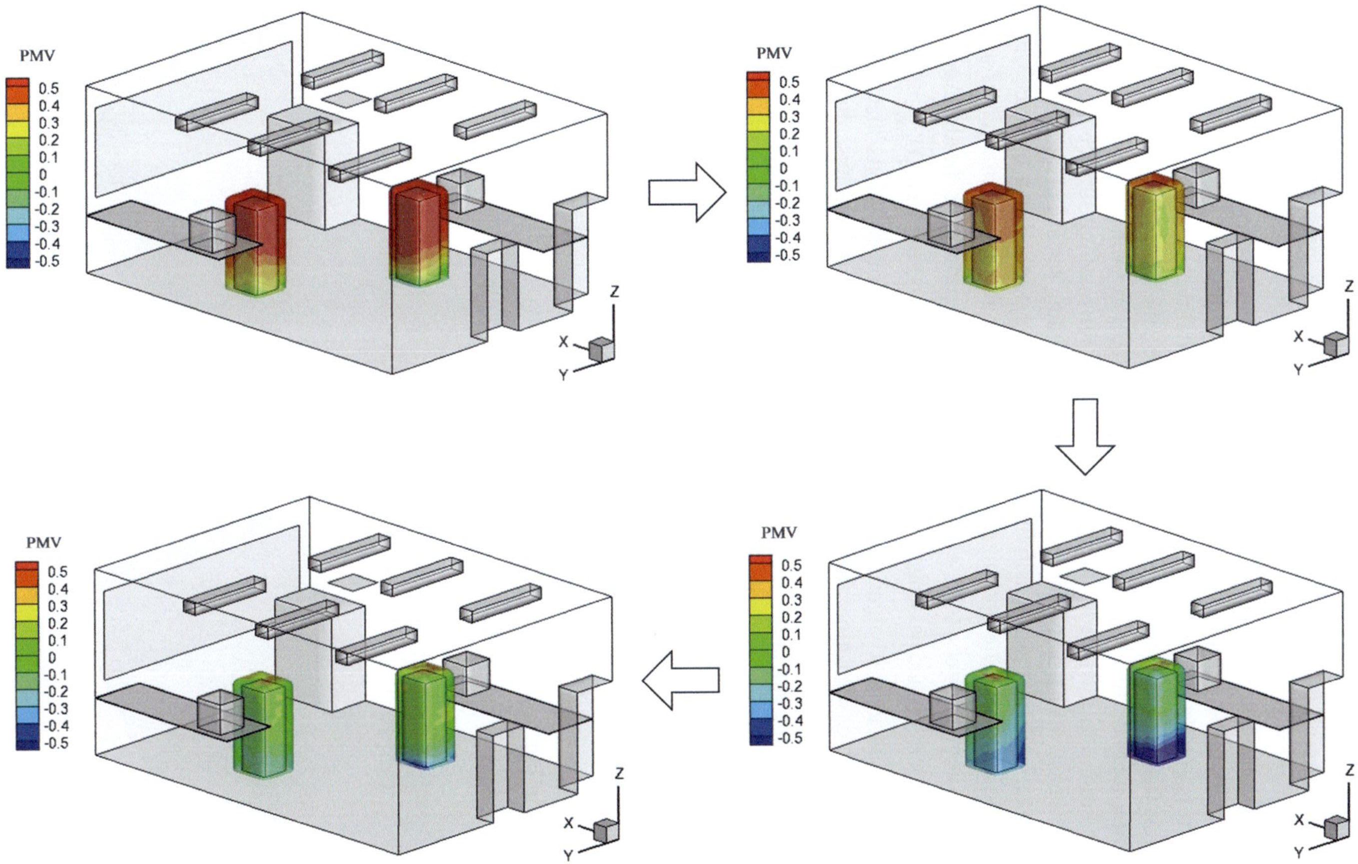

Figure 2.21 Optimization procedure for thermal comfort around occupants

2.3.2.2.2 INVERSE DESIGN OF AN AIRCRAFT CABIN ENVIRONMENT UNDER THE LIMITATION OF THE AIRWORTHINESS STANDARD

Aircraft must satisfy airworthiness standards in order to carry passengers, and we inversely designed the aircraft cabin environment to meet these standards. Fig. 2.22 is a model of half of a five-row aircraft cabin. The ventilation rate was 9.4 L/(s·person). The metabolic rate of each passenger was 75 W/m^2, the clothing insulation value was 0.57 clo, the mechanical work production of the passengers was 0 W/m^2, and the water vapor pressure was 228 Pa. The parameters to be optimized were the inlet air temperature (T_{inlet}: 16–26°C); the inlet flow direction (angle between x-z surface and flow direction: −45° to 45°, where "−" of "−45" refers to the direction point below the x-z surface); the size of the inlet (0.01−0.03 m); and the inlet location (on Wall 2, at a distance of 0-0.36 m from the ceiling). The design domains were the following: for thermal comfort, the surrounding surface at a distance of 0.1 m from a passenger's body; for age of air, the surrounding surface at a distance of 0.05 m from the head; and for draft sensitivity, the surrounding surface at a distance of 0.05 m from the head and feet. The optimization goals were the average thermal comfort around passengers' bodies ($\overline{|PMVc|} < 0.3$); average age of air around passengers' heads ($\overline{Age} < 125$ s); and average velocity around passengers' heads ($\overline{|\mathbf{U}|}_{head} < 0.2$ m/s) and feet ($\overline{|\mathbf{U}|}_{feet} < 0.35$ m/s).

The CFD-based genetic algorithm was able to find several solutions that satisfied all the design objectives after more than 200 solutions had been tried. Because the design variables did not exhibit any relationships with one another, the curve-fitting method could not be used to reduce the calculation effort. We obtained the following ranges of variables that satisfied the objectives according to the multiple solutions: 23°C to 26°C for the temperature; 10° to 45° for the inlet flow direction; and any size and location for the inlet geometry parameters. The best ventilation conditions were 25°C for temperature; (1.0, −0.4, 0) for direction; 0.03 m for inlet size; and 0.24 m down from the ceiling on Wall 2 for inlet location. The corresponding air quality parameters were, respectively, $\overline{|PMVc|} = 0.091$, $\overline{Age} = 109$ s, $\overline{|\mathbf{U}|}_{head} < 0.125$ m/s, and $\overline{|\mathbf{U}|}_{feet} < 0.091$ m/s. The optimization procedure for thermal comfort is illustrated in Fig. 2.23.

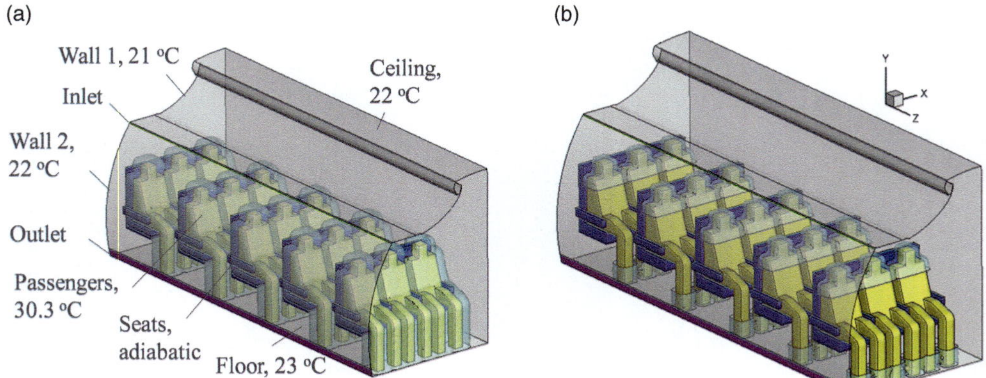

Figure 2.22 Half section of an aircraft cabin with five rows: (a) overview of the cabin model and the surrounding surfaces at a distance of 0.1 m from the passengers' bodies; (b) overview of the head and foot zones

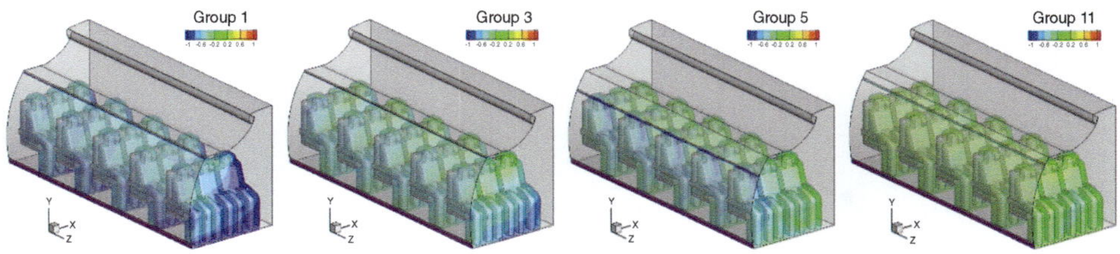

Figure 2.23 Optimization procedure for thermal comfort around occupants

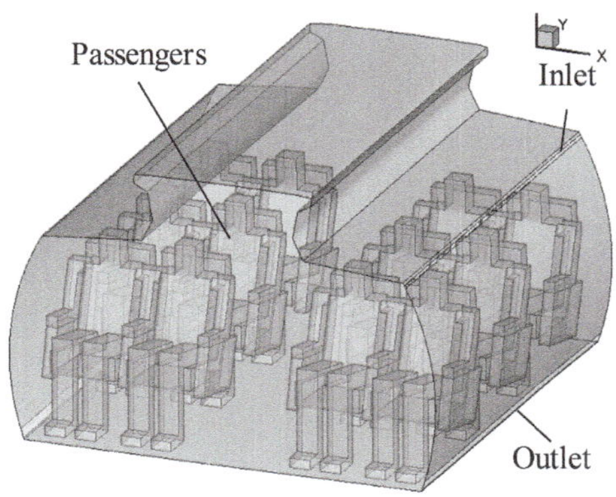

Figure 2.24 Three-row cabin model

2.3.2.2.3 THE CONSTRAINT METHOD

According to the results for the five-row cabin, the inlet size and location had little effect on the design variables, and the influence of direction was limited. Because the ventilation rate and inlet flow temperature in an aircraft cabin may exhibit a similar relationship to that shown in Fig. 2.18, we next set up an aircraft cabin design case with fixed inlet flow direction, inlet location and size, but variable velocity and temperature. The inverse design process used the constraint and optimization methods to identify the inlet parameters.

The cabin model is shown in Fig. 2.24. The cabin had three rows of seats, each occupied by a heated dummy at 75 W. The clothing insulation value was 0.57 clo, the mechanical work production of the passengers was 0 W/m^2, and the water vapor pressure was 228 Pa. During a flight in the summer, the outside surface temperature of the aircraft was assumed to be −46.8°C. The thickness of the thermal insulation material in the aircraft wall was 0.0762 m, with a thermal resistance of 0.035 Wm^{-1}K^{-1}. The heat transfer coefficient at the exterior surface of the cabin was about 100 Wm^{-2}K^{-1}. We set the inlet flow velocity at 0.6–2.7 m/s (with a corresponding ventilation rate of 6–11 h^{-1}) and the inlet air temperature at 11–25°C. The goal was to find the optimal inlet parameters for creating a comfortable and healthful environment around the occupants (i.e., the dummies). The same design objective functions and parameter settings used in the office case (shown in Fig. 2.9) were adopted for this cabin. Approximately 70 trials were required for convergence of the constraint method. Fig. 2.25 shows the optimal PMV = 0 curve in the three-row aircraft cabin. The predicted relationships between inlet flow velocity and the PD and age of air in the monitored zone are presented in Figs. 2.26 and 2.27, respectively.

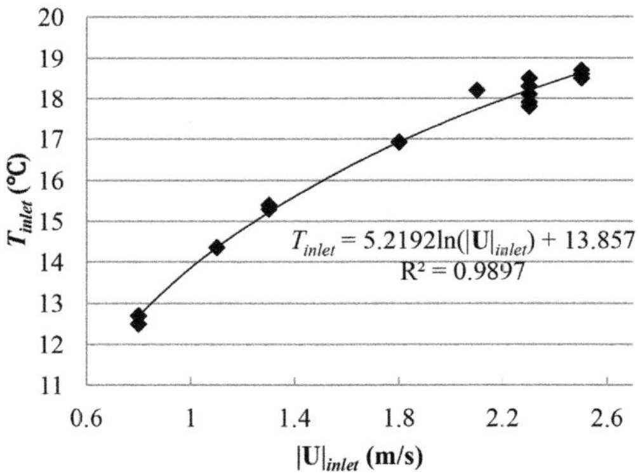

Figure 2.25 Curve of PMV = 0 in the three-row aircraft cabin

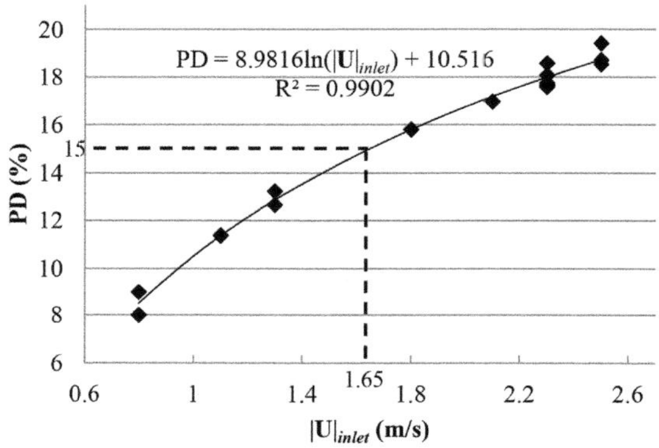

Figure 2.26 Relationship between PD and inlet flow velocity in the three-row aircraft cabin

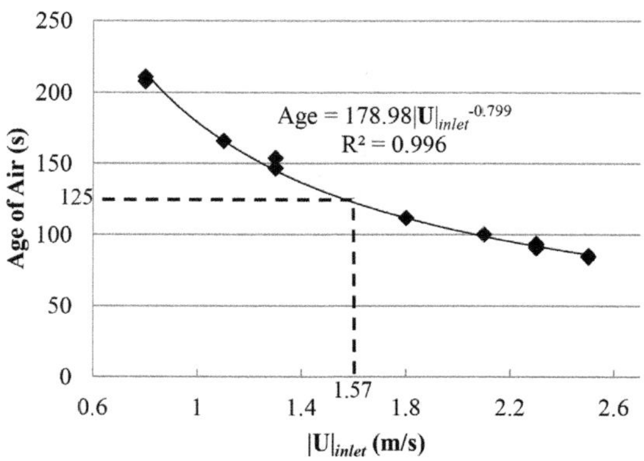

Figure 2.27 Relationship between age of air and inlet flow velocity in the three-row aircraft cabin

The regression equations for the three curves are provided in the figures. Following the procedure of the constraint method, the study identified the optimal inlet flow parameters, as shown on the red line between lower limit P1 (1.57 m/s, 16.2°C) and upper limit P2 (1.65 m/s, 16.5°C) in Fig. 2.28.

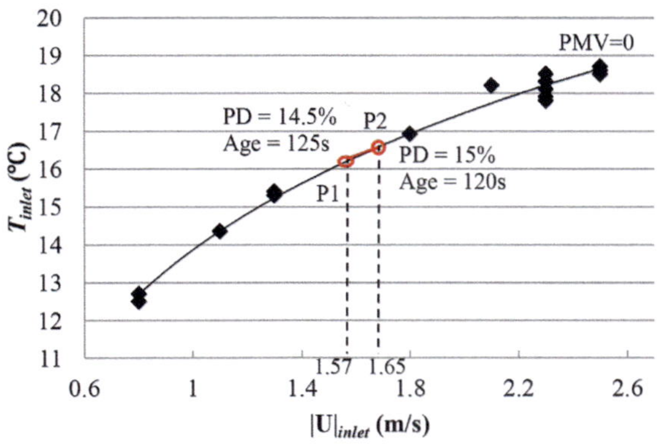

Figure 2.28 Optimal inlet parameters in the three-row aircraft cabin according to the constraint method

2.3.2.2.4 THE OPTIMIZATION METHOD

Next, we employed the optimization method to inversely design the air conditions in the three-row cabin. Figs. 2.29 through 2.31 present the optimal curves for PMV = 0, PD = 15% and age of air = 125 s, using the multi-objective optimization method with a total of approximately 100 trials. The regression equations for these curves are again shown in the figures. The red line between lower limit Q1 (1.5 m/s, 15.9°C) and upper limit Q2 (1.72 m/s, 16.7°C) in Fig. 2.32 represents the optimal inlet solutions for the cabin. These solutions are very similar to those obtained by means of the constraint method (the red line in Fig. 2.28). We used several few points on this red line (as shown in Fig. 2.32) to double check the statuses of PMV, PD, and age of air under these supply conditions and confirmed that all of them satisfied the prescribed design objectives (thus verifying the viability and effectiveness of the developed method).

The inlet flow parameters determined by the traditional calculation method, which considers only the thermal balance and ventilation rate requirement of the whole cabin, were 1.3 m/s for supply air velocity and 19°C for supply air temperature. The age of air in the breathing area was 145 s (according to the equation in Fig. 2.27). It is clear that such parameters could produce hot, stuffy air around the passengers.

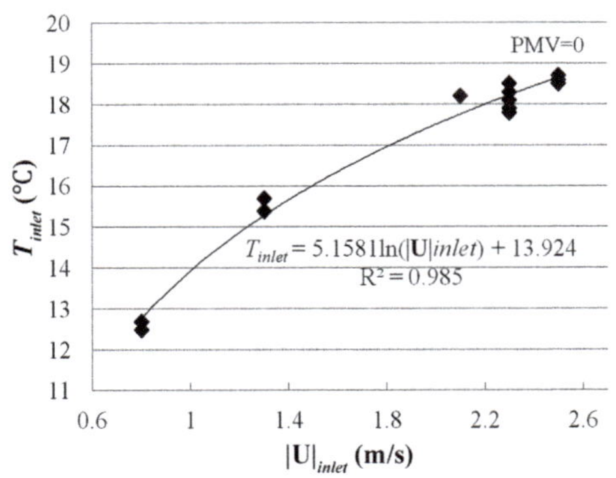

Figure 2.29 Relationship between inlet flow velocity and temperature for PMV = 0

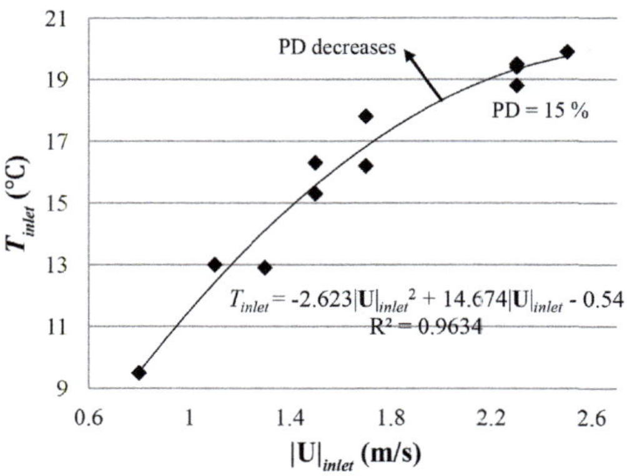

Figure 2.30 Relationship between inlet flow velocity and temperature for PD = 15%

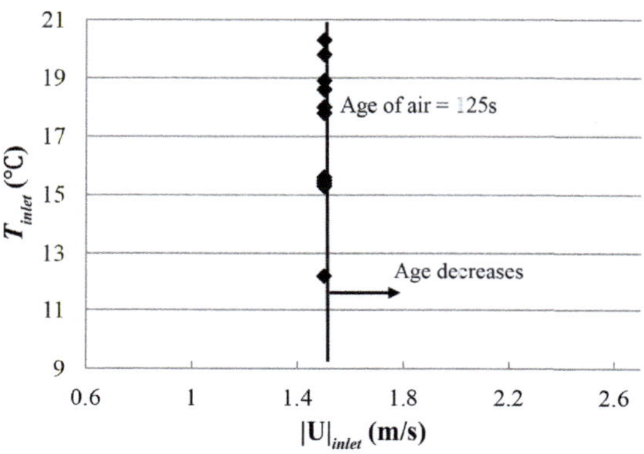

Figure 2.31 Relationship between inlet flow velocity and temperature for age of air = 125 s

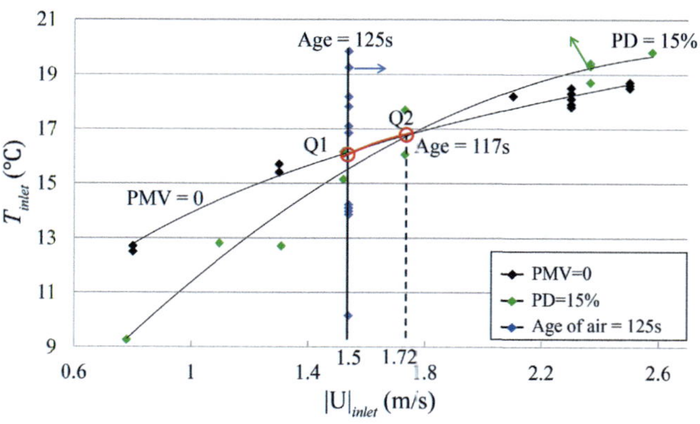

Figure 2.32 Optimal inlet parameters in the three-row aircraft cabin according to the multi-objective optimization method

2.4 Discussion

Optimization algorithms can be generally divided into gradient-free algorithms and gradient-based algorithms. Gradient-free algorithms (the representative method is GA) are developed on the basis of probability theory and they perform well in globally optimal result identification. Gradient-based algorithms (the representative algorithm is adjoint method) are based on the gradient concept, which tends to converge fast. This section compares the prediction performance of these two approaches for indoor environment inverse identification and design.

We first tested the capabilities of the two methods for inverse identification of inlet boundary conditions in the two-dimensional mixing convection case (Blay et al., 1992) as described before. The 2-D model is shown in Fig. 2.33. Experimental conditions and results were obtained from the literature as presented in Section 2.3.1.1.

The first test employed the two methods, respectively, to identify the inlet flow velocity magnitude (assumed to be unknown) from the experimental data (velocity and temperature) at the 10 indoor points. The search range for velocity was 0 to 1 m/s (the actual value being 0.57 m/s), and the accuracy was 0.01 m/s. The group size for GA was 8, the crossover coefficient was 0.8, and the mutation coefficient was 0.1. Here the cost function is defined as

$$F = \frac{1}{3} \times \sqrt{\frac{\Sigma\left(u_i^{'} - u_i\right)^2}{\Sigma u_i^2} + \frac{\Sigma\left(v_i^{'} - v_i\right)^2}{\Sigma v_i^2} + \frac{\Sigma\left(T_i^{'} - T_i\right)^2}{\Sigma T_i^2}} \quad (i = 1, 2, 3\ldots, 10) \tag{2.18}$$

where F is the cost function value, $u_i^{'}$ and $v_i^{'}$ are the simulated velocity at the ten points, T'_i is the simulated temperature at the ten points, and u_i, v_i, and T_i are the measured results. Fig. 2.34 shows that both of the methods were able to identify the actual velocity, but the GA required greater calculation effort than the adjoint method. Thus, the adjoint method performed better in simple cases. However, many indoor environments are much more complex.

The second test assumed that both the inlet flow velocity magnitude and temperature were unknown. The variable search range was 0–1 m/s for velocity and 273.15–303.15 K for temperature. The accuracy levels were 0.01 m/s and 0.5 K. The cost function was the same as

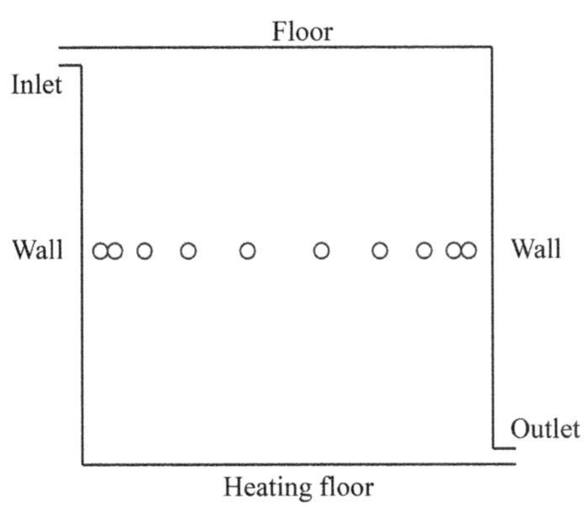

Figure 2.33 Model used for identification of inlet parameters from experimental data at ten indoor points

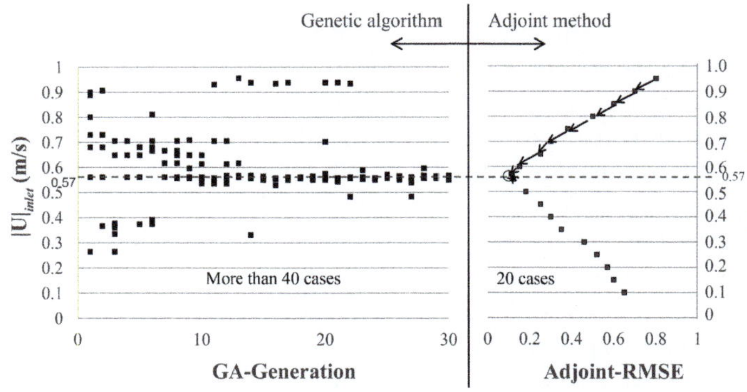

Figure 2.34 Comparison of inlet-flow velocity calculations by the GA and adjoint methods

that in Eq. (2.18). Fig. 2.35 presents the deviation for every coupled velocity and temperature. The GA was always able to find the globally optimal results (0.57 m/s and 287.15 K), while the adjoint method tended to produce different locally optimal results with different initial parameters. This case demonstrates the advantage of the GA over the adjoint method in the identification of globally optimal results.

The study further tests the capabilities of the two methods for indoor environment inverse design, using the same case as shown in Fig. 2.36, with multiple inlet conditions to be determined. The inlet flow velocity magnitude was set from 0 to 5 m/s with an accuracy of 0.5 m/s. The supply air direction could vary from 25° to −25° with an accuracy of 5°, and the air temperature could vary from 10 to 30°C with an accuracy of 1°C. PMV in Zone 1 as shown in Fig. 2.36 was adopted as the optimization cost function. The cost function is expressed as:

$$\overline{\mathrm{PMV}} = \frac{1}{16}\Sigma\left|\mathrm{PMV}_i\right| \quad (i = 1,2,3,\ldots,16) \tag{2.19}$$

where PMV_i is the PMV value at the ith monitoring point. The goal of the optimization was to minimize the value of $\overline{\mathrm{PMV}}$, and the condition for terminating the optimization simulation

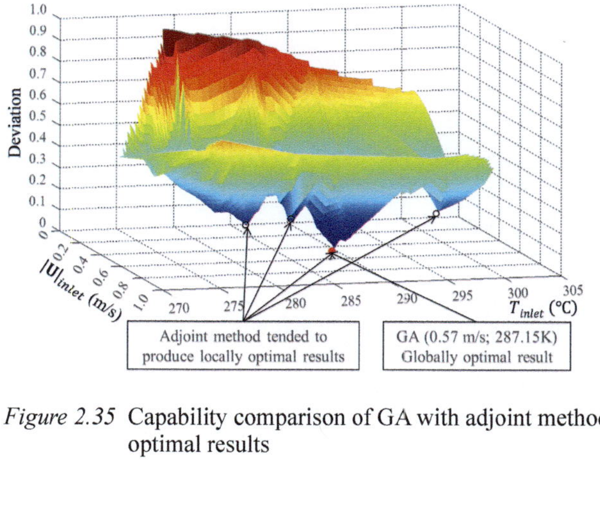

Figure 2.35 Capability comparison of GA with adjoint method in the identification of globally optimal results

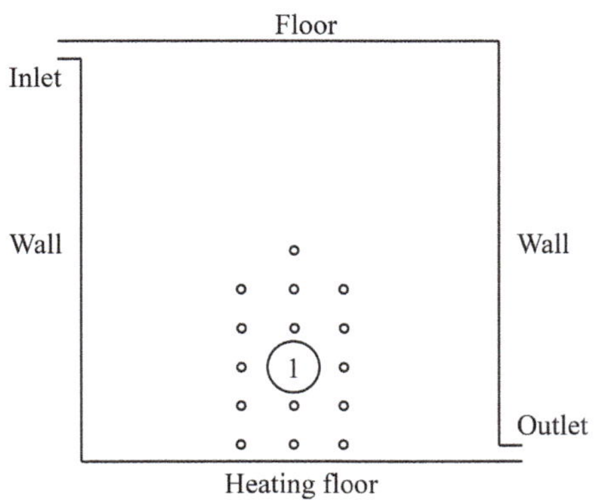

Figure 2.36 Optimization of inlet flow according to the thermal comfort criterion in Zone 1

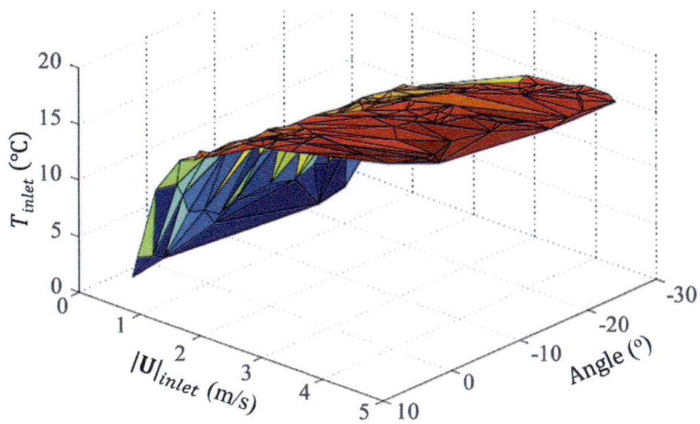

Figure 2.37 Iso-surface formed by the supply air velocity, temperature, and direction values that satisfy the PMV requirement in Zone 1

was set as $\overline{PMV} \leq 0.1$, i.e., the average PMV of Zone 1 was between −0.1 and 0.1. The GA results for different combinations of inlet flow temperature, velocity magnitude, and direction form a surface (Fig. 2.37), whereas the adjoint method can produce only one point on the surface at a time. It implies that the GA can produce multiple solutions from one search process. Results of all these three cases indicate that the genetic algorithm is more appropriate for indoor environment inverse design than the adjoint method, although the advantages of the adjoint method can be taken to accelerate the optimization. The integration of these two methods (and others) will be introduced in Chapter 6.

2.5 Summary

This chapter describes a CFD-based MOGA simulation program that can be used to predict and optimize flow control conditions satisfying multiple design indices in various confined spaces. Indoor environmental quality indicators including PMV, PD, and age of air are used to design a comfortable and healthful indoor environment. We propose two system design/optimization methods: the constraint method and the optimization method. Both methods are superior to the conventional approach.

The constraint method requires less calculation effort than the optimization method, while the multi-objective optimization approach provides a more inclusive range with fewer assumptions. Both methods provide similar results, as evidenced by the very similar final ranges of the optimal inlet parameters. As a result, the constraint method is recommended for most practical applications due to less computing cost.

Our analysis shows that the supply air temperature has a slight effect on the age of air in the cases studied, primarily because of the strong forced convection features in these cases. Therefore, for similar cases, the age of air should be optimized only as a function of supply air velocity (or flow rate). This could largely reduce the simulation effort for the multi-objective optimization method.

For inverse design cases in which the design variables are not strongly related to one another, neither the constraint method nor the optimization method may be suitable. In these situations, the CFD-based genetic algorithm may be the only suitable method for finding for optimal result(s). A significant calculation effort is required in such practices. Therefore, we propose several methods to reduce the calculation effort. These methods are described in Chapter 6

Notes

1 This chapter is partially reproduced from the paper entitled: Inverse prediction and optimization of flow control conditions for confined spaces using a CFD-based genetic algorithm, *Building and Environment*, 64, 77–84. The reprint is with permission from Elsevier.

2 This chapter is partially reproduced from the paper entitled: Inverse design methods for indoor ventilation systems using CFD-based multi-objective genetic algorithm, *Building Simulation*, 7(6), 661–669. The reprint is with permission from Springer.

References

ASHRAE. *ASHRAE Handbook – HVAC applications (SI)*. Atlanta, GA: American Society of Heating, Refrigerating and Air Conditioning Engineers. 2011.

Belegundu AD, Murthy DV, Salagame RR, Contans EW. Multi-objective optimization of laminated ceramic composites using genetic algorithms. *Proceedings of the 5th AIAA/USAF/NASA/ISSMO Symposium on Multidisciplinary Analysis and Optimization.* 1994; 1015–1022.

Blay D, Mergui S, Niculae C. Confined turbulent mixed convection in the presence of a horizontal buoyant wall jet. *Fundamentals of Mixed Convection.* 1992; 213: 65–72.

Campos F, Geremia P, Skaperdas E. Automatic optimization of automotive designs using mesh morphing and CFD. *Proceedings of the 2nd ANSA META International Congress.* 2007; 14–15.

Cieniawski SE, Eheart JW, Ranjithan S. Using genetic algorithms to solve a multiobjective groundwater monitoring problem. *Water Resources Research.* 1995; 31(2):399–409.

Cui W, Ouyang Q, Zhu Y, Hu S. Prediction model of human thermal sensation under low-air-pressure environment. *Proceedings of the 8th International Symposium on Heating, Ventilation and Air Conditioning.* 2014; 329–336.

Cui W, Zhu Y. Systematic study on passengers' thermal comfort under low-air-pressure environment in commercial aircraft cabin. *Presented in the Annual Meeting of the Center for Cabin Air Reformative Environment (973 Project), Chongqing, China.* 2015.

Elliott L, Ingham DB, Kyne AG, Mera NS, Pourkashanian M, Wilson CW. A real coded genetic algorithm for the optimization of reaction rate parameters or chemical kinetic modeling in a perfectly stirred reactor. *Genetic and Evolutionary Computation Conference.* 2002; 138–144.

Elliott L, Ingham DB, Kyne AG, Mera NS, Pourkashanian M, Wilson CW. Incorporation of physical bounds on rate parameters for reaction mechanism optimization using genetic algorithms. *Combust Science Technology.* 2003; 175(4):619–648.

Fanger P O. *Thermal comfort, analysis and applications in environmental engineering*. Copenhagen: Danish Technical Press. 1970.

Fanger PO, Christensen NK. Perception of draught in ventilated spaces. *Ergonomics.* 1986; 29(2): 215–235.

Fanger PO, Melikov AK, Hanzawa H, Ring J. Turbulence and draft. *ASHRAE Journal*. 1989; 31(4): 18–25.

FLUENT 12.1.2 for the ANSYS Release Version 12.1, 2009.

Fong KF, Hanby VI and Chow TT. HVAC system optimization for energy management by evolutionary programming. *Energy and Buildings.* 2006; 38(3):220–231.

Fonseca CM. and Fleming PJ. Multiobjective genetic algorithms. *Proceedings of the IEE colloquium on "Genetic Algorithms for Control Systems Engineering"*. 1993.

Goldberg DE. *Genetic algorithms in search, optimization, and machine learning*. Reading: Addison-Wesley. 1989.

Grefenstette JJ. Optimization of control parameters for genetic algorithms. *IEEE Transactions on Systems, Man, and Cybernetics.* 1986; 122–128.

Gulsen M, Smith AE, Tate DM. A genetic algorithm approach to curve fitting. *International Journal of Production Research*. 1995; 33 (7): 1911–1923.

Hedge A. Predicting sick building syndrome at the individual and aggregate levels. *Environment International.* 1996; 22(1): 3–19.

Holland JH. *Adaptation in natural and artificial systems: An introductory analysis with applications to biology, control, and artificial intelligence*. Ann Arbor: U Michigan Press. 1975.

Horn J, Nafpliotis N, Goldberg DE. A niched Pareto genetic algorithm for multiobjective optimization. *Proceedings of the First IEEE Conference on Evolutionary Computation.* 1994; 82–87.

Huang W, Lam HN. Using genetic algorithms to optimize controller parameters for HVAC systems. *Energy and Buildings*. 1997; 26(3):277–282.

Jameson A. Optimum aerodynamic design using CFD and control theory. *AIAA.* 1995; 1729: 124–131.

Karr CL, Stanley DA, Scheiner BJ. Genetic algorithm applied to least squares curve fitting. *U.S. Bureau of Mines Report of Investigations.* 1991; 9339.

Kato S, Lee JH. Optimization of hybrid air-conditioning system with natural ventilation by GA and CFD. *Proceedings of the 25th AIVC Conference, Ventilation and Retrofitting.* 2004; 15–17.

Kato S, Murakami S. New ventilation efficiency scales based on spatial distribution of contaminant concentration aided by numerical simulation. *ASHRAE Transactions*. 1988; 94(2): 309–30.

Kato S, Pochanart P, Hirokawa J, Kajii Y, Akimoto H, Ozaki Y, Obi K, Katsuno T, Streets DG, Minko NP. The influence of Siberian forest fires on carbon monoxide concentrations at Happo, Japan. *Atmospheric Environment*. 2002; 36(2): 385–390.

Koza JR. Genetic Programming: On The Programming of Computers by Means of Natural Selection. *Cambridge, Massachusetts Institute of Technology Press.* 1992.

Lattès R, Lions JL. *The method of quasi-reversibility: Applications to partial differential equations*. New York: American Elsevier Publishing Company. 1969.

Lu H, Yen GG. Rank-density-based multiobjective genetic algorithm and benchmark test function study. *IEEE Transactions on Evolutionary Computation.* 2003; 7(4):325–343.

Ma Z, Wang S. Supervisory and optimal control of central chiller plants using simplified adaptive models and genetic algorithm. *Applied Energy.* 2011; 88:198–211.

Mahinthakumar GK, Sayeed M. Hybrid genetic algorithm – Local search methods for solving groundwater source identification inverse problems. *Water Resources Planning and Management.* 2005; 131(1):45–57.

Malkawi AM, Srinivasan RS, Yi YK, Choudhary R. Performance-based design evolution: the use of genetic algorithms and CFD. *Proceedings of the Eighth International IBPSA Conference.* 2003; 793–798.

Muñoz AG, Ayala-Ramirez V, Alfaro-Ayala JA, Acosta BMT. Optimization of the transition piece applying genetic algorithms. *Applied Thermal Engineering.* 2011; 31(16):3214–3225.

Nassif N, Kajl S, Sabourin R. Optimization of HVAC control system strategy using two-objective genetic algorithm. *HVAC&R Research.* 2005; 11(3):459–486.

Obayashi S, Takanashi S. Genetic algorithm for aerodynamic inverse optimization problems. *Proceedings of the 1st IEE/IEEE International Conference on Genetic Algorithm in Engineering Systems: Innovations and Applications.* 1995; 7–12.

Ozkol I, Komurgoz G. Determination of the optimum geometry of the heat exchanger body via a genetic algorithm. *Numerical Heat Transfer Part A – Applications.* 2005; 48(3):283–296.

Polifke W, Geng W, Dobbeling K. Optimization of rate coefficients for simplified reaction mechanisms with genetic algorithms. *Combust Flame.* 1998; 113(1):119–135.

Queipo N, Devarakonda R, Humphrey JAC. Genetic algorithms for thermosciences research – Application to the optimized cooling of electronic components. *International Journal of Heat and Mass Transfer.* 1994; 37(6):893–908.

Sakamoto Y, Nagaiwa A, Kobayasi S, Shinozaki T. An optimization method of district heating and cooling plant operation based on genetic algorithm. *ASHRAE Transactions.* 1999; 105(2): 104–115.

Sandberg M, Sjoberg M. The use of moments for assessing air quality in ventilated rooms. *Building and Environment.* 1983; 18(4): 181–97.

Schaffer J. Multiple objective optimization with vector evaluated genetic algorithms. *Proceedings of the International Conference on Genetic Algorithm and Their Applications.* 1985.

Schaumann EJ., Balling RJ, Day K. Genetic algorithms with multiple objectives. *Proceedings of the Seventh AIAA/USAF/NASA/ISSMO Symposium on Multidisciplinary Analysis and Optimization.* 1998; 2114–2123.

Srinivas M, Patnaik LM. Adaptive probabilities of crossover and mutation in genetic algorithms. *IEEE Transactions on Systems, Man, and Cybernetics.* 1994; 24(4):656–667.

Sun Y, Wang X. Optimization of air flow field of the melt blowing slot die via numerical simulation and genetic algorithm. *Journal of Applied Polymer Science.* 2010; 115(3):1540–1545.

Wang SW, Jin X. Q. Model-based optimal control of VAV air-conditioning system using genetic algorithm. *Building and Environment.* 2000; 35:471–478.

West AC, Sherif SA. Optimization of multistage vapour compression systems using genetic algorithms. Part 1: Vapour compression system model. *International Journal of Energy Research.* 2001; 25(9):803–812.

West AC, Sherif SA. Optimization of multistage vapour compression systems using genetic algorithms. Part 2: Application of genetic algorithm and results. *International Journal of Energy Research.* 2001; 25(9):813–824.

Wetter M. GenOpt® Generic Optimization Program User Manual, Version 3.1.0. LBNL–94720. 2011.

Wetter M, Wright, J. Comparison of a generalized pattern search and a genetic algorithm optimization method. *In: Proceedings of the Eighth IBPSA Conference.* 2003; 3:1401–1408.

Wright JA, Loosemore HA, Farmani R. Optimization of building thermal design and control by multi-criterion genetic algorithm. *Energy and Buildings.* 2002; 34(9):959–972.

Xue Y, Zhai ZJ, Chen Q. Inverse prediction and optimization of flow control conditions for confined spaces using a CFD-based genetic algorithm. *Building and Environment.* 2013; 64:77–84.

Yuan X, Chen Q, Glicksman LR, Hu Y, Yang X. Measurements and computations of room airflow with displacement ventilation. *ASHRAE Transactions.* 1999; 105:340.

Zhai Z, Xue Y, Chen Q. Inverse design methods for indoor ventilation systems using CFD-based multi-objective genetic algorithm. *Building Simulation.* 2014; 7(6):661–669.

Zhou L, Haghighat F. Optimization of ventilation system design and operation in office environment, Part I: Methodology. *Building and Environment.* 2009; 44(4):651–656

Zhou L, Haghighat F. Optimization of ventilation systems in office environment, Part II: Results and discussions. *Building and Environment.* 2009; 44(4):657–665.

Zitzler E, Thiele L. Multiobjective evolutionary algorithms: A comparative case study and the strength Pareto approach. *IEEE Transactions on Evolutionary Computation.* 1999; 3(4):257–271.

3 CFD-based adjoint method[1,2,3]

This chapter demonstrates the use of a CFD-based adjoint method to inversely design an indoor environment and to identify optimal solutions that satisfy design objectives. The development of the method for the inverse design of air supply parameters, including inlet location and size and supply temperature and velocity, is discussed in detail. This study implemented the method in OpenFOAM and validated it for the identification of the air supply parameters necessary to achieve the measured air velocity and temperature in an indoor environment. Since the adjoint method can find only the local optima, calculations with different initial inlet air conditions may lead to multiple optimal solutions. When any of the air supply parameters identified as boundary conditions for forward CFD simulations were used, the computed air distribution in the design domain was the same as that used as a design objective. The computing costs did not depend on the number of design variables. Furthermore, this study aimed to design desirable thermal environments in an office and a single-aisle airliner cabin, and used the CFD-based adjoint method to find the optimal design values for the air supply inlet location, inlet size, temperature, and velocity. By setting the region around the occupants as the design domain with a minimum predicted mean vote as the indicator of thermal comfort, this study aimed to determine the corresponding air supply parameters. The results show that it is possible to find the optimal air supply parameters in fewer than 10 design cycles if the initial conditions for the design variables are provided within a reasonable range.

3.1 Introduction

In an enclosed environment, air supply parameters such as inlet location and size and supply temperature and velocity are critical in determining the air distributions. This investigation used the CFD-based adjoint method to optimize these parameters in the design of an indoor environment. The corresponding numerical algorithm, and its implementation in OpenFOAM, are also discussed.

This chapter then describes the validation of the CFD-based adjoint method for inverse design of a built environment. The validation process used known air supply inlet location, inlet size, temperature, and velocity as well as airflow fields from experiments in the existing literature. With the measured air velocity and temperature in the flow domain as the design objective, the adjoint method was used to identify the air supply inlet location, inlet size, temperature, and velocity. If the calculated parameters were the same as the known parameters, the adjoint method would be validated. This study first set only the air supply temperature and velocity as the design variables and then added the inlet location and size.

Next, to demonstrate the performance of the method with an ideal design objective, this study started with desirable air velocity and temperature distributions in a design domain. The adjoint method was employed to find the corresponding air supply inlet location and size, and the supply temperature and velocity. When these parameters had been determined as the boundary conditions for forward CFD simulations, the CFD results could be used to determine whether or not the air velocity and temperature distributions in the design domain were the same as the desired ones.

Finally, this study used the adjoint method for inverse design of two complex thermal environments: an office and an airliner cabin. For complicated cases such as these, it can be very challenging to define the design objectives, obtain accurate results, and reduce computing costs. This chapter describes our effort in these areas.

3.2 CFD-based adjoint method

The CFD-based adjoint method uses the following processes to identify the optimal design variable ξ, which represents the air supply inlet location (x_{in}, y_{in}) and size (h_{in}) and the supply temperature ($\mathbf{U}_{in}$) and velocity (T_{in}) in this study. The method creates a vector, or a 1 × 5 matrix, $\xi = (x_{in}, y_{in}, h_{in}, \mathbf{U}_{in}, T_{in})$, and any mathematical operations involving ξ are matrix operations.

3.2.1 Design objective vs. objective function

The first step is to establish the design objective in the occupied zone of a given space and construct the corresponding objective function $O(p, \mathbf{U}, T)$, which will be minimal with an optimal ξ. The objective function is actually a mathematical explanation of the design objective. One can construct various objective functions for a design objective. For example, the design objective can be the thermal comfort or air quality in the occupied zone of a room. The objective function can then be constructed to make the |PMV| or age of the air very small with appropriate $\mathbf{U} = (U_1, U_2, U_3)$ and T distributions in the occupied zone. Zhou and Haghighat (2009) have provided further examples of design objectives and objective functions. Then, using Ω to represent the design domain, the design objective is to:

$$\text{Min } O(\xi) = \int_{\Omega} f(\mathbf{U},T)d\Omega \tag{3.1}$$

subject to state equations or Navier-Stokes equations for steady-state incompressible Newtonian flow that are denoted by $\mathbf{N} = (N_1, N_2, N_3, N_4, N_5)$:

$$N_1 = \frac{\partial U_i}{\partial x_i} = 0 \tag{3.2}$$

$$N_{i,i=2,3,4} = U_j \frac{\partial U_i}{\partial x_j} + \frac{\partial p}{\partial x_i} - \frac{\partial}{\partial x_j}\left(\nu\left(\frac{\partial U_i}{\partial x_j} + \frac{\partial U_j}{\partial x_i}\right)\right) - \gamma g_i (T - T_{op}) = 0 \tag{3.3}$$

$$N_5 = U_j \frac{\partial T}{\partial x_j} - \frac{\partial}{\partial x_j}\left(\kappa \frac{\partial T}{\partial x_j}\right) = 0 \tag{3.4}$$

where ν is the effective viscosity, κ is the effective conductivity, T_{op} is the operating temperature, γ is the thermal expansion coefficient of the air, and g_i is the ith component of the gravity vector. To facilitate the mathematical operations described in the following sections, the Navier-Stokes equations can be re-written in vector form:

$$N_1 = -\nabla \cdot \mathbf{U} = 0 \tag{3.5}$$

$$(N_2, N_3, N_4)^T = (\mathbf{U} \cdot \nabla)\mathbf{U} + \nabla p - \nabla \cdot (2\nu D(\mathbf{U})) - \gamma \mathbf{g}(T - T_{op}) = 0 \tag{3.6}$$

$$N_5 = \nabla \cdot (\mathbf{U}T) - \nabla \cdot (\kappa \nabla T) = 0 \tag{3.7}$$

where $D(\mathbf{U}) = (\nabla \mathbf{U} + (\nabla \mathbf{U})^T)/2$ is the rate of strain tensor and **g** is the gravity vector.

3.2.2 Principle of the CFD-based adjoint method

The adjoint method computes the derivative of the objective function with respect to the design variables, so that the method can search for the direction that gradually minimizes the objective function. The curve in Fig. 3.1 can be used to represent the relationship between a design variable and the design objective. The adjoint method first initializes a starting point (ξ, $O(\xi)$). The method then computes the gradient and adjusts the design variable in the direction that minimizes the objective function. This procedure is repeated until the minimum has been found. It is possible, of course, that this method will identify only the local optima. However, a reasonable estimate of the ξ can speed up convergence, and thus local optima can be avoided.

With an initialized ξ, the initial $O(\mathbf{U}, T)$ calculated by solving the Navier-Stokes equations is often not the minimum value. The CFD-based adjoint method determines the derivative of the objective function over the design variables ($dO/d\xi$) in order to search for the trend in varying ξ that can lead to a smaller O. Since O is a function of **U** and T but not ξ, the gradient $dO/d\xi$ cannot be directly computed. However, because ξ determines p, **U**, and T, it indirectly determines O. Therefore, the adjoint method introduces a Lagrangian multiplier (p_a, $\mathbf{U}_a$, T_a), which stands for adjoint velocity, temperature, and pressure, respectively, to make $dO/d\xi$ computable. The adjoint variables could be regarded as purely mathematical terms, but they do have physical significance. According to Giles and Pierce (2000), one way of looking at them is that they represent the influence of an arbitrary source term on the function of interest (the Navier-Stokes equations in this case). Our method then uses an augmented objective function:

$$L = O + \int_{\omega} (p_a, \mathbf{U}_a, T_a)\mathbf{N}d\omega \tag{3.8}$$

where ω denotes the computational domain. Since $\mathbf{N} = \mathbf{0}$, L is actually equal to O in Eq. (3.8). Therefore, finding the value of ξ that minimizes O becomes finding the value of

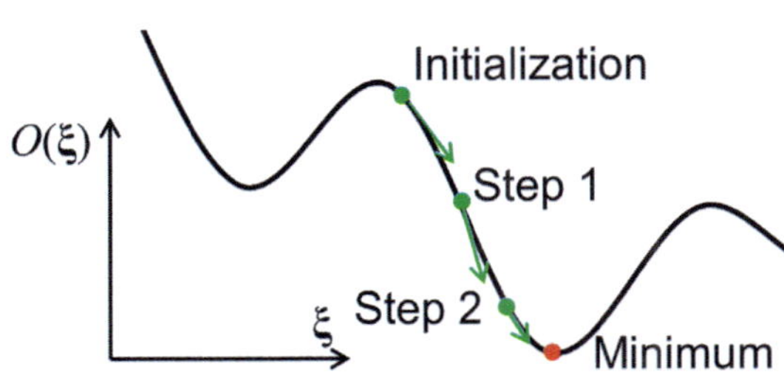

Figure 3.1 Principle of the adjoint method

ξ that minimizes the augmented objective function L. Our goal was to calculate the gradient $dL/d\xi$ instead of $dO/d\xi$. Taking into account the variation in ξ, the variation in L can be expressed as:

$$\delta L = \delta_\xi L + \delta_p L + \delta_\mathbf{U} L + \delta_T L \tag{3.9}$$

which includes the contributions from changes in ξ and the corresponding changes in (p, **U**, T). The state equations must be calculated once for each variable to satisfy $\mathbf{N} = \mathbf{0}$. To find the relationship between the variation in ξ and that in L in order to determine $dL/d\xi$, the (p_a, $\mathbf{U}_a$, T_a) are chosen to satisfy:

$$\delta_p L + \delta_\mathbf{U} L + \delta_T L = 0 \tag{3.10}$$

Eq. (3.10) eliminates the contributions of δp, $\delta\mathbf{U}$, and δT in Eq. (3.9). Then:

$$\delta L = \delta_\xi L \Rightarrow \frac{dL}{d\xi} = \frac{\partial L}{\partial L} \tag{3.11}$$

which is straightforward. Eq. (3.10) must be fulfilled for any (dp, $d\mathbf{U}$, dT) that is entailed by $d\xi$, and this can be accomplished by setting the three partial derivatives on the left-hand side at zero, which creates three independent equations:

$$\delta_p O + \int_\omega (p_a, \mathbf{U}_a, T_a)\delta_p \mathbf{N} d\omega = 0 \tag{3.12}$$

$$\delta_\mathbf{U} O + \int_\omega (p_a, \mathbf{U}_a, T_a)\delta_\mathbf{U} \mathbf{N} d\omega = 0 \tag{3.13}$$

$$\delta_T O + \int_\omega (p_a, \mathbf{U}_a, T_a)\delta_T \mathbf{N} d\omega = 0 \tag{3.14}$$

To make Eqs. (3.12), (3.13), and (3.14) computable, we first expand terms $\delta_p\mathbf{N}$, $\delta_U\mathbf{N}$, and $\delta_T\mathbf{N}$ as follows:

$$\delta_p N_1 = 0 \tag{3.15}$$

$$\delta_p (N_2, N_3, N_4)^T = \nabla \delta p \tag{3.16}$$

$$\delta_p N_5 = 0 \tag{3.17}$$

$$\delta_\mathbf{U} N_1 = -\nabla \cdot \delta \mathbf{U} \tag{3.18}$$

$$\delta_\mathbf{U} (N_2, N_3, N_4)^T = (\delta\mathbf{U}\cdot\nabla)\mathbf{U} + (\mathbf{U}\cdot\nabla)\delta\mathbf{U} - \nabla\cdot(2\nu D(\delta\mathbf{U})) \tag{3.19}$$

$$\delta_\mathbf{U} N_5 = \nabla T \cdot \delta \mathbf{U} \tag{3.20}$$

$$\delta_T N_1 = 0 \tag{3.21}$$

$$\delta_T (N_2, N_3, N_4)^T = -\gamma \mathbf{g} \delta T \tag{3.22}$$

$$\delta_T N_5 = \nabla\cdot(\mathbf{U}\delta T) - \nabla\cdot(\kappa\nabla\delta T) \tag{3.23}$$

The discussion, up to this point has neglected the variation in eddy viscosity ν, which is a valid approach only for laminar flow regimes. For turbulent flows, therefore, we apply a common approximation known as frozen turbulence (Dwight and Brezillon, 2006; Othmer, 2008). Then Eq. (3.12) can be written as:

$$\int_{\Omega} d\Omega \frac{\partial f}{\partial p}\delta p + \int_{\omega} d\omega \mathbf{U}_a \cdot \nabla \delta p = 0 \tag{3.24}$$

The flow domain ω can be decomposed into two parts, boundary Γ and interior Ψ. Integrating by parts yields the following:

$$\begin{aligned}
&\int_{\Omega} d\Omega \frac{\partial f}{\partial p}\delta p + \int_{\omega} d\omega \mathbf{U}_a \cdot \nabla \delta p \\
&= \int_{\Omega} d\Omega \frac{\partial f}{\partial p}\delta p + \int_{\omega} d\omega [\nabla \cdot (\mathbf{U}_a \delta p) - \delta p \nabla \cdot \mathbf{U}_a] \\
&= \int_{\Omega} d\Omega \frac{\partial f}{\partial p}\delta p + \int_{\Gamma} d\Gamma \mathbf{U}_a \cdot \vec{n}\delta p - \int_{\Psi} d\Psi (\nabla \cdot \mathbf{U}_a)\delta p \\
&= \int_{\Gamma} d\Gamma (\mathbf{U}_a \cdot \vec{n} + \frac{\partial f}{\partial p})\delta p - \int_{\Psi} d\Psi (\nabla \cdot \mathbf{U}_a - \frac{\partial f}{\partial p})\delta p = 0
\end{aligned} \tag{3.25}$$

Then we have

$$-\nabla \cdot \mathbf{U}_a + \frac{\partial f}{\partial p} = 0 \tag{3.26}$$

$$\int_{\Gamma} d\Gamma \mathbf{U}_a \cdot \vec{n}\delta p = 0 \tag{3.27}$$

which are the adjoint continuity equation and boundary condition, respectively.

The same integral transformation steps can be performed on Eqs. (3.13) and (3.14) to make them computable. For Eq. (3.13), the transformation is as follows:

$$\begin{aligned}
&\int_{\Omega} d\Omega \frac{\partial f}{\partial \mathbf{U}}\delta \mathbf{U} + \int_{\omega} d\omega p_a(-\nabla \cdot \delta \mathbf{U}) \\
&+\int_{\omega} d\omega \mathbf{U}_a[(\delta \mathbf{U} \cdot \nabla)\mathbf{U} + (\mathbf{U} \cdot \nabla)\delta \mathbf{U} - \nabla \cdot (2\nu D(\delta \mathbf{U}))] + \int_{\omega} d\omega T_a \nabla T \cdot \delta \mathbf{U} = 0
\end{aligned} \tag{3.28}$$

$$\int_{\omega} d\omega p_a \nabla \cdot \delta \mathbf{U} = \int_{\omega} d\omega [\nabla \cdot (p_a \delta \mathbf{U}) - \delta \mathbf{U} \nabla p_a] = \int_{\Gamma} d\Gamma p_a \vec{n} \cdot \delta \mathbf{U} - \int_{\Psi} d\Psi \delta \mathbf{U} \cdot \nabla p_a \tag{3.29}$$

$$\begin{aligned}
&\int_{\omega} d\omega \mathbf{U}_a (\delta \mathbf{U} \cdot \nabla)\mathbf{U} \\
&= \int_{\omega} d\omega [\nabla \cdot ((\mathbf{U}_a \cdot U)\delta \mathbf{U}) - \delta \mathbf{U} \cdot \nabla (\mathbf{U}_a \cdot \mathbf{U})] \\
&= \int_{\Gamma} d\Gamma (\mathbf{U}_a \cdot \mathbf{U})\vec{n} \cdot \delta \mathbf{U} - \int_{\psi} d\psi \delta \mathbf{U} \cdot \nabla (\mathbf{U}_a \cdot \mathbf{U}) \\
&= \int_{\Gamma} d\Gamma (\mathbf{U}_a \cdot \mathbf{U})\vec{n} \cdot \delta \mathbf{U} - \int_{\psi} d\Omega (\nabla \mathbf{U}_a \cdot \mathbf{U}) \cdot \delta \mathbf{U}
\end{aligned} \tag{3.30}$$

$$\begin{aligned}
&\int_{\omega} d\omega \mathbf{U}_a(\mathbf{U}\cdot\nabla)\delta\mathbf{U} \\
&= \int_{\omega} d\omega[\nabla\cdot((\mathbf{U}_a\cdot\delta\mathbf{U})\mathbf{U}) - \mathbf{U}\cdot\nabla(\mathbf{U}_a\cdot\delta\mathbf{U})] \\
&= \int_{\Gamma} d\Gamma(\vec{n}\cdot\mathbf{U})\mathbf{U}_a\cdot\delta\mathbf{U} - \int_{\Psi} d\Psi\mathbf{U}\cdot\nabla(\mathbf{U}_a\cdot\delta\mathbf{U}) \\
&= \int_{\Gamma} d\Gamma(\vec{n}\cdot\mathbf{U})\mathbf{U}_a\cdot\delta\mathbf{U} - \int_{\Psi} d\Psi(\mathbf{U}\cdot\nabla)\mathbf{U}_a\cdot\delta\mathbf{U}
\end{aligned} \tag{3.31}$$

$$\begin{aligned}
&\int_{\omega} d\omega \mathbf{U}_a[-\nabla\cdot(2\nu D(\delta\mathbf{U}))] \\
&= \int_{\Gamma} d\Gamma[2\nu\vec{n}\cdot D(\mathbf{U}_a)\cdot\delta\mathbf{U} - 2\nu\vec{n}\cdot D(\delta\mathbf{U})\cdot\mathbf{U}_a] - \int_{\psi} d\psi\nabla\cdot(2\nu D(\mathbf{U}_a))\cdot\delta\mathbf{U}
\end{aligned} \tag{3.32}$$

$$\begin{aligned}
&\int_{\psi} d\Psi(\nabla p_a - \nabla\mathbf{U}_a\cdot\mathbf{U} - (\mathbf{U}\cdot\nabla)\mathbf{U}_a - \nabla\cdot(2\nu D(\mathbf{U}_a)) + T_a\nabla T + \frac{\partial f}{\partial \mathbf{U}})\delta\mathbf{U} \\
&+\int_{\Gamma} d\Gamma(-p_a\vec{n} + (\mathbf{U}_a\cdot\mathbf{U})\vec{n} + (\vec{n}\cdot\mathbf{U})\mathbf{U}_a + 2\nu\vec{n}\cdot D(\mathbf{U}_a) + T_a\nabla T + \frac{\partial f}{\partial \mathbf{U}})\delta\mathbf{U} \\
&-\int_{\Gamma} d\Gamma[2\nu\vec{n}\cdot D(\delta\mathbf{U})\cdot\mathbf{U}_a] = 0
\end{aligned} \tag{3.33}$$

Then we have:

$$\nabla p_a - \nabla\mathbf{U}_a\cdot\mathbf{U} - (\mathbf{U}\cdot\nabla)\mathbf{U}_a - \nabla\cdot(2\nu D(\mathbf{U}_a)) + T_a\nabla T + \frac{\partial f}{\partial \mathbf{U}} = 0 \tag{3.34}$$

$$\begin{aligned}
&\int_{\Gamma} d\Gamma(-p_a\vec{n} + (\mathbf{U}_a\cdot\mathbf{U})\vec{n} + (\vec{n}\cdot\mathbf{U})\mathbf{U}_a + 2\nu\vec{n}\cdot D(\mathbf{U}_a) + T_a\nabla T + \frac{\partial f}{\partial \mathbf{U}})\delta\mathbf{U} \\
&-\int_{\Gamma} d\Gamma[2\nu\vec{n}\cdot D(\delta\mathbf{U})\cdot\mathbf{U}_a] = 0
\end{aligned} \tag{3.35}$$

which are the adjoint momentum equations and boundary conditions, respectively.

Likewise, Eq. (3.14) can be transformed as follows:

$$\int_{\Omega} d\Omega\frac{\partial f}{\partial T}\delta T + \int_{\omega} d\omega\mathbf{U}_a(-\gamma\mathbf{g}\delta T) + \int_{\omega} d\omega T_a[\nabla\cdot(\mathbf{U}\delta T) - \nabla\cdot(\kappa\nabla\delta T)] = 0 \tag{3.36}$$

$$\begin{aligned}
&\int_{\omega} d\omega T_a\nabla\cdot(\mathbf{U}\delta T) = \int_{\omega} d\omega[\nabla\cdot(T_a\mathbf{U}\delta T) - \mathbf{U}\delta T\nabla T_a] \\
&= \int_{\Gamma} d\Gamma(\vec{n}\cdot T_a\mathbf{U}\delta T) - \int_{\psi} d\psi\mathbf{U}\cdot\nabla T_a\delta T
\end{aligned} \tag{3.37}$$

$$\begin{aligned}
&\int_{\omega} d\omega T_a\nabla\cdot(\kappa\nabla\delta T) \\
&= \int_{\omega} d\omega[\nabla\cdot(\kappa T_a\nabla\delta T) - \kappa\nabla\delta T\cdot\nabla T_a] \\
&= \int_{\omega} d\omega[\nabla\cdot(\kappa(\nabla(T_a\delta T) - \delta T\nabla T_a)) - \kappa\nabla\delta T\cdot\nabla T_a] \\
&= \int_{\omega} d\omega[\nabla\cdot(\kappa\nabla(T_a\delta T)) - \kappa\nabla\delta T\cdot\nabla T_a] - \int_{\omega} d\omega\nabla\cdot(\kappa\delta T\nabla T_a)
\end{aligned} \tag{3.38}$$

$$= \int_{\omega} d\omega[\nabla\cdot(\kappa\nabla(T_a\delta T)) - [\nabla\cdot(\kappa\delta T\nabla T_a) - \kappa\delta T\nabla\cdot\nabla T_a]] - \int_{\omega} d\omega\nabla\cdot(\kappa\delta T\nabla T_a)$$

$$= -\int_{\Gamma} d\Gamma(\kappa\vec{n}\cdot\nabla T_a\delta T) + \int_{\Psi} d\Psi\nabla\cdot(\kappa\nabla T_a)\delta T + \int_{\omega} d\omega[\nabla\cdot(\kappa\nabla(T_a\delta T)) - \nabla\cdot(\kappa\delta T\nabla T_a)]$$

$$= -\int_{\Gamma} d\Gamma(\kappa\vec{n}\cdot\nabla T_a\delta T) + \int_{\Psi} d\Psi\nabla\cdot(\kappa\nabla T_a)\delta T + \int_{\omega} d\omega\nabla\cdot(\kappa T_a\nabla\delta T)$$

$$= \int_{\Psi} d\Psi\nabla\cdot(\kappa\nabla T_a)\delta T + \int_{\Gamma} d\Gamma(\kappa\vec{n}\cdot T_a\nabla\delta T - \kappa\vec{n}\cdot\nabla T_a\delta T)$$

$$\begin{aligned}&\int_{\Psi} d\Psi(-\mathbf{U}\cdot\nabla T_a + \nabla\cdot(\kappa\nabla T_a) - \mathbf{U}_a\gamma\mathbf{g} + \frac{\partial f}{\partial T})\delta T \\ &+\int_{\Gamma} d\Gamma[(\vec{n}\cdot T_a\mathbf{U} + \kappa\vec{n}\cdot\nabla T_a + \frac{\partial f}{\partial T} - \mathbf{U}_a\gamma\mathbf{g})\delta T - \kappa\vec{n}\cdot T_a\nabla\delta T] = 0\end{aligned} \tag{3.39}$$

Then we have:

$$-\mathbf{U}\cdot\nabla T_a + \nabla\cdot(\kappa\nabla T_a) - \mathbf{U}_a\gamma\mathbf{g} + \frac{\partial f}{\partial T} = 0 \tag{3.40}$$

$$\int_{\Gamma} d\Gamma(\vec{n}\cdot T_a\mathbf{U} + \kappa\vec{n}\cdot\nabla T_a + \frac{\partial f}{\partial T} - \mathbf{U}_a\gamma\mathbf{g})\delta T - \int_{\Gamma} d\Gamma(\kappa\vec{n}\cdot T_a\nabla\delta T) = 0 \tag{3.41}$$

which are the adjoint energy equation and boundary conditions, respectively.

3.2.2.1 Adjoint equations

According to the integral transformation steps above, the final form of the adjoint equations is:

$$-\nabla\cdot\mathbf{U}_a = 0 \tag{3.42}$$

$$-\nabla\mathbf{U}_a\cdot\mathbf{U} - (\mathbf{U}\cdot\nabla)\mathbf{U}_a - \nabla\cdot(2\nu D(\mathbf{U}_a)) + \nabla p_a + T_a\nabla T + \frac{\partial f}{\partial\mathbf{U}} = 0 \tag{3.43}$$

$$-\mathbf{U}\cdot\nabla T_a + \nabla\cdot(\kappa\nabla T_a) - \gamma\mathbf{U}_a\cdot\mathbf{g} + \frac{\partial f}{\partial T} = 0 \tag{3.44}$$

Eqs. (3.42), (3.43), and (3.44) have unknown adjoint variables (p_a, $\mathbf{U}_a$, T_a). Since Eq. (3.42) has the same form as the continuity equation in the Navier-Stokes equations, it is called the adjoint continuity equation. Likewise, Eqs. (3.43) and (3.44) are called the adjoint momentum equation and adjoint energy equation, respectively. Previous studies by Jameson (1995) and Jameson et al. (1998) have proven the solvability of the adjoint equations. The solutions for these equations require the definition of the corresponding boundary conditions.

3.2.2.2 Adjoint boundary conditions

According to the integral transformation steps above, on the boundary domain Γ we have:

$$\int_{\Gamma} d\Gamma\mathbf{U}_a\cdot\vec{n}\delta p = 0 \tag{3.45}$$

$$\int_{\Gamma} d\Gamma(-p_a\vec{n} + (\mathbf{U}_a \cdot \mathbf{U})\vec{n} + (\vec{n}\cdot\mathbf{U})\mathbf{U}_a + 2\nu\vec{n}\cdot D(\mathbf{U}_a) + T_a\nabla T + \frac{\partial f}{\partial \mathbf{U}})\delta\mathbf{U} - \int_{\Gamma} d\Gamma[2\nu\vec{n}\cdot D(\delta\mathbf{U})\cdot\mathbf{U}_a] = 0 \tag{3.46}$$

$$\int_{\Gamma} d\Gamma(\vec{n}\cdot T_a\mathbf{U} + \kappa\vec{n}\cdot\nabla T_a + \frac{\partial f}{\partial T} - \mathbf{U}_a\gamma\mathbf{g})\delta T - \int_{\Gamma} d\Gamma(\kappa\vec{n}\cdot T_a\nabla\delta T) = 0 \tag{3.47}$$

3.2.2.2.1 ADJOINT PRESSURE AND VELOCITY BOUNDARY CONDITIONS

The adjoint pressure and velocity boundary conditions can be determined by Eqs. (3.46) and (3.47). Both $\delta\mathbf{U}$ and $\mathbf{U}_a$ are divergenceless fields, and at least one of them vanishes on boundary Γ. Therefore, in Eq. (3.46), we have

$$\int_{\Gamma} d\Gamma 2\nu\vec{n}\cdot[D(\mathbf{U}_a)\cdot\delta\mathbf{U} - D(\delta\mathbf{U})\cdot\mathbf{U}_a] = \int_{\Gamma} d\Gamma\nu[(\vec{n}\cdot\nabla)\mathbf{U}_a\cdot\delta\mathbf{U} - (\vec{n}\cdot\nabla)\delta\mathbf{U}\cdot\mathbf{U}_a] - \int_{\Gamma} d\Gamma\nabla\nu\cdot[(\vec{n}\cdot\mathbf{U}_a)\cdot\delta\mathbf{U} - (\vec{n}\cdot\delta U)\cdot\mathbf{U}_a] \tag{3.48}$$

In Eq. (3.48), $\nabla\nu = 0$ for laminar flows because ν contains only molecular viscosity. For turbulent flow, $\nabla\nu = 0$ at the wall and inlet with homogeneous turbulence. At the outlet, $\nabla\nu$ is generally non-zero. Othmer (2008) assumed that the product of $\nabla\nu$ with differential terms in brackets in Eq. (3.48) to be of second order and therefore neglected the integration with $\nabla\nu$. In this study, we have made the same assumption, and thus Eq. (3.46) can be simplified as:

$$\int_{\Gamma} d\Gamma[\vec{n}(\mathbf{U}_a\cdot\mathbf{U}) + \mathbf{U}_a(\mathbf{U}\cdot\vec{n}) + \nu(\vec{n}\cdot\nabla)\mathbf{U}_a - p_a\vec{n} + \frac{\partial f}{\partial \mathbf{U}}]\cdot\delta\mathbf{U} - \int_{\Gamma} d\Gamma\nu(\vec{n}\cdot\nabla)\delta\mathbf{U}\cdot\mathbf{U}_a = 0 \tag{3.49}$$

Next, we combine Eqs. (3.45) and (3.49) in order to define the adjoint velocity boundary conditions for the inlet, wall, and outlet.

At the wall, the air velocity is fixed at 0 and the pressure is $\nabla p = 0$. Decomposing $\nabla\cdot\delta\mathbf{U} = 0$ into the normal and in-plane directions, we have:

$$\nabla\cdot\delta\mathbf{U} = 0 = (\vec{n}\cdot\nabla)\delta\mathbf{U}_n + \nabla_t\cdot\delta\mathbf{U}_t \tag{3.50}$$

where $\delta\mathbf{U}_n$ is the normal component and ∇_t and $\delta\mathbf{U}_t$ are the in-plane components of ∇ and U, respectively. Since $\delta\mathbf{U} = 0$, $\delta\mathbf{U}_t = 0$ and the first integral of Eq. (3.49) cancels. Hence:

$$(\vec{n}\cdot\nabla)\delta\mathbf{U}_n = 0 \tag{3.51}$$

$$(\vec{n}\cdot\nabla)\delta\mathbf{U} = (\vec{n}\cdot\nabla)\delta\mathbf{U}_t \tag{3.52}$$

Eq. (3.49) can thus be reduced to

$$\int_{\Gamma} d\Gamma\nu(\vec{n}\cdot\nabla)\delta\mathbf{U}_t\cdot\mathbf{U}_{at} = 0 \tag{3.53}$$

where $\mathbf{U}_{at}$ is the in-plane component of $\mathbf{U}_a$. From Eqs. (3.45) and (3.53), we have

$$\mathbf{U}_a\big|_{wall} = 0 \tag{3.54}$$

For the adjoint pressure, Eqs. (3.45) and (3.46) do not impose a boundary condition. In order to ensure the well-pose of the adjoint equation, we set

$$\nabla p_a\big|_{wall} = 0 \tag{3.55}$$

At the outlet, because the boundary conditions for **U** and p are zero velocity gradient and zero pressure, respectively, the second integral of Eq. (3.49) cancels and Eq. (3.45) satisfies with $\delta p = 0$. Therefore:

$$\vec{n}(\mathbf{U}_a \cdot \mathbf{U}) + \mathbf{U}_a(\mathbf{U} \cdot \vec{n}) + \nu(\vec{n} \cdot \nabla)\mathbf{U}_a - p_a\vec{n} + \frac{\partial f}{\partial \mathbf{U}} = 0 \tag{3.56}$$

Again, decomposing Eq. (3.56) into the normal and in-plane directions, we have:

$$\mathbf{U}_a \cdot \mathbf{U} + \mathbf{U}_{an}\mathbf{U}_n + \nu(\vec{n} \cdot \nabla)\mathbf{U}_{an} - p_a + \vec{n} \cdot \frac{\partial f}{\partial \mathbf{U}} = 0 \tag{3.57}$$

$$\mathbf{U}_{at}\mathbf{U}_n + \vec{t} \cdot \frac{\partial f}{\partial \mathbf{U}} = 0 \tag{3.58}$$

where $\mathbf{U}_{an}$ is the normal component of $\mathbf{U}_a$. Therefore, the adjoint pressure boundary condition can be expressed as:

$$p_a = \mathbf{U}_a \cdot \mathbf{U} + \mathbf{U}_{an}\mathbf{U}_n + \nu(\vec{n} \cdot \nabla)\mathbf{U}_{an} + \vec{n} \cdot \frac{\partial f}{\partial \mathbf{U}} \tag{3.59}$$

Moreover, the tangential component of the adjoint velocity boundary condition can be determined from Eq. (3.58). Since

$$(\vec{n} \cdot \nabla)\mathbf{U}_{an} = \nabla \cdot \mathbf{U}_a - \nabla_t \mathbf{U}_{at} = -\nabla_t \mathbf{U}_{at} \tag{3.60}$$

the gradient of $\mathbf{U}_{an}$ can be obtained. The above discussion completes our definition of the adjoint boundary conditions at the outlet.

At the inlet, the air velocity can vary, i.e., $\delta\mathbf{U} \neq 0$, and the inlet air pressure is $\nabla p = 0$. From Eq. (3.45) we have:

$$\mathbf{U}_{an}\big|_{inlet} = 0 \tag{3.61}$$

Then the second term of Eq. (3.49) cancels, and we again have Eqs. (3.57) and (3.58). Thus, the tangential component of the adjoint velocity boundary condition can be determined from Eq. (3.58). Since $\mathbf{U}_{\mathrm{an}} = 0$, we can express the adjoint pressure on the inlet as:

$$p_a\big|_{inlet} = \mathbf{U}_a \cdot \mathbf{U} + \nu(\vec{n} \cdot \nabla)\mathbf{U}_{an} + \vec{n} \cdot \frac{\partial f}{\partial \mathbf{U}} \tag{3.62}$$

In an enclosed environment, the objective function normally is not directly related to the **U** and p on the boundaries. Therefore:

$$\frac{\partial f}{\partial \mathbf{U}} = 0 \tag{3.63}$$

Then, according to Eq. (3.58), we have:

$$\mathbf{U}_{at}\big|_{inlet,outlet} = 0 \tag{3.64}$$

$$\mathbf{U}_{a}\big|_{inlet} = 0 \tag{3.65}$$

Next, because of the continuity equation, we have:

$$\mathbf{U}_{a}\big|_{outlet} = 0 \tag{3.66}$$

In summary, the velocity and pressure boundary conditions for the adjoint equations are:

$$\mathbf{U}_{a}\big|_{\Gamma} = 0 \tag{3.67}$$

$$p_{a}\big|_{inlet,outlet} = \nu(\vec{n}\cdot\nabla)\mathbf{U}_{an} \tag{3.68}$$

$$\nabla p\big|_{wall} = 0 \tag{3.69}$$

3.2.2.2.2 ADJOINT TEMPERATURE BOUNDARY CONDITIONS

The adjoint temperature boundary conditions can be determined by Eq. (3.47). Because the air temperature at the wall is fixed in this study, the first term of Eq. (3.47) cancels, and we have:

$$\int_{\Gamma} d\Gamma(\kappa\vec{n}\cdot T_a\nabla\delta T) = 0 \tag{3.70}$$

Thus

$$T_a\big|_{\text{wall}} = 0 \tag{3.71}$$

At the outlet, the air temperature boundary condition has a zero gradient, and thus the second term of Eq. (3.47) cancels and $\mathbf{U}_a$ is equal to zero on the boundary. In an enclosed environment, the objective function normally does not have a direct relationship with *T*. Therefore:

$$\int_{\Gamma} d\Gamma[T_a(\mathbf{U}\cdot\vec{n}) + \kappa\vec{n}\cdot\nabla T_a]\cdot\delta T = 0 \tag{3.72}$$

$$T_a(\mathbf{U}\cdot\vec{n}) + \kappa\vec{n}\cdot\nabla T_a = 0 \tag{3.73}$$

Eq. (3.73) defines the adjoint temperature boundary condition at the outlet.

At the inlet, the air temperature can vary, so $\delta T \neq 0$ and $\nabla\delta T$ are assumed to be 0. Therefore, the second term in Eq. (3.47) is cancelled, and Eq. (3.73) also expresses the adjoint temperature boundary condition at the inlet.

With the use of the adjoint equations from Section 3.1.2.1 and the corresponding boundary conditions as defined in Section 3.1.2.2, the solutions for the adjoint equations can be obtained. The solutions are then used to compute $dL/d\xi$, which is expressed as:

$$\frac{dL}{d\xi} = \frac{\partial L}{\partial\xi} = \frac{\partial O}{\partial\xi} + \int_{\omega}(p_a, \mathbf{U}_a, T_a)\frac{\partial\mathbf{N}}{\partial\xi}d\omega \tag{3.74}$$

Since the Navier-Stokes equations $\mathbf{N} = \mathbf{0}$ are solved numerically, $\mathbf{N}$ becomes the residuals of the corresponding solution to the Navier-Stokes equations. Thus, $\partial\mathbf{N}/\partial\xi$ actually computes the partial derivative of the solution residuals of the Navier-Stokes equations with respect to the design variables ξ.

3.2.2.3 Adjusting the design variables

The simple steepest descent algorithm (Bryson and Ho, 1975) can be used to adjust ξ as follows:

$$\xi_{new} = \xi_{old} + d\xi = \xi_{old} - \lambda[\frac{\partial O}{\partial \xi} + \int_{\omega} (p_a, \mathbf{U}_a, T_a)\frac{\partial \mathbf{U}}{\partial \xi} d\omega]^T \tag{3.75}$$

where λ is a positive constant. Thus, the variation in L is always negative, and the value of the objective function always decreases. To compute $d\xi$ in Eq. (3.75), the adjoint equations are solved to determine the adjoint variables (p_a, $\mathbf{U}_a$, T_a). Then, as the objective function O in this study is a function only of **U** and T, $\partial O/\partial\xi = 0$. Finally, one needs to evaluate the partial derivative $\partial\mathbf{N}/\partial\xi$ in Eq. (3.75) in order to obtain $d\xi$.

3.2.2.4 Evaluation of the gradient

The evaluation of the partial derivative $\partial\mathbf{N}/\partial\xi$ in Eq. (3.75) requires differentiation of the algorithms that are used to compute **N** (Sokolowski and Zolesio, 1992; Jimack, 1997). In this study, ξ denotes the air supply inlet location, inlet size, temperature, and velocity. Our literature review identified three common methods for evaluating $\partial\mathbf{N}/\partial\xi$: the chain rule of differentiation (Wikipedia, n.d.), the finite difference method (FDM), and the finite element method (FEM). The chain rule of differentiation has been implemented mainly in automatic differentiation software (Griwank, 2000). It allows functions such as **N** to be differentiated with respect to input parameters such as ξ to evaluate the derivative $\partial\mathbf{N}/\partial\xi$ as the output. This technique is powerful, and derivatives of an arbitrary order can be computed automatically. Unfortunately, the automatic differentiation software is not generally efficient, and mature tools of this kind are not available (Schneider, 2006). The FDM and FEM have been implemented primarily by hand coding. Programming the FDM (Gill et al., 1981; Gunzburger, 2003) requires minimal effort, but the method is inefficient, as the time required to evaluate the derivative is proportional to the number of design variables. In later research, Schneider and Jimack (2008) have simply discretized the Navier-Stokes equations by means of the FEM to compute $\partial\mathbf{N}/\partial\xi$. The present study has done the same.

In using the FEM to discretize **N**, we chose the shape functions, S, as the basis functions. Considering the energy equation in the Navier-Stokes equations as an example, the discrete equation determined by the FEM for node j can be expressed as:

$$N_{5,j} = \int_{\omega} \kappa\ \nabla T \cdot \nabla S_j d\omega + \int_{\omega} (\mathbf{U} \cdot\ \nabla T) S_j d\omega \tag{3.76}$$

For the diffusion term in Eq. (3.76), we have:

$$\begin{aligned}
\int_{\omega} \kappa\ \nabla T \cdot \nabla S_j d\omega &= \kappa \sum_{E_j} \int_{\omega_E} \nabla T \cdot \nabla S_j d\omega \\
&= \kappa \sum_{E_j} \int_{\omega_E} \nabla_E T \cdot \nabla_E S_j \left|\det(J_E)\right| d\omega \\
&= \kappa \sum_{E_j} \sum_{k=1}^{e} \int_{\omega_E} \nabla_E [S_{gE(k)} T_{gE(k)}] \cdot \nabla_E S_j \left|\det(J_E)\right| d\omega \\
&= \kappa \sum_{E_j} \sum_{k=1}^{e} \sum_{l=1}^{f} \sigma_l [\nabla_E S_{gE(k)}(q_l) \cdot \nabla_E S_j(q_l)] T_{gE(k)} \left|\det(J_E)\right|
\end{aligned} \tag{3.77}$$

where the outermost sum is over those elements E_j that have node j as a vertex, and e is the number of Lagrange linear shape functions per element in this study. The last step in

Eq. (3.77) is the numerical integration using the Newton-Cotes quadrature for the integration in the prior step, so q_l $(l=1,2,..,f)$ are quadrature points on the master element and σ_l are the quadrature weights. Additionally, $gE(k)$ is the global node number of node k of element E_j, ∇_E denotes the gradient restricting the master element at point q_l, S is the shape function, and J_E is the Jacobian of the element mapping. In the same way, the convection term in Eq. (3.76) can be expressed as:

$$\begin{aligned}&\int_{\omega}(\mathbf{U}\cdot\nabla T)S_j d\omega\\&=\sum_{E_j}\sum_{k=1}^{e}\sum_{l=1}^{f}\{\sigma_l[\sum_{k=1}^{e}S_{gE(k)}(q_l)\mathbf{U}_{gE(k)}]\cdot\nabla_E S_{gE(k)}(q_l)S_j(q_l)\}T_{gE(k)}\left|\det(J_E)\right|\end{aligned}\tag{3.78}$$

Similarly, for the momentum equation, we have:

$$N_{i,j}=\int_{\omega}\nu\ \nabla U_i\cdot\nabla S_j d\omega+\int_{\omega}\mathbf{U}\cdot\ \nabla U_i\cdot S_j d\omega+\int_{\omega}p\vec{i}\cdot\nabla S_j d\omega\tag{3.79}$$

$$\begin{aligned}&\int_{\omega}\nu\ \nabla U_i\cdot\nabla S_j d\omega\\&=\nu\sum_{E_j}\int_{\omega_E}\nabla U_i\cdot\nabla S_j d\omega\\&=\nu\sum_{E_j}\int_{\omega_E}\nabla_E U_i\cdot\nabla_E S_j\left|\det(J_E)\right|d\omega\\&=\nu\sum_{E_j}\sum_{k=1}^{e}\int_{\omega_E}\nabla_E[S_{gE(k)}U_i^{gE(k)}]\cdot\nabla_E S_j\left|\det(J_E)\right|d\omega\\&=\nu\sum_{E_j}\sum_{k=1}^{e}\sum_{l=1}^{f}\sigma_l[\nabla_E S_{gE(k)}(x_l)\cdot\nabla_E S_j(x_l)]U_i^{gE(k)}\left|\det(J_E)\right|\end{aligned}\tag{3.80}$$

$$\begin{aligned}&\int_{\omega}\mathbf{U}\cdot\nabla U_i\cdot S_j d\omega\\&=\sum_{E_j}\int_{\omega_E}\mathbf{U}\cdot\nabla U_i\cdot S_j d\omega\\&=\sum_{E_j}\int_{\omega_E}\mathbf{U}\cdot\nabla_E U_i\cdot S_j\left|\det(J_E)\right|d\omega\\&=\sum_{E_j}\sum_{k=1}^{e}\int_{\omega_E}[\sum_{k=1}^{e}S_{gE(k)}\mathbf{U}^{gE(k)}]\cdot\nabla_E[S_{gE(k)}U_i^{gE(k)}]\cdot S_j\left|\det(J_E)\right|d\omega\\&=\sum_{E_j}\sum_{k=1}^{e}\sum_{l=1}^{f}\{\sigma_l[\sum_{k=1}^{e}S_{gE(k)}(x_l)\mathbf{U}^{gE(k)}]\cdot\nabla_E S_{gE(k)}(x_l)\cdot S_j(x_l)\}U_i^{gE(k)}\left|\det(J_E)\right|\end{aligned}\tag{3.81}$$

$$\begin{aligned}&\int_{\omega}p\ \nabla S_j d\omega\\&=\sum_{E_j}\int_{\omega_E}p\ \nabla S_j d\omega\\&=\sum_{E_j}\int_{\omega_E}p\ \nabla_E S_j\left|\det(J_E)\right|d\omega\\&=\sum_{E_j}\sum_{k=1}^{e}\int_{\omega_E}[S_{gE(k)}p^{gE(k)}]\cdot\nabla_E S_j\left|\det(J_E)\right|d\omega\\&=\sum_{E_j}\sum_{k=1}^{e}\sum_{l=1}^{f}\sigma_l[S_{gE(k)}(x_l)\cdot\nabla_E S_j(x_l)]p^{gE(k)}\left|\det(J_E)\right|\end{aligned}\tag{3.82}$$

For the continuity equation, we have:

$$\begin{aligned}
N_{1,j} &= \int_{\omega} S_j \nabla \cdot \mathbf{U} d\omega \\
&= \sum_{E_j} \int_{\omega_E} S_j \nabla \cdot \mathbf{U} d\omega \\
&= \sum_{E_j} \int_{\omega_E} S_j \nabla_E \cdot \mathbf{U} \left|\det(J_E)\right| d\omega \\
&= \sum_{E_j} \sum_{k=1}^{e} \int_{\omega_E} S_j \nabla_E \cdot [S_{gE(k)} \mathbf{U}^{gE(k)}] \left|\det(J_E)\right| d\omega \\
&= \sum_{E_j} \sum_{k=1}^{e} \sum_{l=1}^{f} \sigma_l S_j(x_l) [\mathbf{U}^{gE(k)} \cdot \nabla_E S_{gE(k)}(x_l)] \left|\det(J_E)\right|
\end{aligned} \tag{3.83}$$

With discretization of the Navier-Stokes equations by the FEM, the $\partial \mathbf{N}/\partial \xi$ can be calculated straightforwardly when ξ denotes the air supply temperature and velocity. When ξ denotes the inlet location and size, this study mathematically sets ξ as the coordinates of the nodes located on the boundary of the air supply diffusers. The discretization of the Navier-Stokes equations by the FEM produces two non-constant parts with respect to the node positions that comprise the derivative $\partial \mathbf{N}/\partial \xi$, which can also be computed directly. In the discretized equations above, the non-constant expressions with respect to the node positions are:

$$\nabla_E S_{gE(k)}(x_l),\ \nabla_E S_j(x_l),\ \left|\det(J_E)\right|$$

The first two terms are the gradient of the shape function, and the third term is the determinant of the element-mapping Jacobian. Thus, the calculation of $\partial \mathbf{N}/\partial \xi$ depends on the evaluation of the derivatives of the upper three expressions over the node positions. When node j and the boundary nodes at ξ share the same element E, the derivatives of the upper three expressions over the node positions can be computed directly. Readers can refer to Schneider and Jimack (2008) for the calculation details, which are relatively straightforward. Otherwise, $\partial \mathbf{N}/\partial \xi$ should be evaluated by the chain rule of differentiation. For example:

$$\frac{\partial N_{5,j}}{\partial \xi} = \frac{\partial N_{5,j}}{\partial \eta} \frac{\partial \eta}{\partial \xi} \tag{3.84}$$

where node j and the node at coordinates $\boldsymbol{\eta}$ share the same element E. Therefore, $\partial N_{5,j}/\partial \boldsymbol{\eta}$ can be directly evaluated, but $\partial \boldsymbol{\eta}/\partial \xi$ cannot be evaluated unless a parametric mesh is used. In the CFD simulation for an indoor environment, the mesh is too fine to allow mapping among all the nodes, and $\partial \boldsymbol{\eta}/\partial \xi$ cannot be evaluated. However, the contributions of the interior elements further away from the air supply diffuser become small very quickly on fine meshes (Schneider, 2006). When node j and ξ do not share the same element E, neglecting $\partial N_{5,j}/\partial \xi$ might be a viable option. Therefore, we have adopted the contributions of the elements that have boundary nodes at ξ as a vertex to evaluate the derivative of the Navier-Stokes equations over the node positions. In summary, $\partial N_{5,j}/\partial \xi$ can be computed by:

$$\partial N_5 / \partial \xi = \sum_{j=1}^{n} \partial N_{5,j} / \partial \xi \tag{3.85}$$

$$\partial S_{5,j} / \partial \xi = 0 \text{ if node } j \text{ and } \xi \text{ do not share the same element } E \tag{3.86}$$

After $\partial \mathbf{N}/\partial \xi$ is calculated, the variation of the node positions on the boundary of the air supply diffusers can be computed to determine the new inlet location and size by Eq. (3.75).

The updated ξ is then used as input for calculating the objective function. The Navier-Stokes equations are solved to determine whether or not the objective function satisfies the convergence criterion. If not, the adjoint equations are solved to obtain new ξ, which can lead to a smaller O. This approach creates a design cycle that can be repeated until the optimal air supply inlet location, inlet size, and supply temperature and velocity are achieved.

3.2.3 Numerical method

This study used Reynolds-averaged Navier-Stokes equations with the re-normalization group (RNG) k-ε model (Yakhot and Orszag, 1986), which has been widely employed in the modeling of indoor airflow (Zhang et al., 2007; Zhai et al., 2007). To reduce the computational load, we "froze" the turbulence (Othmer, 2008; Dwight and Brezillon, 2006), and the turbulent viscosity in the Navier-Stokes equations was re-used for the adjoint diffusion term. Furthermore, to simulate the buoyancy effect in indoor airflow, this study adopted the Boussinesq approximation (Boussinesq, 1903). In solving the Navier-Stokes (NS)/adjoint equations, we used the semi-implicit method for pressure-linked equations (SIMPLE) algorithm (Patankar and Spalding, 1972) to couple the velocity/adjoint velocity and pressure/adjoint pressure. All the equations were implemented in OpenFOAM (2007), which is an open-source CFD software package.

In the design procedure, the NS/adjoint equations were calculated with M_1/M_2 iterations to ensure convergence. The design convergence criteria were (1) $O_{new} < \alpha$ in the first design cycle, where α is a small positive constant set by the designer, and (2) $|O_{new}-O_{old}| < \beta$ after the first design cycle, where O_{old} is the computed objective function in the previous design cycle and β is a small positive constant set by the designer. After obtaining the new air supply inlet location inlet size temperature, and velocity, we used the Gambit journal files (GAMBIT CFD Preprocessor, 1998) to automatically regenerate the corresponding mesh. Fig. 3.2 shows the solution flow chart for the adjoint method.

3.3 Application of the CFD-based adjoint method

3.3.1 Inverse identification problems in built environments

3.3.1.1 Inverse identification of air supply parameters

This study validated the adjoint method by applying it to two two-dimensional-ventilation cases, one with isothermal flow and the other with non-isothermal flow.

3.3.1.1.1 ISOTHERMAL CASE

The isothermal case was taken from Nielsen (1978), who provided detailed experimental data. The case was isothermal because the inlet air temperature and wall temperature were identical. Therefore, we did not consider the energy equation and adjoint energy equation. Fig. 3.3 shows the geometry and dimensions of this case. The figure also shows the mesh used in the calculations. The inlet velocity components were $U_x = 0.455$ m/s and $U_y = 0$, with a turbulence intensity (TI) of 4%.

We first conducted a forward CFD simulation with the RNG k-ε model, assigning the inlet air velocity components as $U_x = 0.455$ m/s and $U_y = 0$, corresponding to a Reynolds number (*Re*) of 5,000. The forward CFD simulation was performed to obtain a flow field in the room, and a section of air velocity field can be used as the design objective. Fig. 3.4 shows

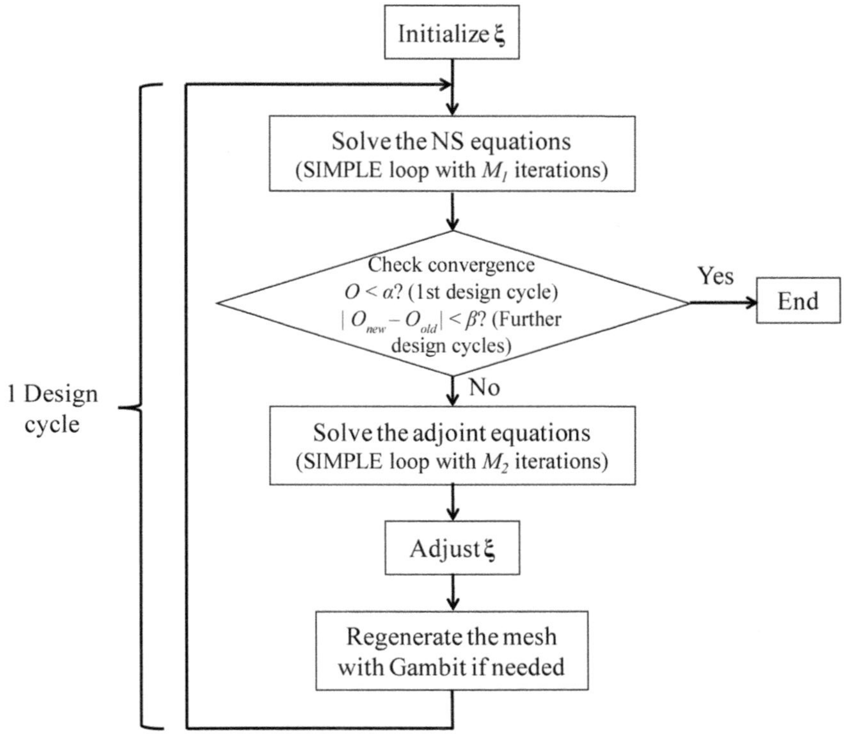

Figure 3.2 Solution flow chart for the adjoint method.

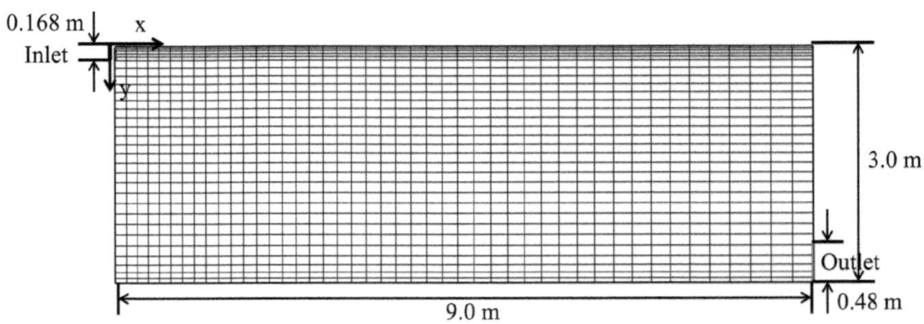

Figure 3.3 Sketch and mesh of the two-dimensional isothermal case

the computed air velocity profile ($\mathbf{U}_0$ in Eq. (3.87)) at the lower half of line $x = 6$ m (design domain Ω), which was selected as the design objective. Thus, the objective function is:

$$O(\mathbf{U}_{inlet}) = \int_{\Omega} \left|\mathbf{U} - \mathbf{U}_0\right|^2 d\Omega \tag{3.87}$$

where $\mathbf{U}_0$ is the air velocity obtained by forward CFD simulation. Next, this study aimed to find the optimal inlet air velocity on the basis of this design objective. As the design objective was produced by the assigned inlet air velocity components of $U_x = 0.455$ m/s and $U_y = 0$, these values are optimal, and the optimized inlet air velocity components may

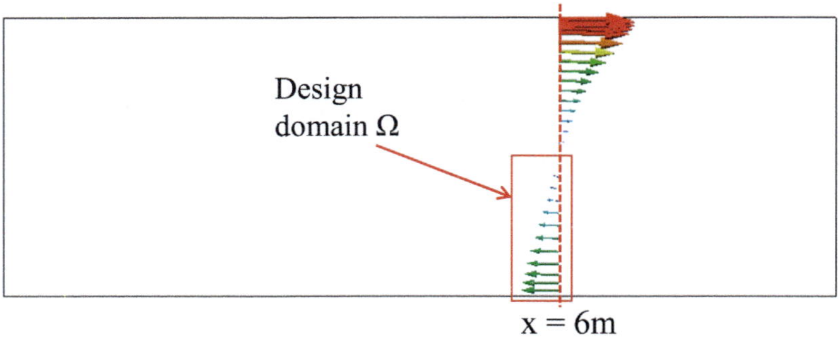

Figure 3.4 Design objective for the two-dimensional isothermal case

be the same as these. However, the solution process may also lead to other values as long as the objective is reached.

Because the isothermal case was simple, it was possible for the adjoint method to produce a converged solution in a small number of iterations. For simplicity, this study used $M_1 = M_2 = 1$ in the calculation. We used four different inlet air velocities as the initial inlet conditions, which were different from the true velocity of $U_x = 0.455$ m/s and $U_y = 0.0$ m/s:

(1) $U_x = 1.0$ m/s and $U_y = 0.0$ m/s
(2) $U_x = 1.0$ m/s and $U_y = 0.1$ m/s
(3) $U_x = 1.0$ m/s and $U_y = -0.1$ m/s
(4) $U_x = 1.0$ m/s and $U_y = 0.3$ m/s

With the initial boundary conditions, the flow and adjoint equations were calculated for 2,000 design cycles with $M_1 = M_2 = 1$. Figs 3.5 and 3.6 show that the first three initial conditions gradually changed to the true result of $U_x = 0.455$ m/s and $U_y = 0.0$ m/s. The error for U_x was only 0.25–1.8%, and the calculated U_y was exactly 0.0 m/s. Unexpectedly, the calculation with an initial velocity of $U_x = 1.0$ m/s and $U_y = 0.3$ m/s led to a final velocity of $U_x = 0.132$ m/s and $U_y = 2.578$ m/s. Furthermore, as illustrated in Fig. 3.7, all the calculations were able to produce a small value of the objective function. Figs 3.5 to 3.7 show that the calculations were unstable in the early design cycles. This instability was caused by the partially converged flow field and adjoint equations, as $M_1 = M_2 = 1$. The calculation was also unstable at design cycle 500, and it was difficult to identify the exact reason.

This study then conducted forward CFD simulations with the inlet air velocities determined by the adjoint method. Fig. 3.8 compares the calculated velocity profiles at $x = 6$ m with the profile computed from the exact inlet air velocity of $U_x = 0.455$ m/s and $U_y = 0$ m/s. Only the calculation with an inlet velocity of $U_x = 0.132$ m/s and $U_y = 2.578$ m/s displayed a slight difference from the other predictions. Therefore, the adjoint method is able to find the optimal inlet air velocity with a small value of the objective function. In other words, the adjoint method is able to inversely find the desired inlet air velocity when the velocity profile in the lower part of x = 6 m was set as the design objective. However, the results indicate that the solution that meets the design objective may not be unique.

3.3.1.1.2 NON-ISOTHERMAL CASE

This investigation also applied the adjoint method to a two-dimensional, non-isothermal case as shown in Fig. 3.9. The dimensions of the flow domain were 1.04 m × 1.04 m, the inlet

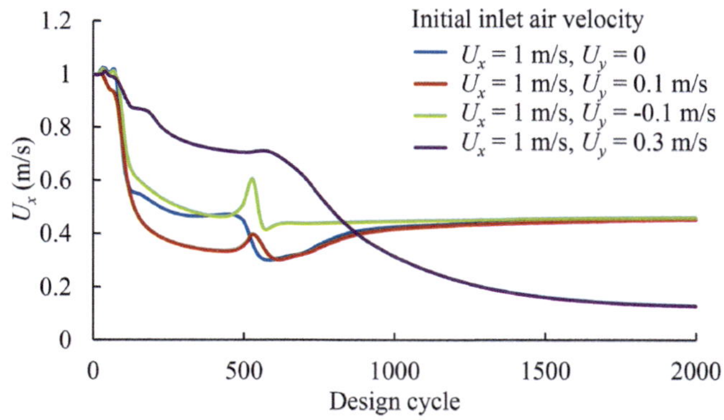

Figure 3.5 Change in U_x with the design cycle

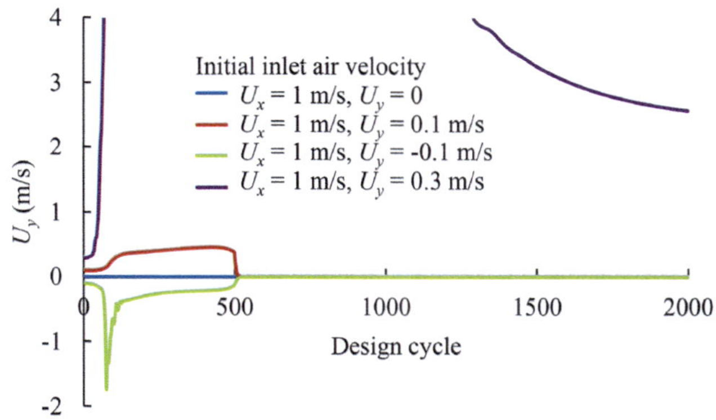

Figure 3.6 Change in U_y with the design cycle

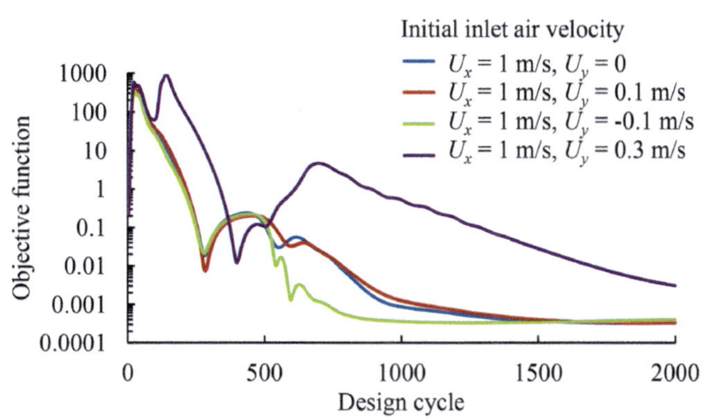

Figure 3.7 Change in the objective function with the design cycle

height was 18 mm, and the outlet height was 24 mm. The inlet air velocity components were $U_x = 0.57$ m/s and $U_y = 0.0$ m/s, and the inlet air temperature (T_{inlet}) was 15°C. The temperature of the walls (T_{wall}) was 15°C, while that of the floor (T_{floor}) was 35°C. Fig. 3.9 also shows the resolution of the mesh used in the numerical simulation. Blay et al. (1992) conducted experimental measurements of the airflow and temperature distributions for this case.

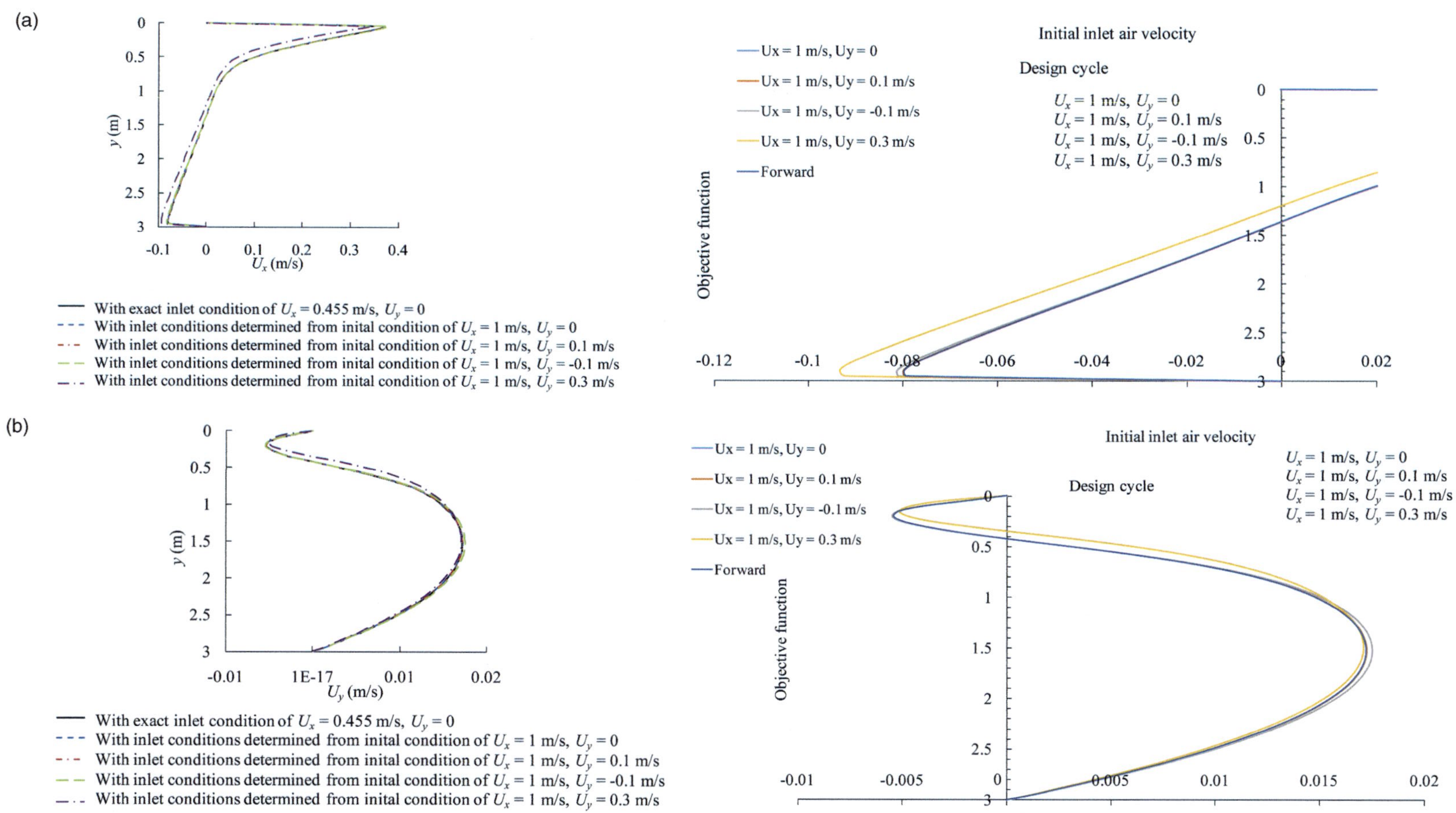

Figure 3.8 Comparison of the design objective (velocity profiles) determined by the adjoint method and that determined from the specified inlet condition of $U_x = 0.455$ m/s, $U_y = 0$ m/s at $x = 6$ m: (a) Comparison of U_x at $x = 6$ m and (b) Comparison of U_y at $x = 6$ m

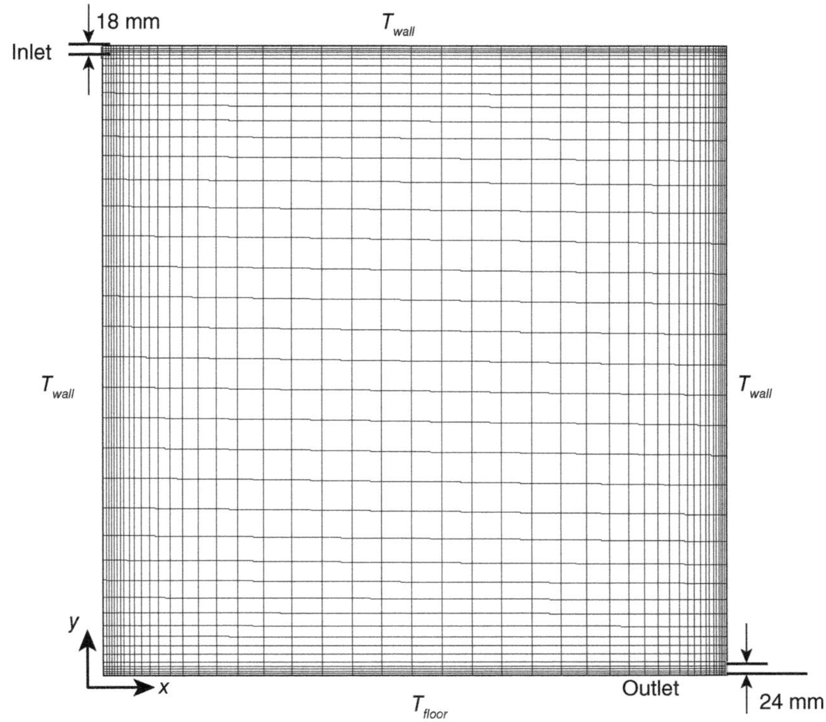

Figure 3.9 Sketch and mesh of the two-dimensional, non-isothermal case

We again conducted a CFD simulation with inlet air velocity of $U_x = 0.57$ m/s and $U_y = 0.0$ and inlet air temperature of 15 °C in order to generate a design objective. The design objective selected for this case was the air velocity profile and/or air temperature profile at mid-cavity (design domain Ω), as indicated by the red line in Fig. 3.10:

Scenario 1: air velocity profile

$$O(\mathbf{U}_{inlet}, T_{inlet}) = \int_{\Omega} |\mathbf{U} - \mathbf{U}_0|^2 \, d\Omega \tag{3.88}$$

Scenario 2: air temperature profile

$$O(\mathbf{U}_{inlet}, T_{inlet}) = \int_{\Omega} (T - T_0)^2 \, d\Omega \tag{3.89}$$

Scenario 3: air velocity and temperature profiles The design variables were the inlet air temperature and velocity. For Scenario 3, to adjust the importance of the velocity error and temperature error in the objective function, we normalized the velocity error by the exact inlet air velocity $U_x = 0.57$ m/s and the temperature error by $(T_{floor} - T_{inlet})$. Thus, we have:

$$O(\mathbf{U}_{inlet}, T_{inlet}) = \frac{1}{U_x^2} \int_{\Omega} |\mathbf{U} - \mathbf{U}_0|^2 \, d\Omega + \frac{1}{(T_{floor} - T_{inlet})^2} \int_{\Omega} (T - T_0)^2 \, d\Omega \tag{3.90}$$

Table 3.1 summarizes the intial inlet air conditions for each scenario:

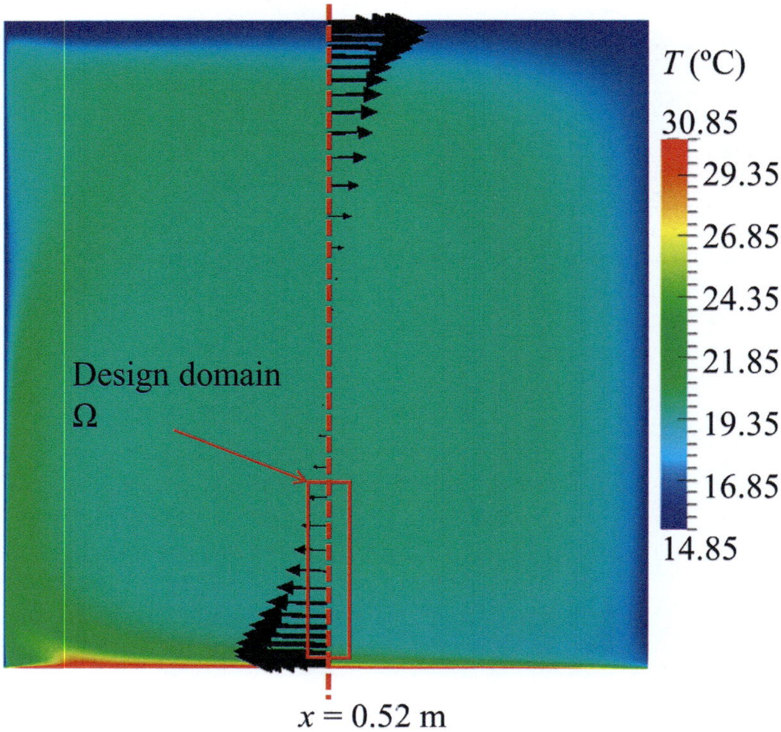

Figure 3.10 Design objective for the two-dimensional, non-isothermal case: velocity and/or temperature profiles at mid-cavity as indicated by the red line

Table 3.1 Initial inlet air conditions for each scenario

	Case	*Initial inlet air temperature (°C)*	*Initial inlet air velocity (m/s)*
Scenario 1	1a	$T = 6.85$	
	1b	$T = 21.85$	
Scenario 2	2a	$T = 6.85$	$U_x = 1.0$,
	2b	$T = 21.85$	$U_y = 0.5$
Scenario 3	3a	$T = 6.85$	
	3b	$T = 21.85$	

Since this was a non-isothermal case, we also used the adjoint method to solve the energy equation. Without iterations ($M_1 = M_2 = 1$), the solution would have diverged as a result of highly significant and sometimes unreasonable changes in the variables. Therefore, we set $M_1 = M_2 > 1$ (as shown in the solution flow chart (Fig. 3.2)) in the calculations in order to achieve a stable and converged solution.

Scenario 1: air velocity as the objective For Scenario 1, this study initialized the inlet velocity and temperature as $U_x = 1$ m/s, $U_y = 0.5$ m/s, and $T = 6.85$°C in Case 1a, and $U_x = 1$ m/s, $U_y = 0.5$ m/s, and $T = 21.85$°C in Case 1b. Furthermore, we used $M_1 = M_2 = 100$

for each design cycle and conducted 100 design cycles. For Case 1a (Fig. 3.11), the objective function was only 10^{-3} at the end of the calculation, indicating that an optimal design had been reached. However, the two initial conditions produced two different optimal inlet conditions. Fig. 3.11(a) shows a final inlet air velocity of $U_x = 0.563$ m/s and $U_y = 0.007$ m/s and a final inlet air temperature of $T = 14.85$°C. The final condition was very close to the true inlet condition of $U_x = 0.57$ m/s, $U_y = 0$ m/s, and $T = 15$°C. However, for Case 1b (Fig. 3.11(b)) the final inlet condition was $U_x = 0.755$ m/s, $U_y = 0.254$ m/s, and $T = 21.85$°C, which was quite different from other final inlet conditions. Because only the velocity field was used as the design objective, it was possible for multiple optimal solutions to exist.

In the non-isothermal case, as in the isothermal case, we conducted forward CFD simulations with final inlet air velocity and temperature determined by the adjoint method. Fig. 3.12 compares the computed velocity profiles at mid-cavity ($x = 0.52$ m) with that determined from the exact inlet condition of $U_x = 0.57$ m/s, $U_y = 0$, and $T = 15$°C. It can be seen that the two final inlet conditions had almost the same velocity profiles as the design objective.

Scenario 2: air temperature as the objective Since it is much harder to obtain a converged result after incorporating the energy equation into the adjoint method, M_1 and M_2 were set at 5,000. In Scenario 2, this study again initialized the inlet velocity and temperature as $U_x = 1$ m/s, $U_y = 0.5$ m/s, and $T = 6.85$°C for the first case (Case 2a) and $U_x = 1$ m/s, $U_y = 0.5$ m/s, and $T = 21.85$°C for the second case (Case 2b). Fig. 3.13 shows that the objective function decreased gradually to about 0.02 in 60 design cycles. The corresponding average temperature difference between the temperature design objective and the calculated

(a)

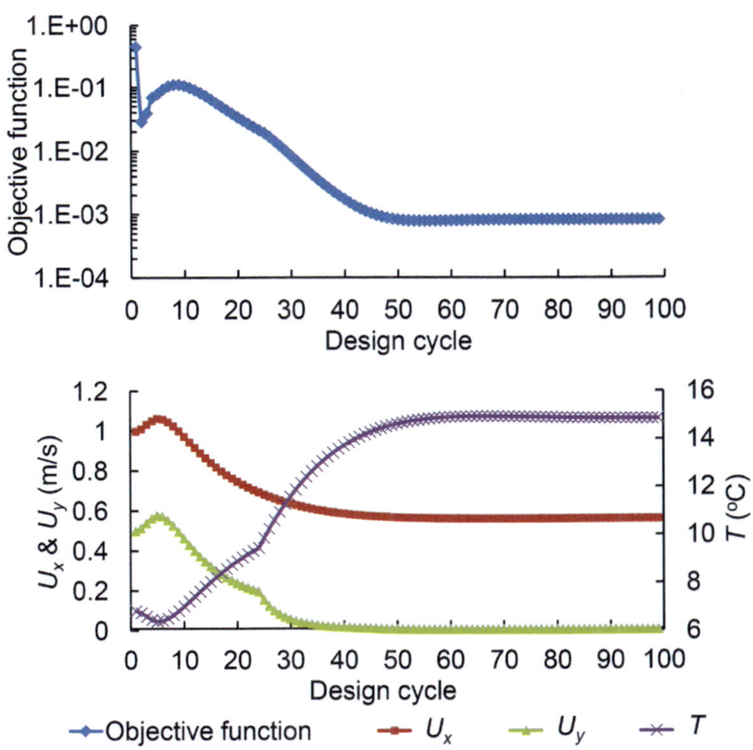

Figure 3.11 caption on p.78 (continued)

(continued)

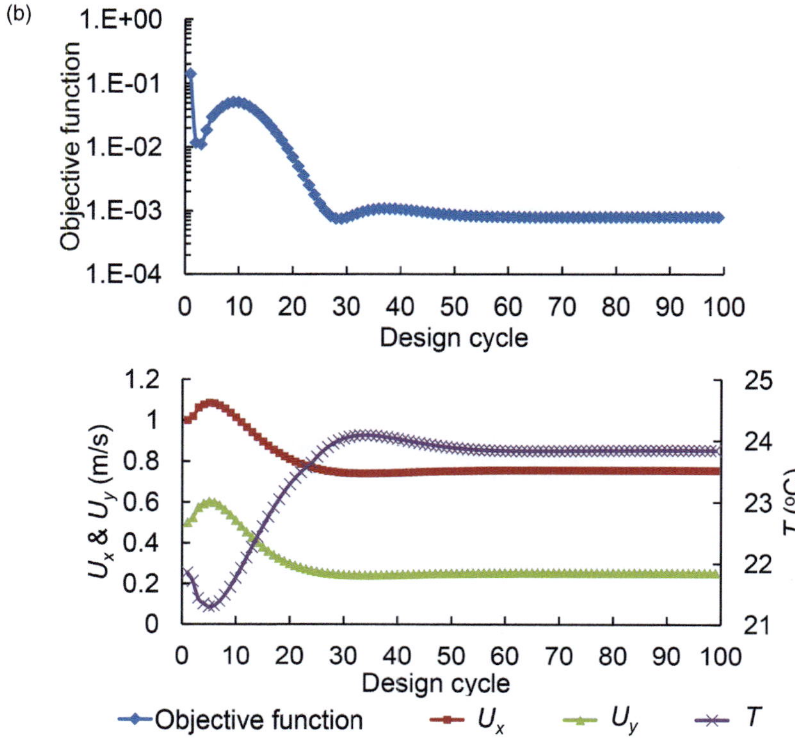

Figure 3.11 Changes in objective function and in U_x, U_y, and T at the inlet with the number of design cycles, with partial flow field as design objective: (a) Case 1a with initial inlet condition of $U_x = 1$ m/s, $U_y = 0.5$ m/s, and $T = 6.85$°C and (b) Case 1b with initial inlet condition of $U_x = 1$ m/s, $U_y = 0.5$ m/s, and $T = 21.85$°C

temperature distribution in the lower part of the mid-cavity was about 0.01 K. However, the optimal inlet air conditions were $U_x = 0.324$ m/s, $U_y = 0.477$ m/s, and $T = 12.44$°C for Case 2a and $U_x = 0.870$ m/s, $U_y = 0.482$ m/s, and $T = 16.47$°C for Case 2b, which are quite different from the true values of $U_x = 0.57$ m/s, $U_y = 0$ m/s, and $T = 15$°C. We also tried some other initial conditions, and the calculation always approached optimal solutions that differed from the true value, although the objective function became as small as 10^{-2}. Because different inlet conditions can lead to similar temperature distributions in the lower part of the mid-cavity, multiple optimal solutions were not a complete surprise to us.

Scenario 3: air velocity and temperature as the objective The design objective in Scenario 3 was to satisfy the air velocity and temperature profile in the lower part of the mid-cavity. This study again used an initial inlet condition of $U_x = 1$ m/s, $U_y = 0.5$ m/s, and $T = 6.85$°C in Case 3a, and a condition of $U_x = 1$ m/s, $U_y = 0.5$ m/s, and $T = 21.8$ 5°C in Case 3b. M_1 and M_2 were set at 1,000, which is a trade-off between the convergence of the calculations and the computing effort. For each condition, we calculated no more than 400 design cycles. As shown in Fig. 3.14, the final thermo-fluid boundary condition at the inlet for Case 3a was $U_x = 0.528$ m/s, $U_y = 0.109$ m/s, and $T = 14.63$°C, and for Case 3b it was $U_x = 0.537$, $U_y = 0.084$ m/s, and $T = 14.68$°C. The two final inlet conditions agreed well with the true condition at the inlet, which was $U_x = 0.57$ m/s, $U_y = 0$ m/s, and $T = 15$°C. It seems that

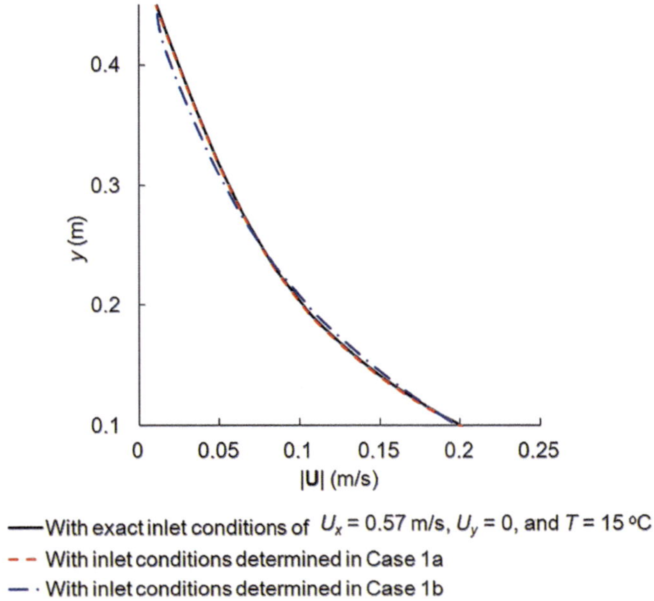

Figure 3.12 Comparison of the design objective (velocity profiles) determined by the adjoint method and that determined from the specified inlet condition of $U_x = 0.57$ m/s, $U_y = 0$ m/s, and $T = 15$°C at $x = 0.52$ m

(a)

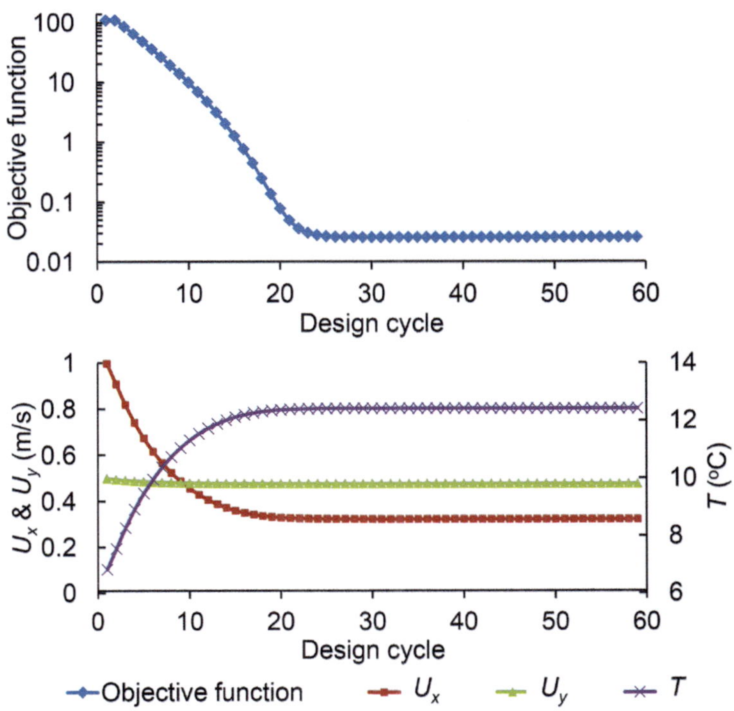

Figure 3.13 caption on p.80 (continued)

(continued)

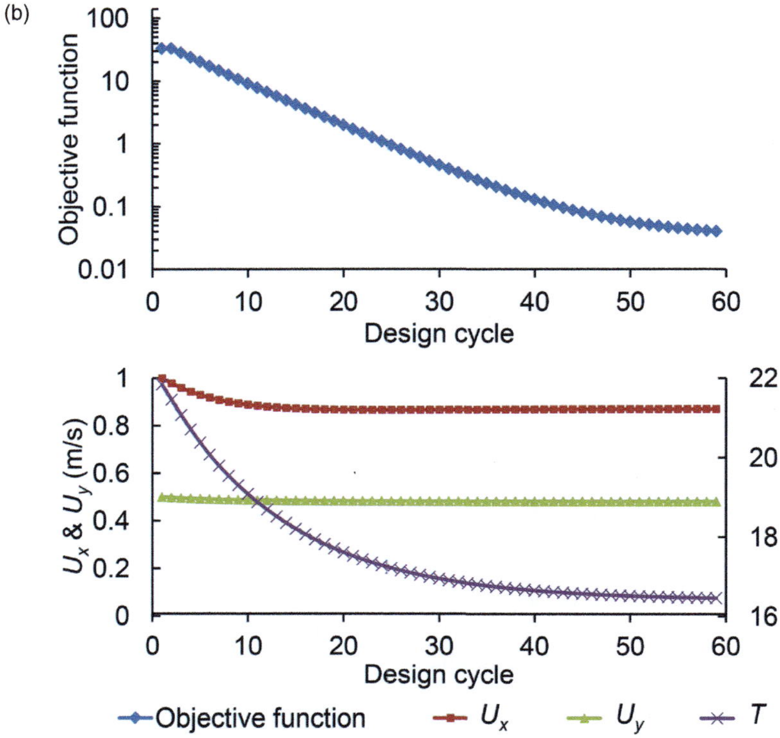

Figure 3.13 Changes in U_x, U_y, and T at the inlet and objective function with the number of design cycles, with partial temperature field as design objective: (a) Case 2a with initial inlet condition of $U_x = 1$ m/s, $U_y = 0.5$ m/s, and $T = 6.85$°C and (b) Case 2b with initial inlet condition of $U_x = 1$ m/s, $U_y = 0.5$ m/s, and $T = 21.85$°C

the optimal solution was unique when both the air velocity and temperature profiles were specified as the design objective.

Fig. 3.14 also shows that the objective function for air velocity became 10^{-3}, and that for air temperature became 10^{-2}. The corresponding average velocity in the lower part of the mid-cavity differed by only 0.003 m/s from the design objective, while the average air temperature differed by only 0.01 K from the design objective.

3.3.1.1.3 DISCUSSION

In the isothermal case, within 1 iteration/cycle × 2,000 cycles, the adjoint method was able to find the optimal inlet air velocity. The calculation time was roughly twice that required by forward CFD simulation for 4,000 iterations (no internal iteration) in solving the flow. However, in the third scenario of the non-isothermal case, the adjoint method was also able to find the optimal inlet conditions in 1,000 iterations/cycle × 400 design cycles. A more complex flow can increase the computing effort by several orders of magnitude.

In addition, the more design objectives there are, such as in Scenario 3, the lower the likelihood that multiple solutions will occur. It is possible that if too many design objectives are specified, then no solution can be obtained.

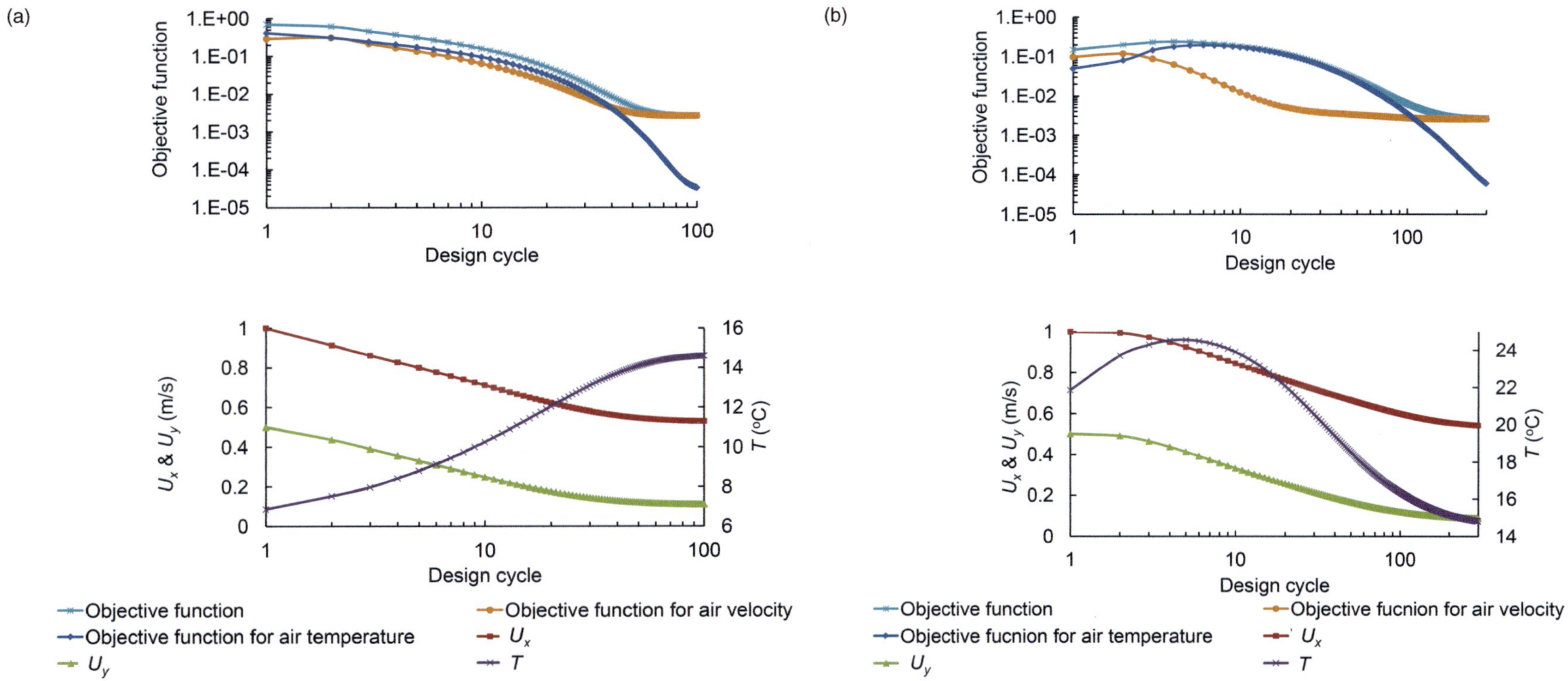

Figure 3.14 Changes in objection function and in U_x, U_y, and T at the inlet with the number of design cycles, with partial flow and temperature fields as design objective: (a) Case 3a with initial inlet condition of $U_x = 1$ m/s, $U_y = 0.5$ m/s, and $T = 6.85$°C and (b) Case 3b with initial inlet condition of $U_x = 1$ m/s, $U_y = 0.5$ m/s, and $T = 21.85$°C

Because this study aimed to verify the applicability of the adjoint method to the optimal design of an indoor environment, the only design variables were the inlet air velocity and temperature. This would seem to be a waste of the adjoint method's capability to inversely identify a large number of design variables (Jameson, 1988; Jameson 1995). However, the design variables in this study were of two different types (velocity and temperature), in contrast to the application of the method in other areas. This study verified the inverse identification of design variables of different types.

On the basis of the results obtained in this study, we expanded our investigation to two more design variables, inlet location and size. We also considered three-dimensional cases such as an aircraft cabin and investigated the setup of design objectives in real engineering cases and the influence of design objectives on the solution procedure.

3.3.1.2 Inverse identification of air supply location, size, temperature, and velocity

3.3.1.2.1 VALIDATION WITH AN ATTAINABLE DESIGN OBJECTIVE

To validate the CFD-based adjoint method, this study used the inverse identification of air supply inlet location, inlet size, temperature, and velocity for a two-dimensional, non-isothermal case with experimental data from Blay et al. (1992). As shown in Fig. 3.15(a), the dimensions of the flow domain were $H \times H = 1.04\ \text{m} \times 1.04\ \text{m}$, the inlet location was $y_{in}/H = 0.991$ (where y_{in} is the y coordinate of the center point of the inlet), the inlet height was $h_{in} = 18$ mm, and the outlet height was 24 mm. In the experiment, the inlet air velocity was $U_x = 0.57$ m/s, $U_y = 0.0$ m/s, and the inlet air temperature (T_{in}) was 15°C. The temperature of the walls (T_{wall}) was 15°C, while that of the floor (T_{floor}) was 35.5°C in order to form a thermal plume. This case is representative of airflows in enclosed environments, as it includes major indoor flow characteristics such as jets and thermal plumes. Furthermore, it was not desirable to use a very complex case for this investigation because of the inherent uncertainties in such a case. Instead, this simple two-dimensional case, which nevertheless incorporated complex flow features, was more desirable for validation.

In their experiment, Blay et al. (1992) had measured the air velocity and temperature on the red dashed lines shown in Fig. 3.15(a). Our inverse simulation thus set the measured air velocity and temperature as a design objective for the purpose of simplification. This was because any complex design objective in an enclosed environment, such as PMV or the age of air, would be based on flow and/or temperature fields. Then the objective function for this case could be constructed as:

$$O(\xi) = \varpi_1 \int_{\Omega} \left|\mathbf{U} - \mathbf{U}_0\right|^2 d\Omega + \varpi_2 \int_{\Omega} (T - T_0)^2 d\Omega \tag{3.91}$$

where ϖ_1 and ϖ_2 were chosen to adjust the relative importance of the two integrals and Ω denotes the design domain represented by the red dashed lines in Fig. 3.15(a). This design objective was clearly attainable because the measured data from the experiment with the corresponding boundary conditions were used in the inverse identification. The boundary conditions for the air supply inlet were velocity $U_x = 0.57$ m/s and $U_y = 0$, temperature $T = 15$°C, location $y_{in}/H = 0.991$, and size $h_{in} = 0.018$ m. Thus, to be successful, the inverse simulation had to find the air supply parameters. Fig. 3.9 shows the resolution of the mesh when the inlet was located at $y_{in}/H = 0.991$. The mesh resolution at the inlet, outlet, and near-wall region was finer because of the high gradients of velocity and temperature in those locations. This principle was applied to regenerate the mesh whenever the air supply inlet location and size changed in the calculation.

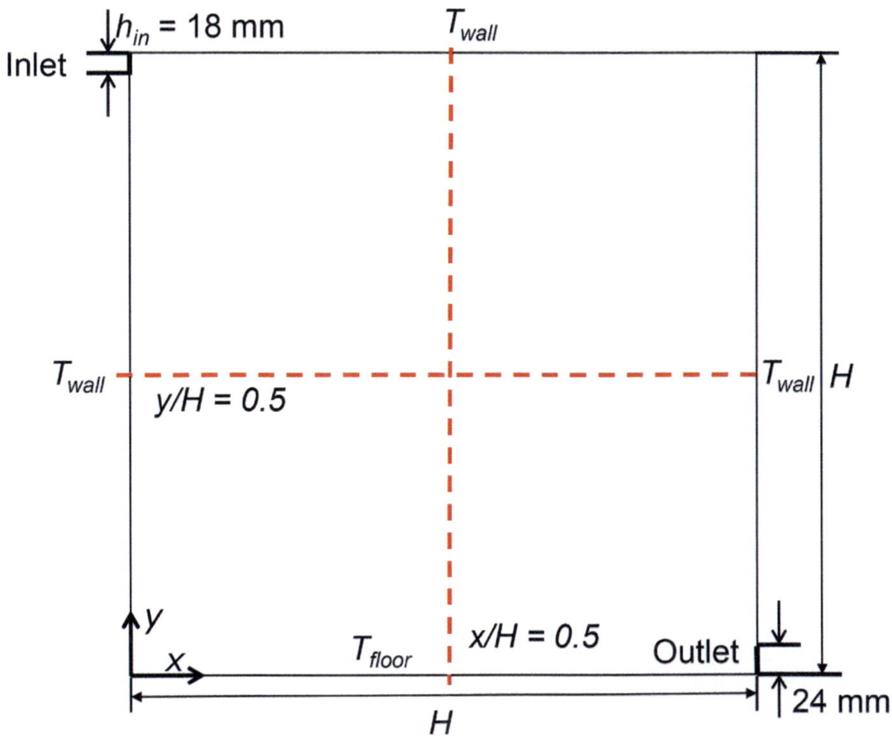

Figure 3.15 Sketch of the two-dimensional, non-isothermal case

For inverse identification of the air supply inlet location, inlet size, temperature, and velocity, a number of design variables were added one by one to test the robustness of the method:

- Scenario 1-1: Inverse identification of the air supply inlet location.
- Scenario 1-2: Inverse identification of the inlet location and supply temperature and velocity.
- Scenario 1-3: Inverse identification of the inlet location, inlet size, and supply temperature and velocity.

Scenario 1-1 used the air supply inlet location as a design variable. The inlet size was fixed at $h_{in} = 0.018$ m, and the supply temperature and velocity were fixed at $U_x = 0.57$ m/s, $U_y = 0$ m/s, and $T_{in} = 15$°C, respectively. Therefore, the optimal air supply location should be $y_{in}/H = 0.991$. Scenario 1-2 assumed that the air supply parameters were unknown. However, the airflow rate was constrained by:

$$(\mathbf{U}_{in} \cdot \vec{n})h_{in} = Q \tag{3.92}$$

where $Q = 0.01026$ m^2/s in this study. Scenario 1-3 assumed that the inlet size was a design variable. For all the scenarios, this study set $\alpha = \beta = 0.01$, $\varpi_1 = \varpi_2 = 1$, and $M_1 = M_2 = 7{,}000$. Table 3.2 summarizes the thermo-fluid conditions for the three scenarios.

Fig. 3.16(a) shows that in all the inverse calculations, variation in the objective function $|O_{new} - O_{old}|$ was smaller than 10^{-2}, and thus the convergence criterion was met. In this

Table 3.2 Summary of the studied scenarios.

Scenario	*Design variables*	*Fixed boundary conditions*	*Initial conditions*
1-1	y_{in}	$U_{x,\,in} = 0.57$ m/s, $U_{y,\,in} = 0$ m/s, $T_{in} = 15$°C, $h_{in} = 0.018$ m, $T_{wall} = 15$°C, $T_{floor} = 35.5$°C	$y_{in}/H = 0.5$
1-2	y_{in}, $U_{y,\,in}$, T_{in}	$Q = 0.01026$ m²/s, $h_{in} = 0.018$ m, $T_{wall} = 15$°C, $T_{floor} = 35.5$°C	$y_{in}/H = 0.5$, $U_{y,\,in} = 0.2$ m/s, $T_{in} = 10$°C
1-3	y_{in}, $U_{x,\,in}$, $U_{y,\,in}$, T_{in}, h_{in}	$Q = 0.01026$ m²/s, $T_{wall} = 15$°C, $T_{floor} = 35.5$°C	$y_{in}/H = 0.5$, $h_{in} = 0.0342$ m, $U_{x,\,in} = 0.3$ m/s, $U_{y,\,in} = 0.2$ m/s, $T_{in} = 10$°C

figure, the number of design cycles is the number of times that the design variables have been updated. Thus, this method was able to inversely identify the locally optimal design variables. The convergence speed for the three scenarios was similar. The increase in the number of design variables did not increase the computational effort, which is a strong feature of the adjoint method. Although all the calculations produced results that were similar to the design objective, there were significant differences in the optimal air supply inlet location, inlet size, temperature, and velocity that were obtained in the three scenarios, as shown in Fig. 3.16(b). It indicates the existence of multiple local optima. The possible reason was that the different combinations of air supply parameters were able to produce similar air distributions in the design domain.

There was a sudden jump in the objective function for Scenario 1-3 at design cycle 16. This jump was caused by a sudden change in the flow feature, which appears at design cycle 17 in Fig. 3.17(c). As depicted in Fig. 3.18, the temperature profiles also changed greatly. The objective function for Scenario 1-3 continued to fluctuate after design cycle 20. The reason for this fluctuation was that the inlet was positioned close to the upper wall and its location fluctuated. The fluctuation in the inlet location (Fig. 3.17(b)) did not strongly affect the flow field, but it did influence the heat transfer between the jet and the upper wall. As shown in Fig. 3.18, the velocity profiles did not change so significant as the temperature

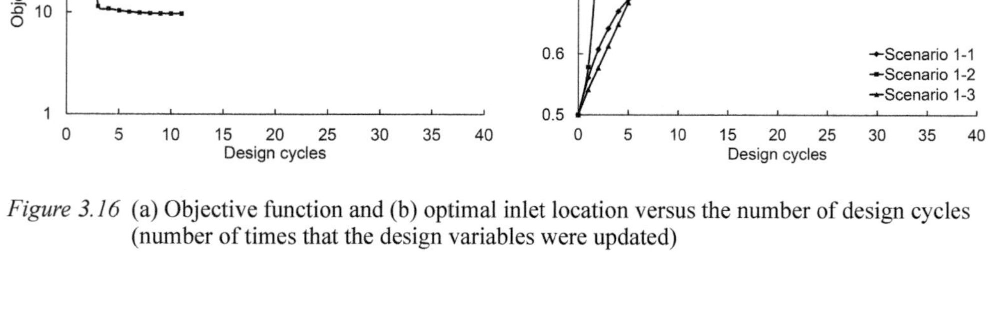

Figure 3.16 (a) Objective function and (b) optimal inlet location versus the number of design cycles (number of times that the design variables were updated)

profiles from design cycle 20 to design cycle 21. Fig. 3.17(d) shows the optimal flow field at design cycle 40 for Scenario 1-3. It differs significantly from the flow field for the initial condition, shown in Fig. 3.17(a).

When the air supply inlet size was set as a design variable in Scenario 1-3, the air supply velocity in the x direction was also a design variable because of the requirement of a constrained air flow rate. The results computed by the CFD-based adjoint method provided simultaneous variations in inlet size h_{in} and air supply velocity in the x direction $U_{x,\, in}$. When both of these parameters change, it is difficult or impossible to maintain a constant air flow rate. Therefore, this study varied $U_{x,\, in}$ and used Eq. (3.92) to determine the air supply size h_{in}. The results in Fig. 3.18 indicate that this approach is effective for the inverse identification of air supply inlet size in an enclosed environment.

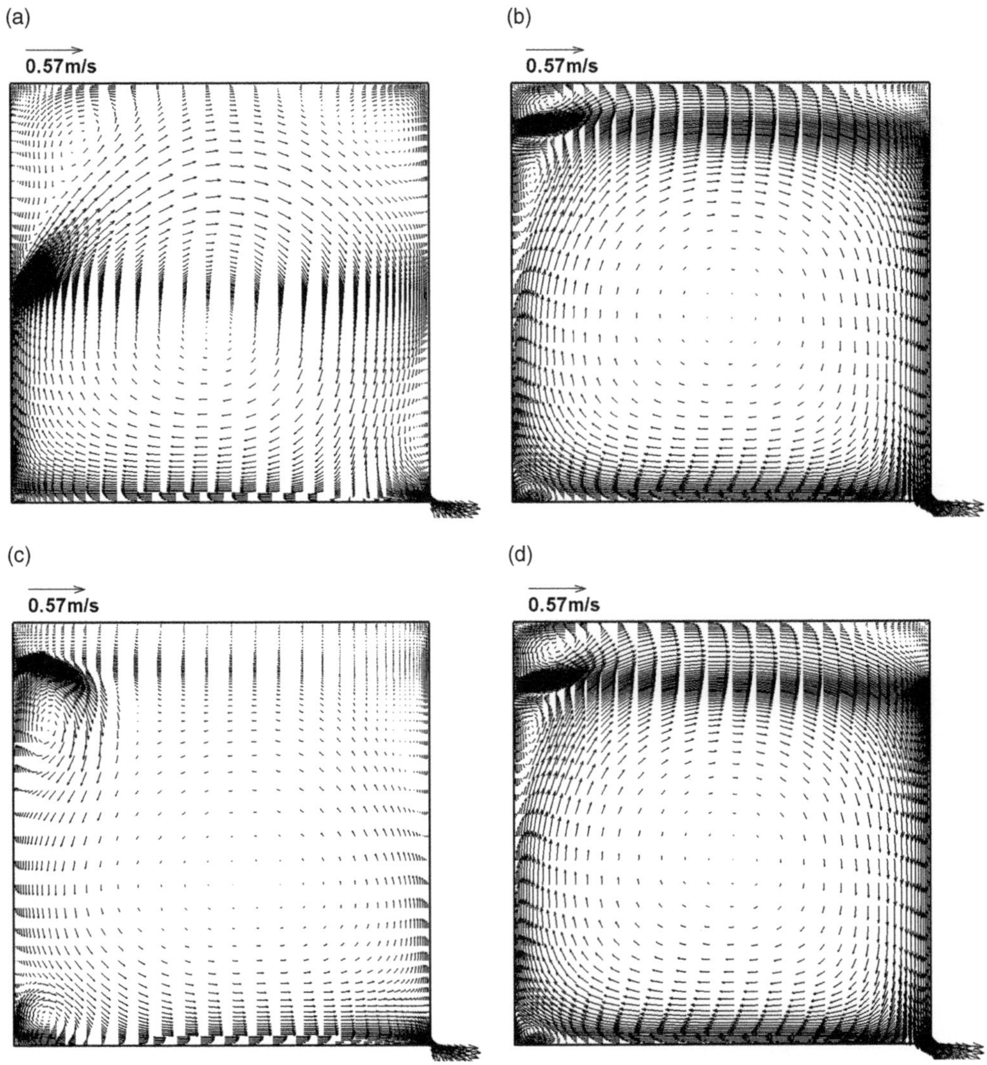

Figure 3.17 Computed flow fields at (a) the initial condition, (b) design cycle 16, (c) design cycle 17, and (d) design cycle 40 for Scenario 1-3 (where the design cycle is the number of times that the design variables were updated)

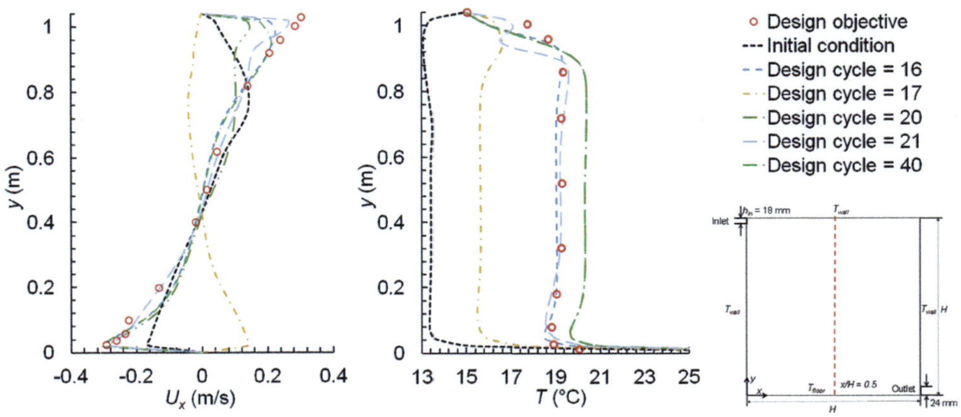

Figure 3.18 Comparison of the velocity and temperature profiles determined by the adjoint method and the design objective at $x/H = 0.5$ for Scenario 1-3

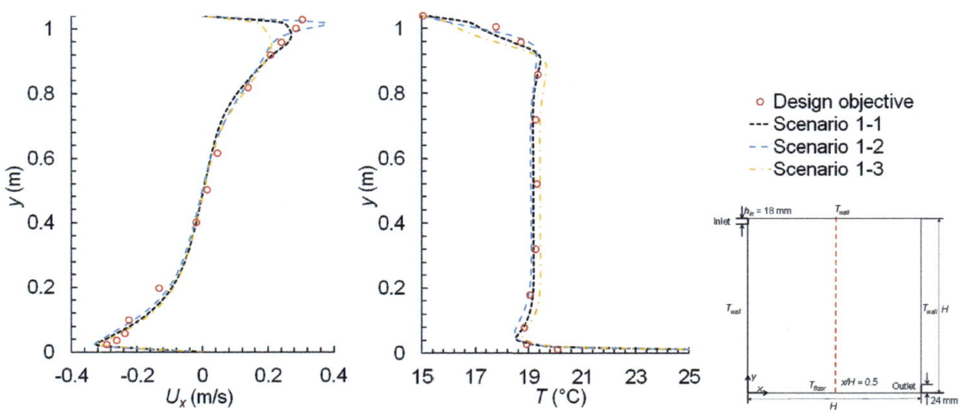

Figure 3.19 Comparison of the optimal velocity and temperature profiles determined by the adjoint method with the design objective at $x/H = 0.5$

The air supply inlet location, inlet size temperature, and velocity obtained by the CFD-based adjoint method were then used as inputs for forward CFD simulations to calculate the air distribution in the cavity. Fig. 3.19 shows the resulting velocity and temperature profiles at $x/H = 0.5$. A comparison of the profiles with the design objective (Blay et al., 1992) at $x/H = 0.5$ reveals that only the velocity profiles near the top wall differed slightly from the design objective. This finding indicates that multiple solutions for the air supply parameters can be obtained by means of the adjoint method, and all the solutions can lead to similar design objectives.

3.3.1.2.2 DEMONSTRATION WITH AN IDEAL DESIGN OBJECTIVE

The second case demonstrated the use of the CFD-based adjoint method to obtain air supply inlet location, inlet size, temperature, and velocity with an ideal design objective in a design domain. On the basis of the two-dimensional, non-isothermal case discussed in Section 3.1, we set a desired air velocity of $\mathbf{U}_0 = (0.1, 0)$ m/s and desired air temperature of $T_0 = 22$°C in the design domain Ω (the red dashed line from $y = 0.05$ m to $y = 0.5$ m in Fig. 3.20) as the design objective. The design domain was analogous to the distance from the ankle to the head of an occupant in the middle of the occupied zone in a room. Eq. (3.91) was again used as the objective function.

This study conducted the optimization for three scenarios. The first two scenarios were constrained by airflow rate according to Eq. (3.92) and had two different, arbitrary sets of initial air supply parameters. For example, in Scenario 2-1 the initial inlet location was in the middle of the left wall ($y_{in}/H = 0.5$), while in Scenario 2-2 the inlet was located in the lower part of the left wall ($y_{in}/H = 0.025$). In Scenario 2-3, the airflow rate was not constrained; rather, it was set arbitrarily. For all the scenarios, we set $\alpha = \beta = 0.1$, $\varpi_1 = 100$, $\varpi_2 = 1$, and $M_1 = M_2 = 7{,}000$. Table 3.3 summarizes the information for the three scenarios.

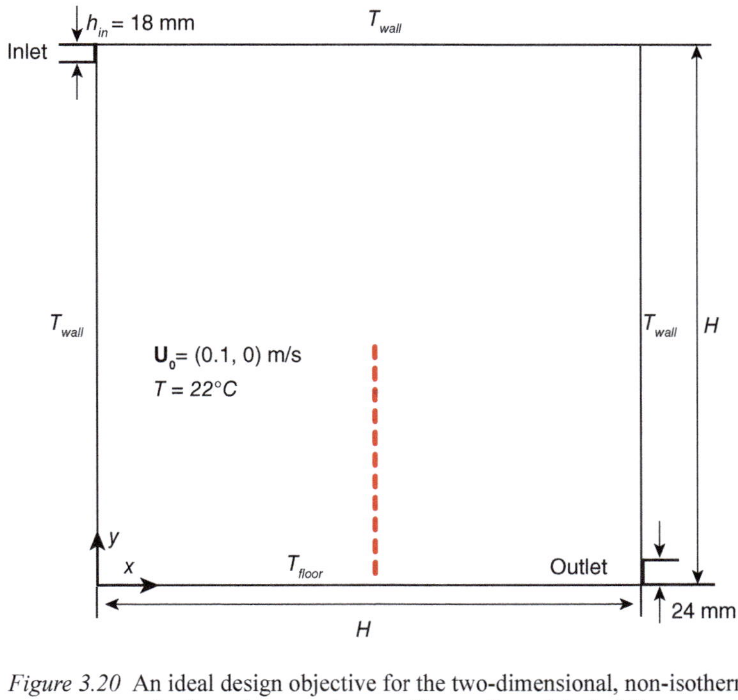

Figure 3.20 An ideal design objective for the two-dimensional, non-isothermal case: the air velocity and temperature profiles on the red dashed line

Table 3.3 Summary of the optimized scenarios

Scenario	*Design variables*	*Fixed values*	*Initial air supply conditions*
2-1		$Q = 0.01026$ m²/s, $T_{wall} = 15$°C, $T_{floor} = 35.5$°C	$y_{in}/H = 0.5$, $h_{in} = 0.0513$ m, $U_{x,in} = 0.2$ m/s, $U_{y,in} = -0.2$ m/s, $T_{in} = 10$°C
2-2	y_{in}, $U_{x,in}$, $U_{y,in}$, T_{in}, h_{in}		$y_{in}/H = 0.025$, $h_{in} = 0.0513$ m, $U_{x,in} = 0.2$ m/s, $U_{y,in} = 0$ m/s, $T_{in} = 10$°C
2-3		$T_{wall} = 15$°C, $T_{floor} = 35.5$°C	$y_{in}/H = 0.049$, $h_{in} = 0.1026$ m, $U_{x,in} = 0.1$ m/s, $U_{y,in} = 0$ m/s, $T_{in} = 20$°C

Fig. 3.21(a) shows that in all three scenarios, the variation in the objective function $|O_{new}-O_{old}|$ was smaller than 10^{-1} after about 50 design cycles. This means that convergence was achieved although the optimal air supply locations were different, as shown in Fig. 3.21(b). Therefore, this method was able to inversely identify locally optimal air supply parameters. In Scenario 2-1, there was a sudden jump in the objective function at the 19th design cycle because of a rapid change in the flow pattern, as depicted in Fig. 3.22. The flow in the lower part of the cavity at design cycle 18 was clockwise, and it became counter-clockwise at design cycle 19. The optimization process dramatically changed the flow fields. Fig. 3.23 shows that the change in the flow features also caused the temperature profiles to differ greatly from design cycle 18 to design cycle 19. Furthermore, the optimal velocity and temperature profiles for Scenario 2-1 at design cycle 46 differed significantly from the initial condition.

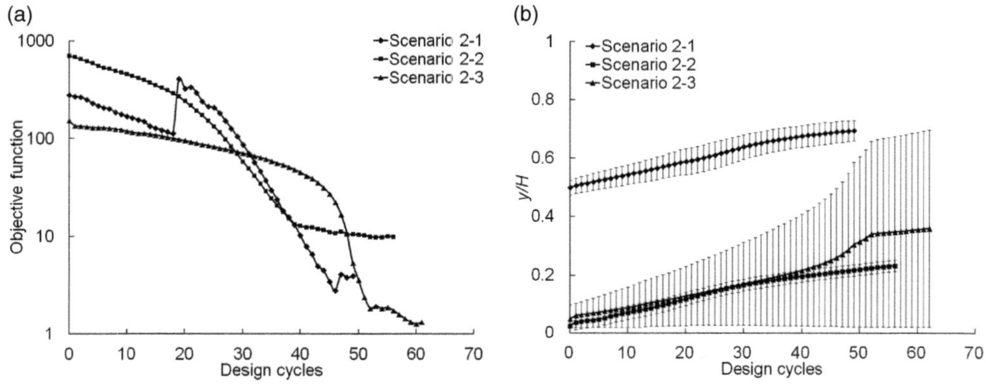

Figure 3.21 (a) Objective function and (b) inlet location (solid lines) and size (vertical bars) determined by the adjoint method versus the design cycle (the number of times that the design variables were updated)

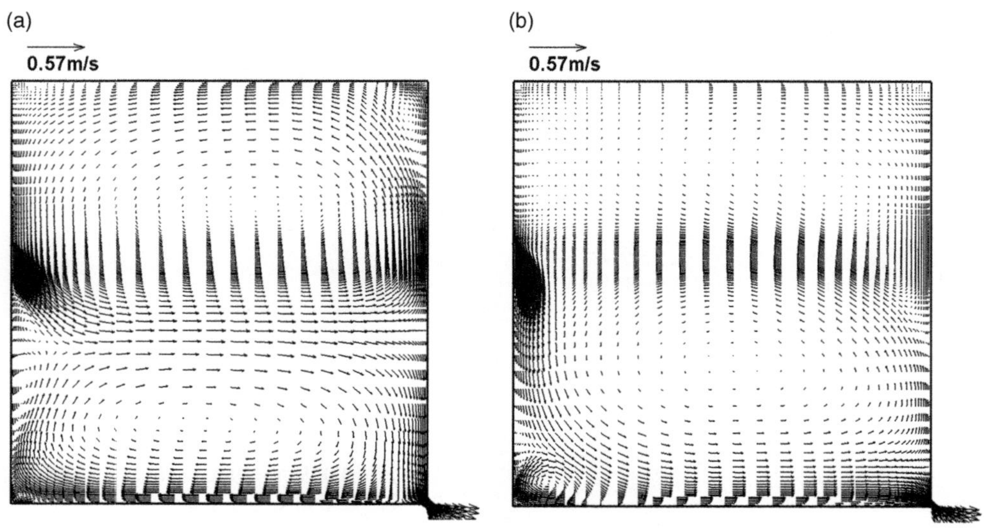

Figure 3.22 Computed flow fields at (a) design cycle 18 and (b) design cycle 19 for Scenario 2-1 (where the design cycle is the number of times that the design variables were updated)

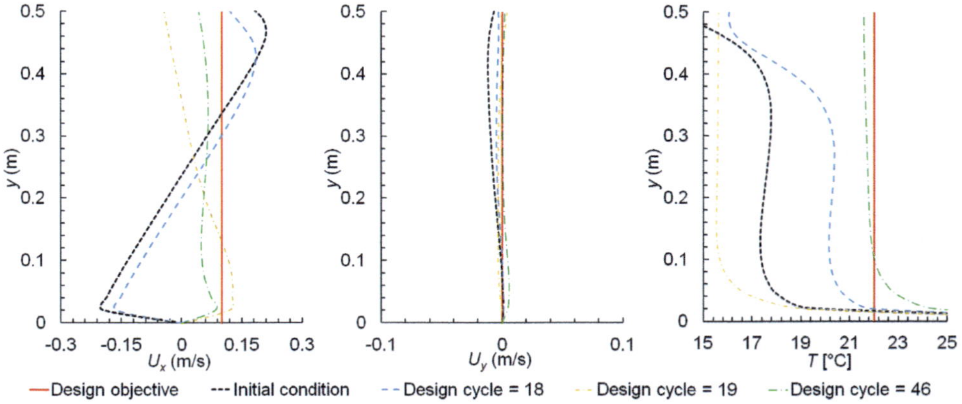

Figure 3.23 Comparison of the velocity and temperature profiles determined by CFD using air supply inlet location, inlet size, temperature, and velocity obtained by the adjoint method, with the design objective on the design domain for Scenario 2-1

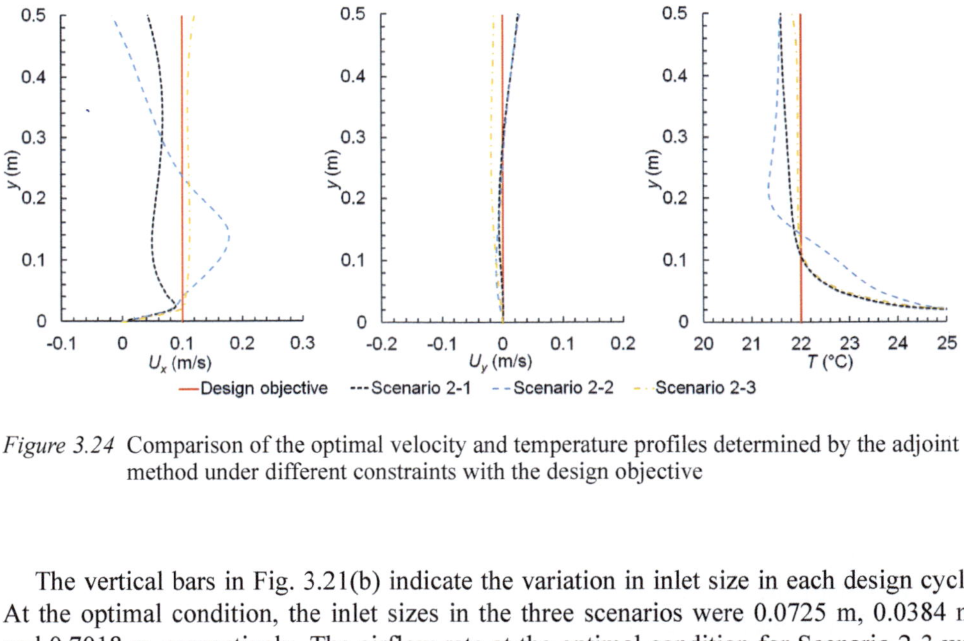

Figure 3.24 Comparison of the optimal velocity and temperature profiles determined by the adjoint method under different constraints with the design objective

The vertical bars in Fig. 3.21(b) indicate the variation in inlet size in each design cycle. At the optimal condition, the inlet sizes in the three scenarios were 0.0725 m, 0.0384 m, and 0.7018 m, respectively. The airflow rate at the optimal condition for Scenario 2-3 was 0.07199 m^2/s, which was about seven times higher than the constrained airflow rates for Scenarios 2-1 and 2-2. Fig. 3.24 shows the velocity and temperature profiles calculated by the forward CFD simulation, with the air supply parameters obtained by the adjoint method as inputs. When the airflow rate was constrained, the calculated velocity and temperature profiles deviated significantly from the desired design objective. Only the air velocity and temperature profiles for Scenario 2-3 were similar to those set in the objective function. Thus, the desired design objective can be obtained only when there is sufficient freedom to choose the corresponding boundary conditions.

3.3.1.2.3 DISCUSSION

The calculations in the previous two sections converged, but they led to different values for the objective functions. For the inverse identification problems with an attainable design objective in Section 3.3.1.2.1, the objective function was expected to be zero if the experimental data contained no errors and the predicted air distributions by CFD simulation were the same as the experimental data. However, errors in the experimental data were inevitable, because of the uncertainties contributed by the measuring instruments and the approximations in the turbulence models used in the simulation. Therefore, the air velocity and temperature profiles predicted by CFD simulation in the design procedures differed from the design objective. These differences explain the relatively large final objective functions in Section 3.3.1.2.1. For the inverse identification problems with an ideal design objective in Section 3.3.1.2.2, the final objective functions for Scenarios 2-1 and 2-2 differed because of the existence of multiple solutions. Furthermore, the final objective function for Scenario 2-3 was much smaller than that in Section 3.3.1.2.1. The possible reasons were a smaller design domain and more freedom in the choice of the corresponding boundary conditions in Scenario 2-3.

Application of the CFD-based adjoint method to the two-dimensional cases can yield multiple optimal solutions even with a single design variable. Multiple trials are required to obtain globally optimal solutions, which is time-consuming. To improve the computational efficiency, one may first take advantage of the CFD-based genetic algorithm (GA) method to determine the initial air supply inlet location, temperature and velocity. This process may lead to multiple solutions. Next, optimization with the CFD-based adjoint method can be conducted with these pre-determined initial air supply parameters to determine whether or not they lead to a unique solution. Theoretically, the combined CFD-based GA and adjoint method would be faster than the pure CFD-based GA. The combined method would also find the global optima, thus compensating for the shortcomings of the pure CFD-based adjoint method. Xue and Zhai (2014) recently tried to accelerate the GA method by integrating it with the adjoint method. This new integrated method is advantageous when the number of variables for optimization is large. In addition, our future research will attempt to solve the NS/adjoint equations by means of fast fluid dynamics (Zuo and Chen, 2009), which could further improve the computational efficiency.

The primary simplifications in this study were to the design variables and the adjoint equations of the turbulence models. For example, we assumed that the inlet air velocity was uniform. The simplifications to the design variables made it easier to demonstrate the strength of the method, although the adjoint method can also handle more complex cases. This study also simplified the turbulence models used in the adjoint equations by "freezing" the turbulence level in order to reduce the computing time. According to Dwight and Brezillon (2006), the accuracy of turbulence approximations is case dependent. Fortunately, the simplification seemed acceptable in our case.

3.3.2 Optimal design of the built environment

3.3.2.1 Inverse design of thermal comfort and indoor air quality in buildings

This section describes the process of defining a desirable design objective, determining the design parameters, and using the CFD-based method in the inverse design of the thermal environment in an office.

3.3.2.1.1 RESEARCH METHOD

Design objective In an enclosed environment such as an office, the desired thermal comfort level can be determined by means of the predicted mean vote (PMV) index (Fanger, 1970). The value of PMV can be calculated by:

$$\begin{aligned}\text{PMV} = {} & (0.303e^{-0.036m} + 0.028)\{m - W - 3.05 \times 10^{-3} \times [5733 - 6.99(m - W) - p_w] \\ & - 0.42 \times [(m - W) - 58.15] - 1.7 \times 10^{-5} m(5867 - p) - 0.0014m(34 - T) \\ & - 3.96 \times 10^{-8} f_{cl} \times [(T_{cl} + 273)^4 - (T_r + 273)^4] - f_{cl} h_c (T_{cl} - T)\}\end{aligned} \tag{3.93}$$

where

$$\begin{aligned}T_{cl} = {} & 35.7 - 0.028(m - W) - I_{cl}\{3.96 \times 10^{-8} f_{cl} \times \\ & [(T_{cl} + 273)^4 - (T_r + 273)^4] + f_{cl} h_c (T_{cl} - T)\}\end{aligned} \tag{3.94}$$

$$h_c = \begin{cases} 2.38(T_{cl} - T)^{0.25} \text{ if } 2.38(T_{cl} - T)^{0.25} > 12.1\sqrt{|\mathbf{U}|} \\ 12.1\sqrt{|\mathbf{U}|} \text{ if } 2.38(T_{cl} - T)^{0.25} < 12.1\sqrt{|\mathbf{U}|} \end{cases} \tag{3.95}$$

$$f = \begin{cases} 1.00 + 1.290 I_{cl} & \text{for } I_{cl} \le 0.078 m^2\,°C/W \\ 1.05 + 0.645 I_{cl} & \text{for } I_{cl} > 0.078 m^2\,°C/W \end{cases} \tag{3.96}$$

In these equations, W (W) is the external work accomplished, m (W/m^2) the metabolic activity, I_{cl} (clo) the thermal resistance of the passengers' clothing, T (°C) the air temperature, T_r (°C) the mean radiant temperature, T_{cl} (°C) the clothing temperature, **U** (m/s) the air velocity, and p_w (Pa) the relative humidity or partial water vapor pressure.

The design objective of this study was to achieve the most thermally comfortable indoor environment, where PMV should approach zero. Thus, the corresponding objective function was constructed as:

$$O(\xi) = \int_{\Omega} (PMV)^2 d\Omega \Big/ \int_{\Omega} d\Omega \tag{3.97}$$

where Ω is the design domain, which is normally the area around the occupants, and ξ is a vector that denotes a series of design variables.

Design parameters In an office, many of the parameters used to calculate PMV are known. For example, this study set the metabolic activity level to 75 W for passengers in a relaxed state and the thermal resistance of the passengers' clothing (clo) to 0.57 clo for summer (ASHRAE Handbook, 2005). The external work accomplished was assumed to be 0 W. The partial water vapor pressure (p_w) of ambient air was set to 1580 Pa, which corresponds to an average relative humidity of 40%. For simplicity, this study assumed the mean radiation temperature to be the same as the air temperature in the cabin. Then the PMV becomes a function only of air temperature and velocity.

The air temperature and velocity fields in an office are determined primarily by design variables such as the geometry of the interior furnishings, internal thermal loads, air supply velocity and temperature from the diffusers, inlet size and location, etc. For the purpose of demonstrating the adjoint method, this study set only the air supply velocity and temperature from the diffusers as the design variables. It should be noted that the computing time required by the adjoint method does not depend on the number of variables.

When PMV is used to evaluate the thermal environment in an office, with the above assumptions, we have the following variables and equations for inverse design of the indoor environment:

- State variables: air velocity **U** (vector), air pressure p, and air temperature T;
- Design variables ξ: inlet air velocity $\mathbf{U}_{\text{in}}$ (vector) and inlet air temperature T_{in}, where $\xi = (\mathbf{U}_{\text{in}}, T_{in})$;
- State equations for the airflow in the domain, which are Navier-Stokes (NS) equations denoted by $\mathbf{N} = (N_1, N_2, N_3, N_4, N_5)$.

This investigation applied the re-normalization group (RNG) k-ε model (Yakhot and Orszag, 1986), which is widely used in indoor airflow modeling (Zhang et al., 2007) to simulate turbulence in the state equations. The Boussinesq approximation (Boussinesq, 1903) was applied to simulate the buoyancy effect, while the air density was assumed to be constant. The design goal was to minimize the objective function with state variables subjected to the state equations.

Adjoint equations The final form of the adjoint equations in this study can be written as:

$$-\nabla \cdot \mathbf{U}_a = 0 \tag{3.98}$$

$$-\nabla \mathbf{U}_a \cdot \mathbf{U} - (\mathbf{U} \cdot \nabla)\mathbf{U}_a - \nabla \cdot (2\nu D(\mathbf{U}_a)) + \nabla p_a + T_a \nabla T + \mathbf{S}_{\mathbf{Ua}} = 0 \tag{3.99}$$

$$-\mathbf{U} \cdot \nabla T_a - \nabla \cdot (\kappa \nabla T_a) + B = 0 \tag{3.100}$$

$$\mathbf{S}_{\mathbf{Ua}} = \begin{cases} 2 \times \text{PMV} \times \dfrac{\partial \text{PMV}}{\partial \mathbf{U}} & \text{for domain } \Omega \\ \mathbf{0} \text{ for domain } \omega \setminus \Omega & \end{cases} \tag{3.101}$$

$$S_{Ta} = \begin{cases} 2 \times \text{PMV} \times \dfrac{\partial \text{PMV}}{\partial T} & \text{for domain } \Omega \\ 0 \text{ for domain } \omega \setminus \Omega & \end{cases} \tag{3.102}$$

The source terms $\mathbf{S}_{\mathbf{Ua}}$ and S_{Ta} in the adjoint equations are determined by the results of the state equations.

In each design cycle in this study, the NS/adjoint equations were calculated with 10,000 iterations to ensure convergence. The design convergence criteria were (1) $O < 0.01$ in the first design cycle and (2) $|O_i - O_{i-1}| < 0.01$ in the ith design cycle (i > 1). The O_{i-1} is the computed objective function in the previous design cycle.

3.3.2.1.2 CASE SETUP

This investigation conducted an inverse design of the thermal environment in an occupied office with displacement ventilation (Yuan et al., 1999) as shown in Fig. 3.25. The office was simulated by an experimental chamber with dimensions of 5.16 m × 3.65 m × 2.43 m, with a displacement diffuser on the side wall supplying air at a rate of 4 air changes per hour. The supply air temperature was 17°C. There were two occupants, two computers, two tables, two boxes, and six lamps in the room as shown in Fig. 3.25(a). For the detailed sizes, locations, and heat release of these items as well as other thermo-fluid boundary conditions, one can refer to Yuan et al. (1999). The experiment measured the air temperature and air velocity along nine vertical poles distributed in the stream-wise center plane and the cross-sectional center plane as shown in Fig. 3.25(b). This figure also shows the mesh that we selected on the basis of our grid independence test. The design domain Ω was set as the area 0.1 m away from the occupants.

We first used thermo-fluid boundary conditions from the experiment by Yuan et al. (1999) to validate the forward CFD simulation in OpenFOAM. The simulation predicted the air velocity and temperature in the office. Fig. 3.26 compares these parameters at two typical positions. The air velocity was normalized by the air supply velocity $|\mathbf{U}_{in}|$ = 0.09 m/s, and the air temperature was normalized by using $\theta = (T - T_{in})/(T_{exhaust} - T_{in})$, where the air supply temperature T_{in} = 17°C and the air exhaust temperate $T_{exhaust}$ = 26.4°C. We found good agreement between the experimental data and the numerical simulation results.

When a ventilation system is designed for an office, the thermo-fluid boundary conditions must be sufficient for removal of the cooling load and compliance with air quality regulations. Therefore, this investigation constrained the air flow rate by fixing the air supply velocity in the x direction. For the inverse design, we initialized the design variables as $\mathbf{U}_{in}$ = (0.09, 0, 0) m/s and T_{in} = 15°C.

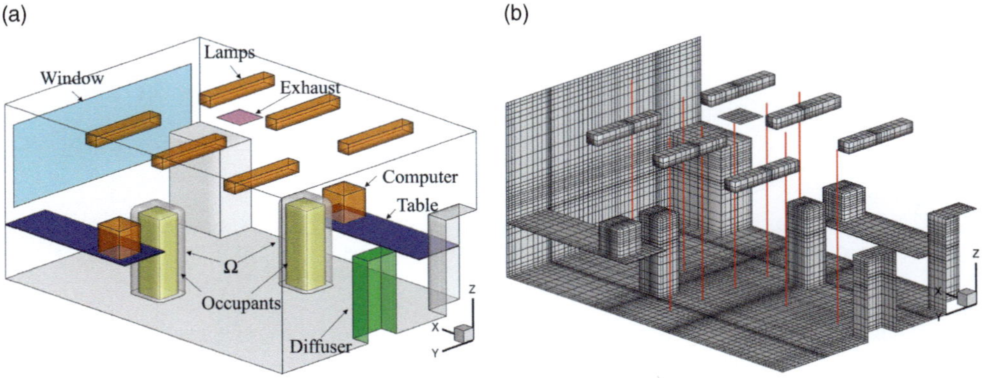

Figure 3.25 (a) Schematic and (b) mesh of displacement ventilation in an occupied office (where the red vertical lines represent measurement locations for air velocity and temperature)

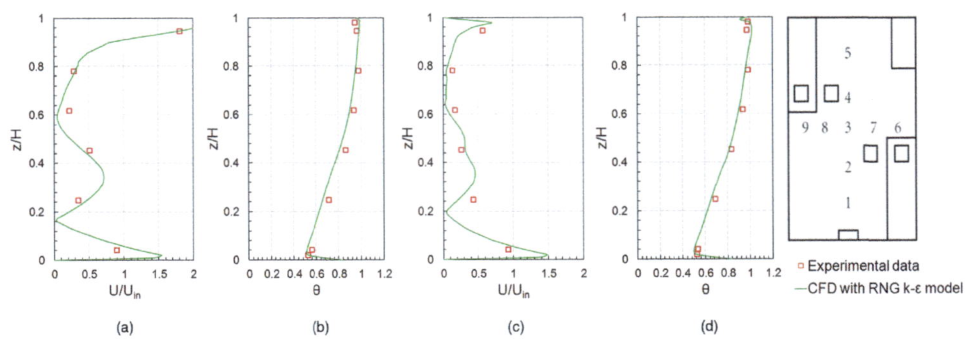

Figure 3.26 Comparison of air velocity and temperature at position 3 ((a) and (b)) and position 8 ((c) and (d))

3.3.2.1.3 RESULTS

This study used the CFD-based adjoint method to find the air supply velocity and temperature for the diffusers that would provide the most comfortable thermal environment in the office. Starting with the initial air supply conditions, Fig. 3.27 shows the changes in the objective function and design variables within 12 design cycles. The objective function and design variables all varied rapidly in the first three design cycles and then became relatively stable in subsequent cycles. The inverse design converged within nine cycles, but the simulation continued until the 12th design cycle had been completed. At the end of the optimization process, the objective function was only 0.08 and the air supply conditions were $U_{in,\ y}$ = -0.01 m/s, $U_{in,\ z}$ = 0.05 m/s, and T_{in} = 290.6 K. The corresponding average of the absolute PMV in the design domain was 0.25. Fig. 3.28 also shows that the thermal comfort level in this office was greatly improved after nine design cycles.

3.3.2.2 Inverse design of thermal comfort in aircraft cabins

This section describes the process of defining a desirable design objective, determining the design parameters, and using the CFD-based method in the inverse design of the thermal environment in an airliner cabin.

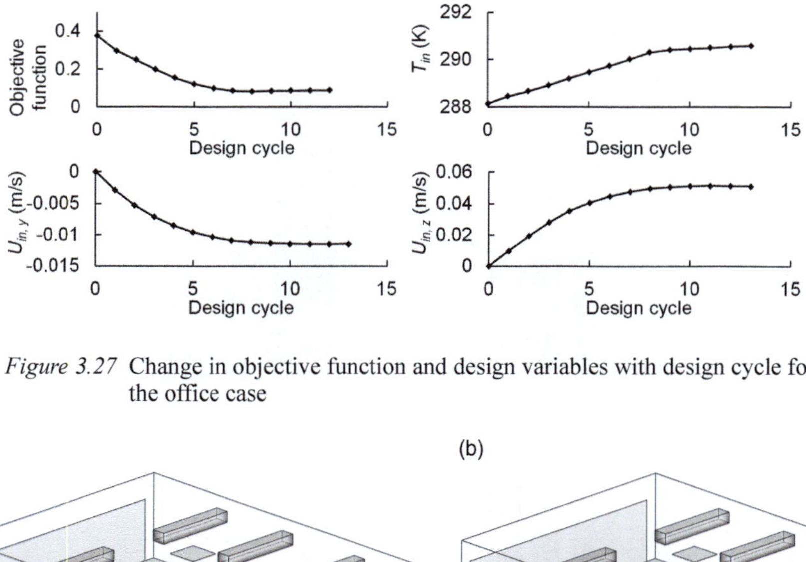

Figure 3.27 Change in objective function and design variables with design cycle for the office case

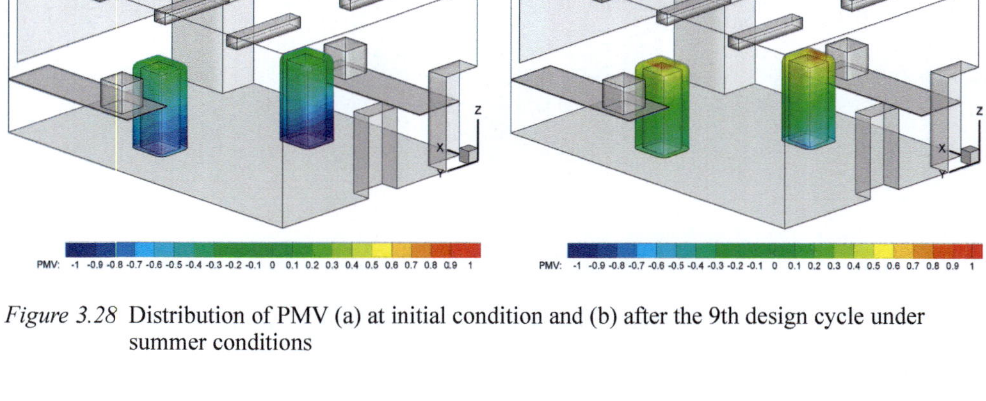

Figure 3.28 Distribution of PMV (a) at initial condition and (b) after the 9th design cycle under summer conditions

3.3.2.2.1 RESEARCH METHOD

Design objective In an enclosed environment such as a building, the desired thermal comfort level can be determined by means of the predicted mean vote (PMV) index (Fanger, 1970). Because an air cabin environment has a lower pressure than that at sea level, which will decrease human body heat loss through convection and increase it through evaporation (Cui et al., 2014), direct use of PMV may not be appropriate. This study used a modified predicted mean vote for air cabins (PMVc) developed by Cui et al. (2015) that accounts for low pressure in a cruising aircraft cabin. PMVc can be determined from PMV as shown in Eqs. (3.103) and (3.104) for the summer and winter seasons, respectively:

$$PMVc(\text{summer}) = -0.0758PMV^2 + 0.6757PMV - 0.1262 \tag{3.103}$$

$$PMVc(\text{winter}) = -0.0696PMV^2 + 0.6906PMV - 0.1369 \tag{3.104}$$

As with the PMV scale, the PMVc scale ranges from −3 (cold) to 0 (neutral) to 3 (hot). The value of PMV can be calculated by Eqs. (3.93) to (3.96). The design objective of this study was to achieve the most thermally comfortable cabin environment, where PMVc should approach zero. Thus, the corresponding objective function was constructed as:

$$O(\xi) = \int_{\Omega} (PMVc)^2 d\Omega \Big/ \int_{\Omega} d\Omega \tag{3.105}$$

Design parameters In an aircraft cabin, many of the parameters used to calculate PMVc are known. For example, this study set the metabolic activity level to 75 W for passengers in a relaxed state (in-flight) and the thermal resistance of the passengers' clothing (clo) to 0.57 clo for summer and 1.01 clo for winter (ASHRAE Handbook, 2005). The external work accomplished was assumed to be 0 W. Considering the low relative humidity and low pressure situation in an aircraft cabin (Nagda and Hodgson, 2001; Park et al., 2011), we set the partial water vapor pressure (p_w) of ambient air to 221.8 Pa, which corresponds to an average relative humidity of 13% inside a cruising aircraft. For simplicity, this study again assumed the mean radiation temperature to be the same as the air temperature in the cabin. Then the PMV becomes a function only of air temperature and velocity.

The air temperature and velocity fields in an aircraft cabin are determined primarily by design variables such as the geometry of the cabin interior furnishings, internal thermal loads, air supply velocity and temperature from the diffusers and gaspers, inlet size and location, etc. For the purpose of demonstrating the adjoint method, this study set only the air supply location, size, velocity, and temperature from the diffusers as the design variables. It should be noted that the computing time required by the adjoint method does not depend on the number of variables.

When PMVc is used to evaluate the thermal environment in the air cabin, with the above assumptions, we have the following variables and equations for inverse design of the air cabin environment:

- State variables: air velocity **U** (vector), air pressure p, and air temperature T;
- Design variables ξ: inlet location y_{in}, inlet size l_{in}, inlet air velocity $\mathbf{U}_{\mathrm{in}}$ (vector), and inlet air temperature T_{in}, where $\xi = (y_{in}, l_{in}, \mathbf{U}_{\mathrm{in}}, T_{in})$;
- State equations for the airflow in the domain, which are Navier-Stokes equations denoted by $\mathbf{N} = (N_1, N_2, N_3, N_4, N_5)$:

This investigation applied the re-normalization group (RNG) k-ε model (Yakhot and Orszag, 1986), which is widely used for indoor airflow modeling (Zhang et al., 2007), to simulate turbulence in the Navier-Stokes equations. The Boussinesq approximation (Boussinesq, 1903) was applied to simulate the buoyancy effect, while the air density was assumed to be constant. The design goal was to minimize the objective function with state variables subjected to the Navier-Stokes equations.

Adjoint equations The final form of the adjoint equations is expressed by Eqs. (3.98) to (3.100) with:

$$\mathbf{A} = \begin{cases} 2 \times \text{PMV}_c \times \dfrac{\partial \text{PMV}_c}{\partial \mathbf{U}} & \text{for domain } \Omega \\ \mathbf{0} \text{ for domain } \omega \setminus \Omega \end{cases} \tag{3.106}$$

$$B = \begin{cases} 2 \times \text{PMV}_c \times \dfrac{\partial \text{PMV}_c}{\partial T} & \text{for domain } \Omega \\ 0 \text{ for domain } \omega \setminus \Omega \end{cases} \tag{3.107}$$

In each design cycle, the Navier-Stokes/adjoint equations were calculated with 10,000 iterations to ensure convergence. The design convergence criteria were (1) $O < 0.01$ in the first design cycle and (2) $|O_i - O_{i-1}| < 0.01$ in the ith design cycle ($i > 1$). The O_{i-1} is the computed objective function in the previous design cycle. Since the air supply location and size may vary in the design procedure, this study used Gambit journal files (GAMBIT CFD Preprocessor, 1998) to automatically regenerate the corresponding mesh. We did not conduct the grid independence test specifically, and the mesh was generated according to the evaluation of grid independence in an aircraft cabin in our previous study (Liu et al., 2013).

3.3.2.2.2 CASE SETUP

This investigation conducted an inverse design of the air environment in a single-aisle, fully-occupied air cabin in order to find the optimal air supply conditions for the diffusers. The gaspers were assumed to be closed. To decrease the computing load, this study performed calculations for only half of a one-row aircraft cabin using periodic and symmetric boundary conditions, as illustrated in Fig. 3.29. This cabin model involved simplifications such as neglecting the under-seat structures. According to a numerical and experimental study of the air distribution in a functional aircraft cabin (Liu et al., 2012; Liu et al., 2013), these simplifications would have little effect on the flow pattern. For half of one row of the cabin, the generated mesh had 0.7 million unstructured cells. Fig. 3.29(a) shows the conventional ventilation system in an aircraft cabin, which creates mixed ventilation. The inlet and outlet were located in the upper and lower parts of the side wall, respectively. This study also considered displacement ventilation (Fig. 3.29(b)) according to the work of Zhang and Chen (2007). The inlet and outlet of the cabin with displacement ventilation were located in the aisle and in the upper part of the side wall, respectively. The design domain Ω was set as the area 0.1 m away from the passengers. To avoid longitudinal flow, we fixed U_z as 0 at the inlet in all calculations.

When a ventilation system is designed for an aircraft cabin, as in the case of the office above, the thermo-fluid boundary conditions must be sufficient for removal of the cooling load and compliance with air quality regulations. ANSI/ASHRAE Standard 161 (2007) requires a minimum ventilation rate of 9.4 L/s per passenger. This standard also specifies that the operative temperature during flight and on the ground should be between 18.3°C and 23.9°C. In accordance with these standards, and with the assumption that 75% of the total heat dissipated by a passenger is convective (Liu et al., 2013), the inlet air temperature in this study ranged from 13.3°C to 18.9°C at an airflow rate of 9.4 L/s.

Under these constraints, we initialized five design variables as shown in Table 3.4. To avoid local optima, the air supply location and temperature were initialized to their median values. For example, the inlet air temperature ranged from 13.3°C to 18.9°C, and T_{in} was initialized to 16.1°C. The inlet location y_{in} was in the range of 0–y_{mix} (0.409 m) for mixed ventilation and 0–y_{dis} (0.28 m) for displacement ventilation. For mixed ventilation, the lower bound was 10 cm higher than the passenger's head to avoid the draft effect. For displacement ventilation, the upper bound was the width of the aisle. The initial inlet locations were then set to $y_{mix}/2$ and $y_{dis}/2$. We set the initial inlet air velocity as $|\mathbf{U}_{in}| = 1.5$ m/s and its direction as normal to the inlet surface, according to our previous measurements in a functional aircraft cabin (Liu et al., 2012). Since the air flow rate was constrained, the initial inlet size was determined accordingly.

Table 3.5 lists the thermal boundary conditions that do not vary during the optimization. The temperatures of the surfaces shown in Table 3.5 were obtained from data measured on a flight at cruising altitude by use of an infrared thermometer (Zhang and Chen, 2007).

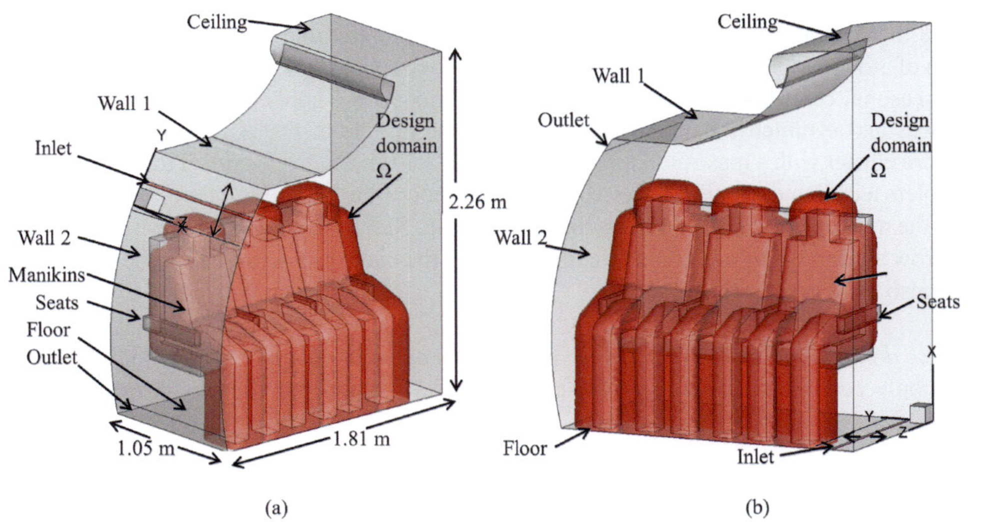

Figure 3.29 Schematic of the computational domain and design domain for the one-row aircraft cabin: (a) mixed ventilation and (b) displacement ventilation

Table 3.4 Ranges and initialized values for the five design variables

Design variable	y_{in} *[m]*		l_{in} *[m]*	$U_{in,x}$ *[m/s]*	$U_{in,y}$ *[m/s]*	T_{in} *[°C]*
	Mixed ventilation	*Displacement ventilation*				
Range	0–y_{mix}	0–y_{dis}	N/A	N/A	N/A	13.3–18.9
Initialized value	$y_{mix}/2$	$y_{dis}/2$	0.0188	1.5	0	16.1

Table 3.5 Thermal boundary conditions that were fixed during the optimization

Surface	*T*	*Surface*	*T*
Wall 1	21°C	Outlet	Zero gradient
Wall 2	22°C	Seats	Adiabatic
Ceiling	22°C	Floor	23°C
Passengers	30.3°C		

In order to address a more realistic scenario, this study also considered the inverse design of the thermal environment for mixed ventilation in a five-row aircraft cabin, as shown in Fig. 3.30. We used symmetric boundary conditions to reduce the computational domain by half. The number of mesh cells for the five-row aircraft cabin was about five times that for the one-row aircraft cabin. The front and back surfaces were set as walls to allow consideration of the end-wall effect. The other setup details and parameters were the same as those for the one-row aircraft cabin.

3.3.2.2.3 RESULTS

This study used the CFD-based adjoint method to find the air supply conditions for the diffusers that would provide the most comfortable thermal environment in the fully occupied cabin of a single-aisle aircraft. For a one-row cabin, we considered two ventilation modes and two seasonal conditions. Periodic boundary conditions were applied on the front and back surfaces and symmetric boundary conditions in the middle of the aisle. With the use of a 16-core cluster with a maximum turbo frequency of 3.3 GHz for each core, it took one hour to complete a design cycle. For a five-row cabin, to reduce the computing time, we considered only the mixed ventilation and the summer condition. Since the number of mesh cells for the five-row aircraft cabin was about five times that for the one-row aircraft cabin, the computing time also increased by five times.

Inverse design of the mixed ventilation system in the one-row aircraft cabin With the initial air supply conditions as listed in Table 3.4, Fig. 3.31 shows the changes in the objective

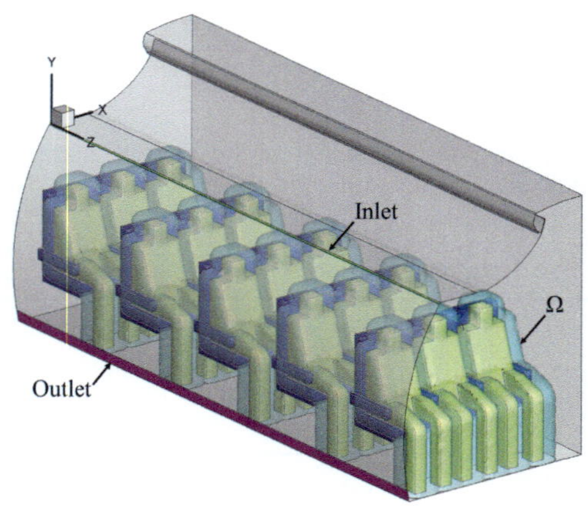

Figure 3.30 Schematic of the computational domain and design domain for the five-row aircraft cabin with mixed ventilation

function and design variables within 20 design cycles. The objective function and design variables all changed rapidly in the first two design cycles and then became relatively stable in subsequent cycles. The inverse design converged within seven cycles (seven hours' computing time), but the simulation continued until 20 cycles had been completed, as specified. At the end of the optimization process, the objective functions were only 0.16 and 0.06 for summer and winter conditions, respectively. The corresponding averages of the absolute PMVc in the design domain were 0.40 and 0.24, respectively. The air supply inlet size and velocity varied less than the inlet location and temperature. This difference indicates that the thermal comfort level was more sensitive to the air supply inlet location and temperature. Furthermore, the optimal air supply location, size, and velocity were almost the same for summer and winter, whereas a variation between seasons was observed in the air supply temperature. This variation can be accounted for by the thermal resistance of the passengers' clothing, which changed according to the season.

To further demonstrate the sensitivity of the thermal comfort level in the aircraft cabin to the design variables, Fig. 3.32 shows the distribution of air velocity and temperature in the first two design cycles and the last design cycle under summer conditions. The flow and temperature fields changed dramatically during the design process. At the low initial air supply temperature, the air flowed directly toward the passengers at head level, which caused the value of PMVc to be less than −1, as shown in Fig. 3.32(a). After optimization, the air supply temperature increased, and the air supply location was raised. In the second design cycle (Fig. 3.32(c)), the supplied air did not flow directly toward the passengers at head level and the cabin air temperature increased, which greatly improved the thermal comfort level in the cabin. This change in the flow field was due to the inertial force from the inlet jets, which is comparable to the buoyancy force from the thermal plumes in the aircraft cabin (Liu et al., 2013). After the 20th design cycle, the inlet location was close to the top of the

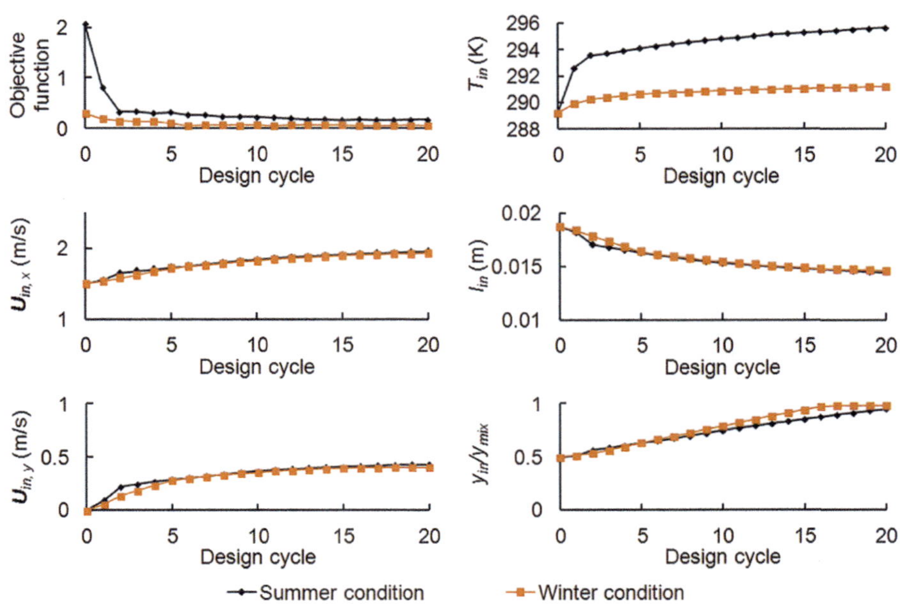

Figure 3.31 Change in objective function and five design variables with design cycle for the case of mixed ventilation

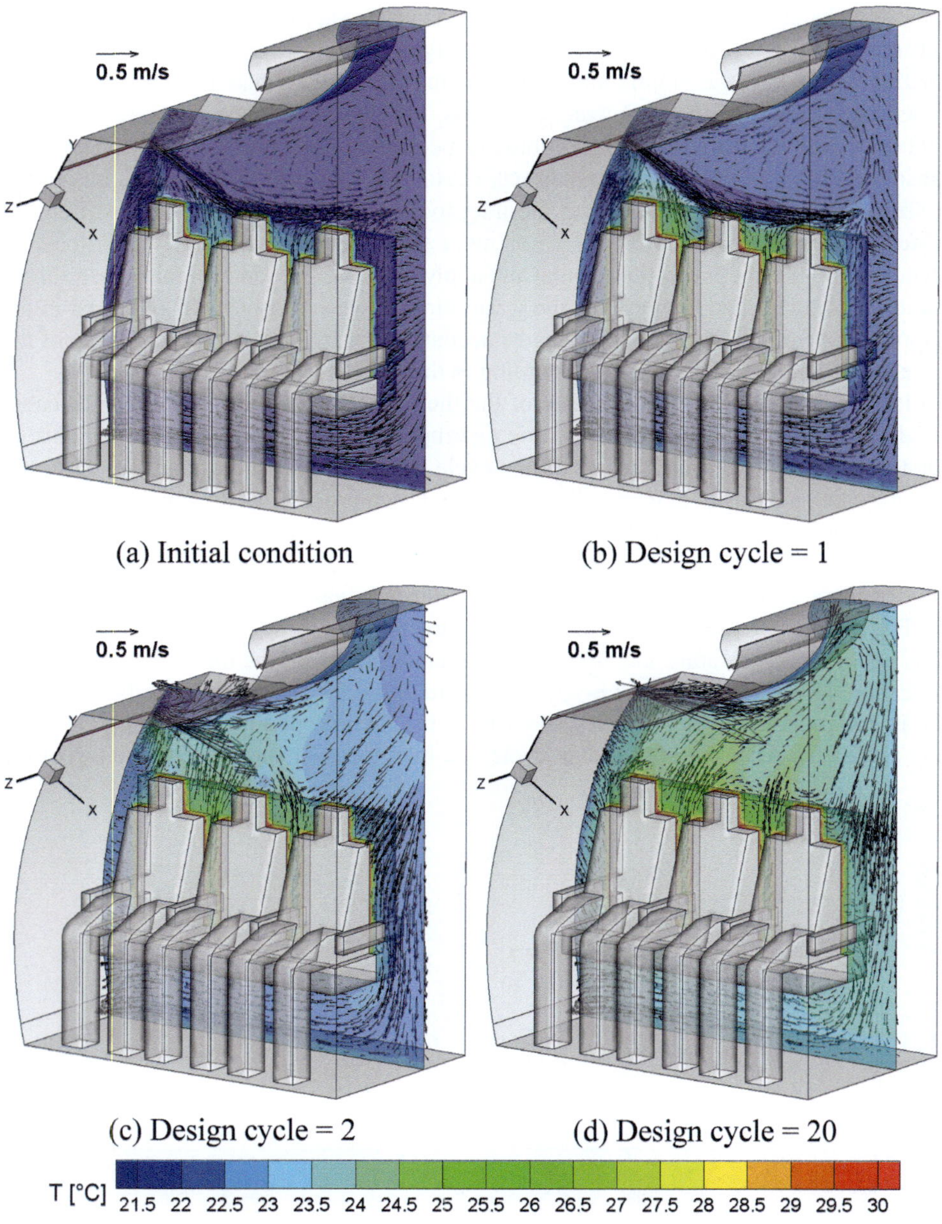

Figure 3.32 Computed airflow and temperature fields (a) at the initial condition and after (b) the 1st design cycle, (c) the 2nd design cycle, and (d) the 20th design cycle under summer conditions

side wall, and the PMVc value in the design domain was between −0.4 and 0.2, as shown in Fig. 3.33(d). The thermal comfort level in this aircraft cabin was very sensitive to the air supply location and temperature, and was greatly improved after only two design cycles, as shown in Fig. 3.33. These results demonstrate the efficiency of the CFD-based adjoint method in the inverse design of an enclosed air environment.

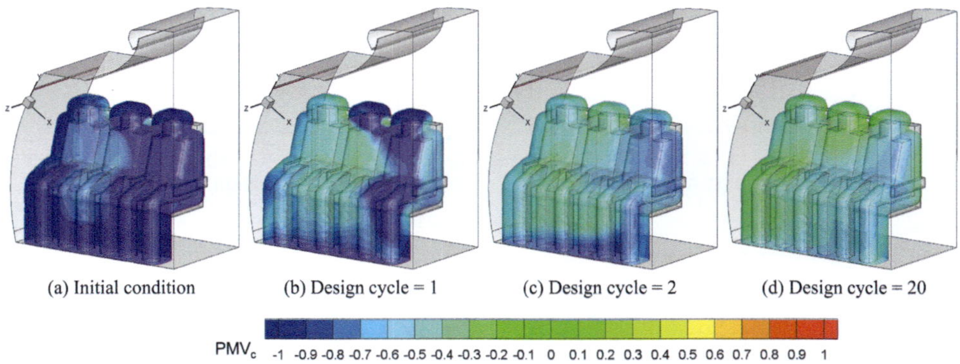

Figure 3.33 Distribution of PMVc (a) at the initial condition and after (b) the 1st design cycle, (c) the 2nd design cycle, and (d) the 20th design cycle under summer conditions

Inverse design of the displacement ventilation system in the one-row aircraft cabin For the displacement ventilation system, this study again followed the inverse design procedure, using the initial air supply conditions in Table 3.4. Fig. 3.34 shows the change in the objective function and the five design variables within 20 design cycles. As in the case with mixed ventilation, the objective function and design variables changed rapidly in the first five design cycles and then became stable. The optimization converged in the eighth design cycle (after eight hours' computing time), in a manner that was similar to but somewhat slower than that of the mixed ventilation case. At the end of the inverse design process, the objective functions were again very small, 0.03 for summer and 0.01 for winter, and led to PMVc values in the design domain of 0.17 and 0.12, respectively. The thermal comfort level for displacement ventilation was better than that for mixed ventilation. The optimal air supply inlet size and velocity were almost the same in both cases, but the air supply location and temperature were different. The thermal comfort level was more sensitive to the air supply location and temperature than to the other design variables. The optimal air supply temperature for displacement ventilation was about 2 K higher than that for mixed ventilation, and this difference has an implication for energy efficiency.

This study also considered the distribution of PMVc under summer conditions, as shown in Fig. 3.35. Note that the inverse design process improved the thermal comfort level in this aircraft cabin after five design cycles. After the 20th design cycle (Fig. 3.35(d)), the thermal comfort level was better than that with the mixed ventilation system (Fig. 3.33(d)). Furthermore, the displacement ventilation system was able to provide a similar thermal comfort level for all three passengers, whereas the mixed ventilation system could not.

Inverse design of the mixed ventilation system in the five-row aircraft cabin With the initial air supply conditions for mixed ventilation as listed in Table 3.4, Fig. 3.36 shows the changes in the objective function and air supply temperature within 20 design cycles. Here we have not shown the other design variables because their changes were very small. The inverse design converged within 15 design cycles, but the simulation continued until 20 cycles had been completed, as specified. At the end of the optimization process, the objective function was only 0.18, which is very close to the condition in the one-row aircraft cabin. The air supply location, size, and velocity exhibited little variation. The optimal air supply temperature was 295.5 K, which is also very close to that in the one-row aircraft cabin. This result confirms

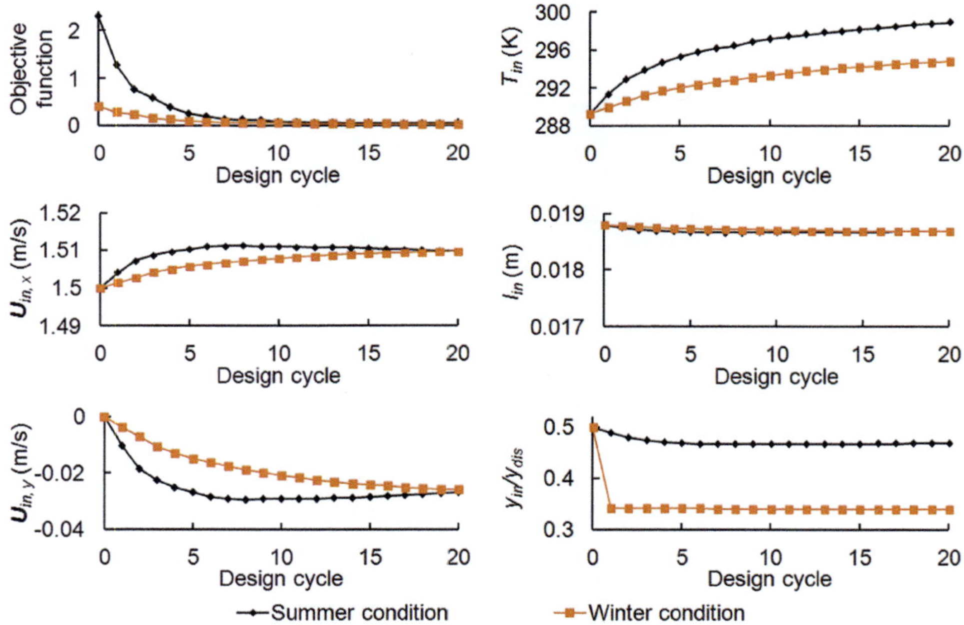

Figure 3.34 Change in objective function and the five design variables with design cycle for the case of displacement ventilation

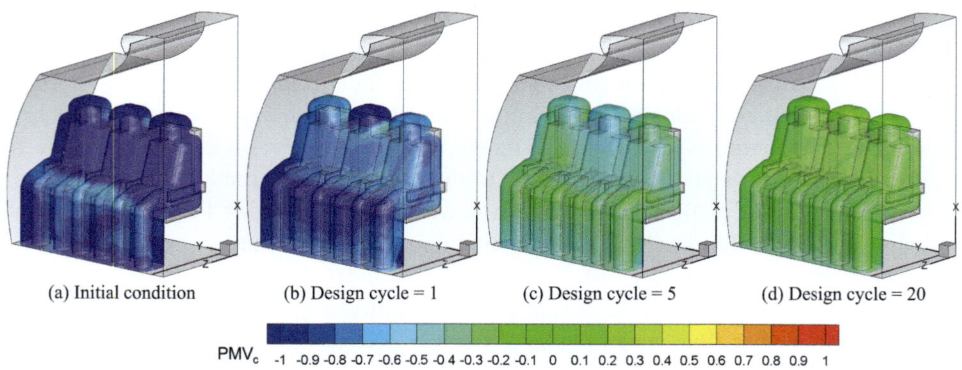

Figure 3.35 Distribution of PMVc (a) at the initial condition and after (b) the 1st design cycle, (c) the 5th design cycle, and (d) the 20th design cycle under summer conditions.

that the thermal comfort level was more sensitive to the air supply temperature than to the other design variables.

Fig. 3.37 shows the distribution of PMVc versus the number of design cycles in the five-row aircraft cabin. It is clear that the thermal comfort level greatly improved within the first 10 design cycles. Therefore, the CFD-based adjoint method can be very efficient in the inverse design of an enclosed air environment.

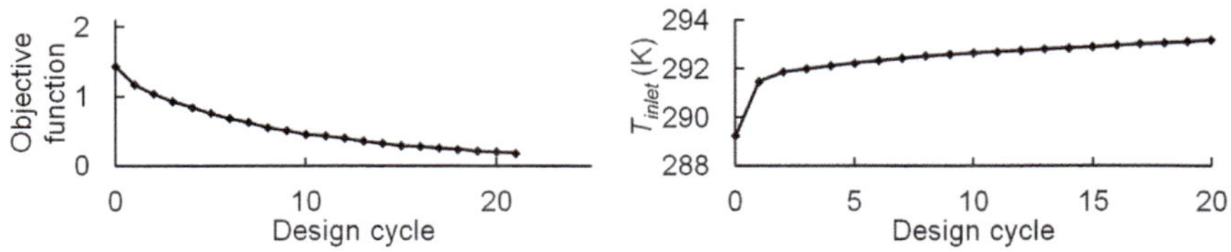

Figure 3.36 Change in objective function and air supply temperature with design cycle for the case of mixed ventilation in the five-row aircraft cabin

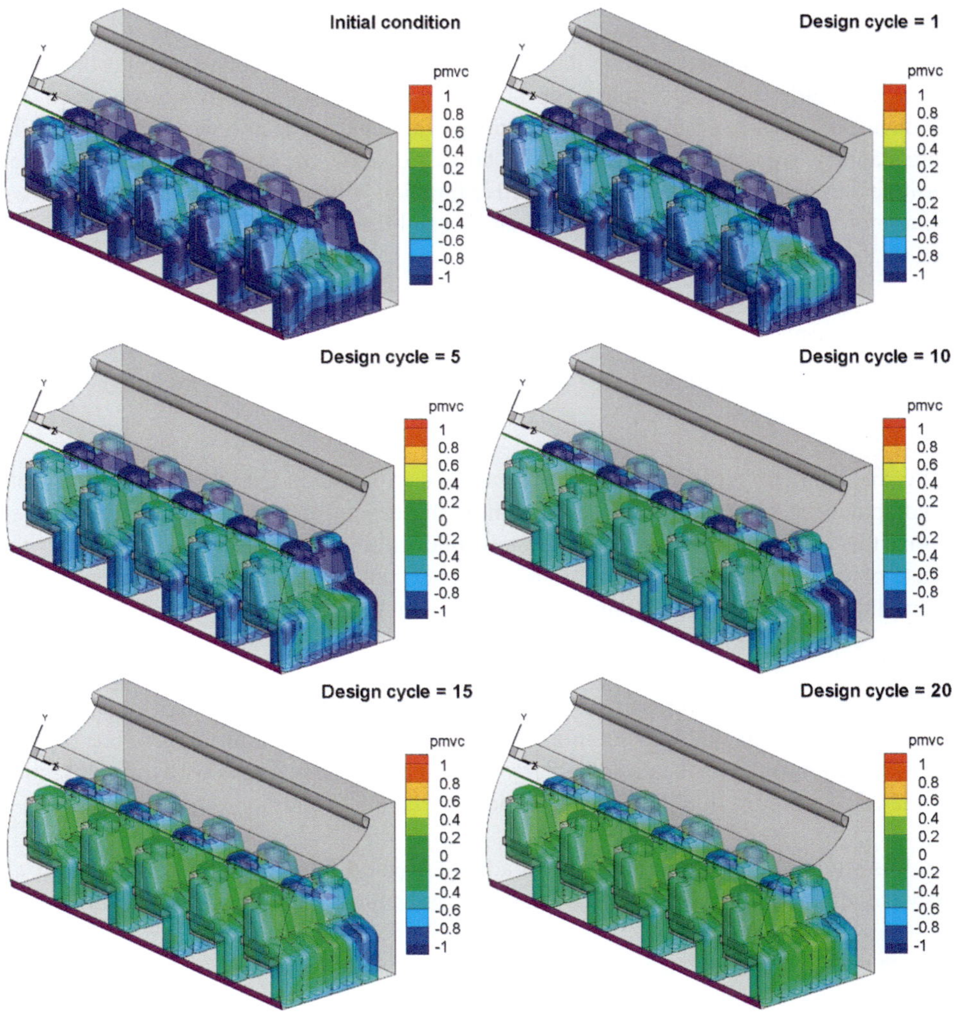

Figure 3.37 Distribution of PMVc at the initial condition and after the 1st, 5th 10th, 15th, and 20th design cycles under summer conditions

3.3.2.2.4 DISCUSSION

This study conducted an inverse design process that identified the five design variables in a single calculation. The beauty of the adjoint method is that its computing load does not depend on the number of design variables. The optimal PMVc distribution obtained by this method is acceptable because the initialized design variables are within a reasonable range. However, the potential for further improvements in the thermal environment of this aircraft cabin is unknown because the method can identify only local optima. Identifying the global optima would require multiple calculations or integration of the method with a gradient-free algorithm. Searching for global optima by means of a gradient-free algorithm, such as a genetic algorithm, would entail a dramatic increase in the computing cost with an increase in the number of design variables.

We obtained a converged result within seven to eight design cycles. The difference in convergence speed between the two ventilation systems may have been due to the difference between the initial estimates of the design variables and the optimal values. Furthermore, because this study used the simple steepest descent algorithm, the convergence speed was dependent on λ in Eq. (5.16), which determines the changes in the design variables. To ensure stability, we applied a constant λ. As far as we know, the quasi-Newton method can determine the second derivative of the objective function with respect to the design variables, and therefore the changes in the design variables can be automatically calculated in order to speed up the convergence process.

3.4 Conclusions

This investigation first validated the CFD-based adjoint method for determining the optimal air supply parameters:

- For the isothermal case, this study used a velocity profile in the lower part of the room computed by a forward CFD simulation as the design objective. The results showed that this adjoint method could accurately find the optimal inlet air velocity with the use of various initial inlet conditions. However, the calculations with different initial conditions led to different optimal conditions, and this outcome implies the existence of multiple solutions.
- For the non-isothermal case, when only the air velocity profile or only the air temperature profile in the lower part of the mid-cavity was used as the design objective, the calculation led to multiple solutions. When the air velocity and temperature profiles were used together as the design objective, each of the calculations with different initial inlet air conditions found a unique solution for the inlet condition.

Next, this study validated the CFD-based adjoint method for inverse identification of air supply parameters, i.e., inlet location, inlet size, and supply temperature and velocity:

- With the use of measured air velocity and temperature at several locations in a ventilated cavity, the CFD-based adjoint method was able to identify the air supply parameters that led to the same air velocity and temperature values in those locations. The results showed that different air supply parameters could result in the same air velocity and temperature in those locations. This finding implies that multiple solutions exist and can be identified by the method.

- When different air supply inlet locations, inlet sizes, temperatures, and velocities were used as design variables, the adjoint method was able to simultaneously identify the locally optimal values for these variables. The computing costs did not vary with the number of design variables.
- This study has successfully demonstrated that it is possible to set desirable design objectives in the flow domain and to obtain the air supply parameters (inlet location, inlet size, temperature, and velocity) needed for achieving the design objective. The initial boundary conditions can be set arbitrarily. When different constraints are used, the calculated air supply parameters can also be very different.

Finally, this investigation used the CFD-based adjoint method to find the optimal air supply conditions for improving thermal comfort in two spaces. The PMV and PMVc indices were used to evaluate thermal comfort in an office and an aircraft cabin, respectively. For the aircraft cabin, this study conducted an inverse design of mixed and displacement ventilation systems under summer and winter conditions. The computed results led us to the following conclusions:

- The CFD-based adjoint method can be used to design a ventilation system that improves thermal comfort in a real indoor environment. With reasonable estimates of the initial design variables, the optimal variables can be identified in a single calculation.
- The optimization process for the displacement ventilation system resulted in a more uniform thermal comfort level than did the process for the mixed ventilation system. The air temperature with the displacement ventilation system was lower than that with the mixed ventilation system, and this difference has an implication for energy efficiency.
- The airflow pattern and thermal comfort level in the cabin were sensitive to the air supply location and temperature. This sensitivity occurred because the inertial force from the inlet jets was comparable to the buoyancy force from the thermal plumes. The optimal air supply temperature differed under summer and winter conditions because of the varying thermal resistance of the passengers' clothing.

Notes

1 This chapter is partially reproduced from the paper entitled Optimal air distribution design in enclosed spaces using an adjoint method, *Inverse Problems in Science and Engineering*, 23(5):760–779. The reprint is with permission from Taylor & Francis.

2 This chapter is partially reproduced from the paper entitled Optimization of air supply location, size, and parameters in enclosed environments using a computational fluid dynamics-based adjoint method, *Journal of Building Performance Simulation*, 9(2):149–161. The reprint is with permission from Taylor & Francis.

3 This chapter is partially reproduced from the paper entitled Inverse design of the thermal environment in an airliner cabin by use of the CFD-based adjoint method, *Energy and Buildings*, 104: 147–155. The reprint is with permission from Elsevier.

References

ASHRAE. *ASHRAE Handbook, 2005*. Thermal comfort chapter, Fundamentals volume of the *ASHRAE Handbook*. Atlanta, GA: ASHRAE, Inc.

ASHRAE, 2007. Standard 161-2007, Air quality within commercial aircraft. Atlanta, GA: ASHRAE, Inc.

Blay D, Mergui S, Niculae C. Confined turbulent mixed convection in the presence of a horizontal buoyant wall jet. *Fundamentals of Mixed Convection, ASME HTD*. 1992;213:65–72.

Boussinesq J. *Theorie analytique de la chaleur*. Paris: Gauthier-Villars. 1903.

Bryson AE, Ho YC. *Applied optimal control*. Washington, DC: Hemisphere. 1975.

Cui W, Ouyang Q, Zhu Y, Hu S. Prediction model of human thermal sensation under low-air-pressure environment. *Proceedings of the 8th International Symposium on Heating, Ventilation and Air Conditioning*, 329–336. Berlin Heidelberg: Springer. 2014.

Cui W, Zhu Y. Systematic study on passengers' thermal comfort under low-air-pressure environment in commercial aircraft cabin. *Presented in the Annual Meeting of the Center for Cabin Air Reformative Environment (973 Project), Chongqing, China*. 2015.

Dwight RP, Brezillon J. Effect of various approximations of the discrete adjoint on gradient-based optimization. *Proceedings of the 44th AIAA Aerospace Sciences Meeting and Exhibit, Reno, NV*. 2006.

Fanger PO. *Thermal comfort, analysis and applications in environmental engineering*. Copenhagen: Danish Technical Press. 1970.

GAMBIT CFD Preprocessor. *User's guide*. Lebanon, NH: FLUENT Inc. 1998.

Giles MB, Pierce NA. An introduction to the adjoint approach to design. *Flow, Turbulence and Combustion*. 2000;65(3–4):393–415.

Gill PE, Murray W, Wright MH. *Practical optimization*. London: Academic Press. 1981.

Griwank A. *Evaluating derivatives, principles and techniques of algorithmic differentiation*. Philadelphia, PA: SIAM. 2000.

Gunzburger MD. *Perspectives in flow control and optimization*. Philadelphia, PA: SIAM. 2003.

Jameson, A. Aerodynamic design via control theory. *Journal of Scientific Computing*. 1988, 3: 233–260.

Jameson A. Optimum aerodynamic design using CFD and control theory. *AIAA paper*. 1995;1729: 124–131.

Jameson A, Martinelli L, Pierce NA. Optimum aerodynamic design using the Navier–Stokes equations. *Theoretical and Computational Fluid Dynamics*. 1998;10(1–4):213–237.

Jimack PK. An optimal finite element mesh for elastostatic structural analysis problems. *Computers and Structures*. 1997;64(1):197–208.

Liu W, Wen J, Chao J, Yin W, Shen C, Lai D, Lin CH, Liu J, Sun H, Chen Q. Accurate and high-resolution boundary conditions and flow fields in the first-class cabin of an MD-82 commercial airliner. *Atmospheric Environment*. 2012;56:33–44.

Liu W, Wen J, Lin CH, Liu J, Long Z, Chen Q. Evaluation of various categories of turbulence models for predicting air distribution in an airliner cabin. *Building and Environment*. 2013;65:118–31.

Nagda N, Hodgson M. Low relative humidity and aircraft cabin air quality. *Indoor Air*. 2001;11(3): 200–214.

Nielsen PV, Restivo A, Whitelaw JH. The velocity characteristics of ventilated rooms. *Journal of Fluids Engineering*. 1978;100(3):291–8.

OpenFOAM. The Open Source CFD Toolbox. http://www.opencfd.co.uk/openfoam.html. 2007.

Othmer C. A continuous adjoint formulation for the computation of topological and surface sensitivities of ducted flows. *International Journal for Numerical Methods in Fluids*. 2008;58:861–877.

Park S, Hellwig RT, Grün G, Holm A. Local and overall thermal comfort in an aircraft cabin and their interrelations. *Building and Environment*. 2011;46(5):1056–64.

Patankar SV, Spalding DB. A calculation procedure for heat, mass and momentum transfer in three-dimensional parabolic flows. *International Journal of Heat and Mass Transfer*. 1972; 15(10):1787–806.

Schneider R. Applications of the discrete adjoint method in computational fluid dynamics. Doctoral dissertation, The University of Leeds. 2006.

Schneider R, Jimack PK. On the evaluation of finite element sensitivities to nodal coordinates. *Electronic Transactions on Numerical Analysis*. 2008;32:134–144.

Sokolowski J, Zolesio JP. *Introduction to shape optimization*. Berlin Heidelberg: Springer. 1992.

Wikipedia, Http://en.wikipedia.org/wiki/Automatic_differentiation. n.d.

Xue Y, Zhai Z. Comparison and integration of generic algorithm and adjoint algorithm for optimizing indoor environments. *Proceedings of the 13th International Conference on Indoor Air Quality and Climate (Indoor Air 2014)*. 2014.

Yakhot V, Orszag SA. Renormalization group analysis of turbulence. *Journal of Scientific Computing*. 1986;1(1):3–51.

Yuan X, Chen Q, Glicksman LR, Hu Y, Yang X. Measurements and computations of room airflow with displacement ventilation. *ASHRAE Transactions*. 1999;105:340–352.

Zhai Z, Zhang Z, Zhang W, Chen Q. Evaluation of various turbulence models in predicting airflow and turbulence in enclosed environments by CFD: Part 1—Summary of prevalent turbulence models. *HVAC&R Research*. 2007;13(6):853–70.

Zhang T, Chen Q. Novel air distribution systems for commercial aircraft cabins. *Building and Environment*. 2007;42(4):1675–1684.

Zhang Z, Zhang W, Zhai Z, Chen Q. Evaluation of various turbulence models in predicting airflow and turbulence in enclosed environments by CFD: Part 2—Comparison with experimental data from literature. *HVAC&R Research*. 2007;13(6):871–86.

Zhou L, Haghighat F. Optimization of ventilation system design and operation in office environment, Part I: Methodology. *Building and Environment*. 2009;44(4):651–656.

Zuo, W, Chen, Q. Real time or faster-than-real-time simulation of airflow in buildings. *Indoor Air*. 2009;19(1):33–44.

4 CFD-based artificial neural network method[1,2]

The CFD-based artificial neural network (ANN) can be used to establish the mapping relationship between the design objectives and the design variables in the inverse design of enclosed environment. The values of design variables and objectives of training samples are obtained by the sampling method and the CFD process, respectively. Then the training samples are used to train the ANN. When the design objectives are set as the input of the ANN and the design variables are set as the output of the ANN, ANN can directly obtain the design result according to the requirement of design objectives. But the prediction error of the ANN will affect the accuracy of design results. When the design variables are set as the input of the ANN and the design objectives are set as the output of the ANN, ANN plays as a surrogate model of CFD to reduce the computational time in the inverse design. And a correction process can be used to avoid the effect of prediction error on the design result. But more computational cost will be involved by the correction process. This chapter details the principles of ANN, including the structural parameters of the ANN, the method used to select training samples, the data processing method, and the training algorithm. Four cases are used to test the application of the CFD-based ANN in the inverse design of enclosed environment.

4.1 Introduction

An artificial neural network (ANN) is a highly sophisticated paradigm that borrows features from human and animal brains to enable recognition of patterns within data. It learns to solve problems by developing a memory capable of associating a large number of input patterns with a resulting set of outputs or effects (Singh et al., 2004). Such networks have been applied widely for predicting and forecasting in areas such as finance, information, medicine, water resources, and environmental science (Khashman, 2010; Liang and Du, 2008; Maier and Dandy, 2000; Mjalli et al., 2007).

In the inverse design of an indoor environment, the design variables (air supply location, size, or other parameters) are obtained from the critical design objectives (air distributions). The most direct approach is to establish the relationship between the design objectives and the design variables. However, because this relationship is usually non-linear in an indoor environment, it is difficult to obtain. An artificial neural network provides a very powerful non-linear computational platform with the ability to train and adapt itself to different inputs and outputs (Mazhar et al., 2013; Yi and Malkawi, 2011). In principle, an ANN can approximate any non-linear dependence between the design variables and objectives (Balabin et al., 2011). Zhang and You (2014a) used an ANN to design an indoor environment. They established the relationship between thermal comfort and the inlet boundary conditions of the environment. The boundary conditions were identified from the required thermal comfort level. However, it is difficult to solve a multi-solution problem by this method. In such a problem, there is a one-to-many inverse mapping relationship, which makes it difficult for the training process to converge. Therefore, only one solution should be obtained in the design stage.

A multi-solution problem can be solved by the indirect inverse design method or non-gradient-based optimization algorithm, such as the Genetic Algorithm. However, the computational costs are usually high because of the large number of CFD processes involved in the design process. In this case, an ANN can serve as a surrogate for the CFD model, mimicking the behavior of the CFD model in buildings. After being trained well, the ANN can be used in place of CFD during the design process to predict the design objectives. The time requirement for the ANN is significantly less than for CFD, and therefore the computational cost is lower. ANNs have been used to predict indoor velocity, temperature, thermal comfort, energy cost, and contaminant source. Ayata et al. (2007) and Ozbalta et al. (2012) concluded that the ANN approach was an efficient tool for predicting indoor air velocity, indoor temperature, and relative humidity. The R^2 values between predicted and actual values of indoor temperature and indoor relative humidity are as high as 0.94 and 0.96, respectively. Liang and Du (2005) used an ANN to predict thermal comfort with six design variables. They found that the ANN could maintain the indoor comfort level within the desired range under both heating and cooling modes. Qin et al. (2012) found that an ANN could predict dynamic energy consumption and thermal-environment parameters reasonably well. Bastani et al. (2012) used an ANN to predict indoor contaminant source location with an accuracy of 90% when three or more monitoring sensors were used.

ANNs have the potential to provide better, quicker, and more practical predictions than any of the traditional methods. Thomas and Soleimani-Mohseni (2007) compared the prediction accuracy of ANNs and linear models. Many different inputs were used, such as the outdoor and indoor temperature, heating power, wall temperatures, ventilation flow rate, and duration of sun radiation. The results showed that ANN models predicted temperature more accurately than linear models. Neto and Fiorelli (2008) compared an ANN model and a detailed design and simulation software program as forecasting tools for the energy demand of an HVAC system in a building. The energy consumption predicted by the ANN model agreed fairly well with the actual value.

In this chapter, the principles of the ANN method are first presented, including structural parameters, the method used to select training samples, the data processing method, and the training algorithm, all of which can affect ANN performance. Next, inverse design methods based on CFD and the ANN are presented, including direct and indirect methods. Finally, the CFD-based ANN method is used to solve the problem of inverse identification and optimal design of an indoor environment.

4.2 Principles of the CFD-based ANN method

4.2.1 Structure of the ANN

The basic unit of an artificial neural network is the neuron, and an ANN is a massively parallel distributed processor made up of simple processing neurons (Saijal et al., 2011). The structure of a single neuron can be seen in Fig. 4.1, where $NE(in)_{aa}$ represents the aa^{th} input of the neuron, ϖ_{aa} represents the aa^{th} weight, *bi* represents the bias, *tran* represents a transfer function, and $NE(out)$ represents the output of the neuron.

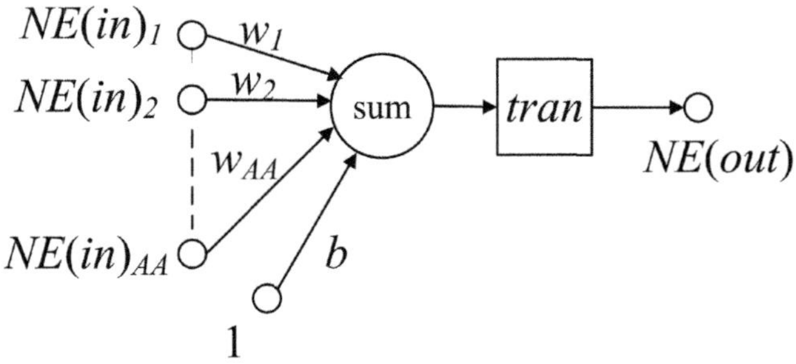

Figure 4.1 Structure of a neuron

The output of the neuron is generated by the following equation:

$$NE(out) = tran\left[\sum_{aa=1}^{AA} \varpi_{aa} NE(in)_{aa} - bi\right] \tag{4.1}$$

In the design of an indoor environment, an ANN is typically used to establish the non-linear relationship between the design variables and design objectives. Therefore, nonlinearity is necessary in the ANN. This nonlinearity is introduced into the network by transfer functions. The *logsig*, *tansig*, and *purelin* transfer functions are the most common choices and their shapes are shown in Fig. 4.2. These functions can have any value from plus infinity to minus infinity. The *logsig* transfer function compresses the output into the range from 0 to 1, the *tansig* transfer function compresses the output into the range from −1 to 1, and the output range of the *purelin* transfer function is from plus infinity to minus infinity.

The basic structure of an ANN consists of one input layer, one output layer, and a minimum of one hidden layer, as shown in Fig. 4.3. The input layer consists of nodes (neurons) that receive input (IN_1, IN_2 . . . IN_{BB}) from the external environment. The hidden layer consists of nodes that receive output from the input layer or the previous hidden layer, transform this output, and pass it on to the next hidden layer or output layer. The outputs of the output layer (OUT_1, OUT_2 . . . OUT_{CC}) are the ANN results. An error function can be established from the ANN results and the desired values (DA_1, DA_2 . . . DA_{CC}). The "backpropagation" method can then be used to adjust the weights and biases. The error function and backpropagation method will be described in detail in Section 4.2.4.

The structure of an ANN determines the number of layers and the number of nodes in each layer. When an ANN is used in the inverse design of an indoor environment, the input and

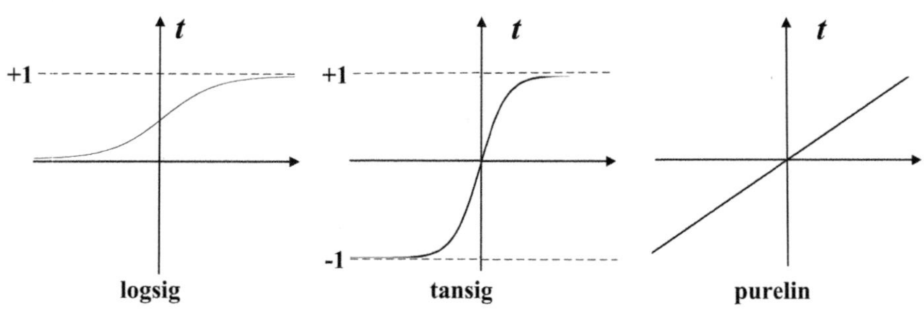

Figure 4.2 Typical transfer functions

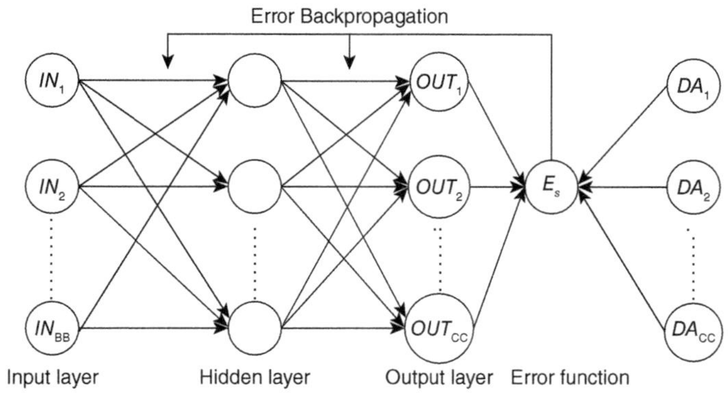

Figure 4.3 Structure of a three-layer ANN

output are the design variables and objectives, respectively. The numbers of nodes in the input and output layer are the same as the numbers of design variables and objectives. The number of hidden layers should be chosen in accordance with the input and output sizes (Vukovic et al., 2010). Smaller networks usually have greater generalization ability. The advantages of larger networks are that they tend to learn quickly and are more able to avoid local minima in the error surface. However, their computational costs are high, and they generally require a large number of training samples to achieve good generalization ability (Maier and Dandy, 2000). It has been proved that an ANN with one hidden layer can approximate any function if sufficient degrees of freedom are provided. Three-layer ANNs are commonly used in practice, particularly in engineering applications (Lu and Viljanen, 2009).

There is no way to determine the optimal number of nodes in the hidden layer without training several networks and estimating the generalization error for each. In order to ensure good generalization ability, various empirical relationships between the number of training samples and the number of connection weights have been suggested (Maier and Dandy, 2000). In most ANN applications, however, the number of nodes in the hidden layer has been determined by experience because the empirical relationships are not universal.

4.2.2 Sampling method for the ANN

To establish the relationship between the design variables and objectives, a number of samples must be generated in order to train the ANN. In the design of an indoor environment, each sample is a CFD case with design variables (boundary conditions) and design objectives (air distributions). When a CFD-based ANN is used in the design of a building environment, the values of the design variables are obtained by means of a sampling method. The values of the design objectives can then be obtained by CFD according to the values of the design variables. Generally, the number of samples is the same as the number of CFD processes. At the present time, most of the building models treated by CFD are complicated, and the simulations require a lot of time. Because of computing capacity limitations, it is usually impractical to generate a large number of samples. The sampling method is an important factor in reducing the computing cost.

The quality of the training samples has a significant impact on ANN performance. The training data should cover the range of inputs for which the network will be used. Multilayer networks can be trained to generalize well within the range of inputs. However, they do not have the ability to accurately extrapolate beyond this range (Demuth et al., 2008). The range of design variables should be confirmed before sampling. The distribution of samples is also regarded as an important index for the training samples. If the key points in the design space are included in the training samples, a nonlinear relationship can be mimicked closely by the ANN. In an indoor environment, many design variables may be involved, and each variable may have many potential values, resulting in a vast number of individuals in the design space. For example, the design variables may include velocity and temperature. If the velocity varies from 1 to 10 m/s with an interval of 0.1 m/s and the temperature varies from 20 to 30°C with an interval of 0.1°C, then there are 100 potential values for each design variable and 10^4 (100 × 100) individuals in the design space. Only a few of these individuals can be selected by the sampling method to represent the distribution of samples and train the ANN. Therefore, the sampling method is important to the performance of the network.

There are two main types of sampling method that can be used to generate training samples for an ANN, probability sampling and nonprobability sampling. Probability sampling is a method in which every individual has a known probability of being selected as part of the sample. The primary types of probability sampling are simple random sampling, systematic sampling, and stratified sampling. Nonprobability sampling is a method in which some individuals in the population have no chance of being selected, or in which the probability of selection cannot be accurately determined. When an ANN is used in the design of an indoor environment, the ranges of design variables for sampling are the same as those of real variables in buildings, and every potential value of the design variable has a chance of being sampled. Therefore, probability sampling should be used here. The simple random sampling, systematic sampling, and stratified sampling methods are presented in the following subsections.

4.2.2.1 Simple random sampling method

Simple random sampling is the most general sampling method. It is a basic type of sampling and can be used as a component of other, more complex methods. It is based on the mathematical principle that a truly random sample (if sufficiently large) will be representative of the target population. The simple random sample has two key properties:

(1) Unbiased. Individuals are chosen randomly from the population, and all individuals have an equal chance of being selected.
(2) Independent. The entire process of sampling is performed in a single step, and each subject is selected independently of the other individuals in the population.

Simple random sampling consists of the following steps:

(a) The populations are coded from 1 to N_p.
(b) A random number "N_r" between 1 and N_p is generated. Random number generation is simple with the use of a computer program, such as the "rand" function in MATLAB.
(c) The population coded with the random number "N_r" is selected as a sample. Steps (a) and (b) are repeated to generate N_s samples.

Although simple random sampling is a commonly used method, it can be extremely time consuming and difficult to organize. It requires a complete list of all members of the target population so that the sample is a real representation of the larger group. All the individuals in the population have the same chance of becoming part of the sample. Furthermore, all possible combinations of the sample are equally likely to occur. Even though the sample is intended to represent the entire population, small minority populations are sometimes missed because the sample size is simply not large enough.

4.2.2.2 Systematic sampling method

Because simple random sampling can be inefficient and time-consuming, statisticians have turned to other methods, such as systematic sampling. Systematic sampling is a statistical method in which individuals are selected from an ordered sampling frame. The most common form of systematic sampling is an equal-probability method. In this approach, the individuals of the population are put into a list, and then every individual in the list is chosen for inclusion in the sample. Progression through the list is treated circularly, with a return to the top once the end of the list is reached. Systematic sampling consists of the following steps:

(a) The populations are coded from 1 to N_p.
(b) The sampling interval, I_s, is determined by dividing the population size N_p by the sample size N_s and rounding to the nearest integer.
(c) To ensure against any possible human bias in this method, the simple random sampling method is used to generate the code for the first sample (N_f).
(d) Subsequent samples are selected according to a specific rule, for example, $N_{dd} = N_f + (dd- 1) \times I_s$.

Systematic sampling allows the researcher to easily yet systematically add N individuals to the random selection of subjects. Another advantage of this method is that the population is evenly sampled. The biggest disadvantage of systematic sampling is that the process of selecting the sample can interact with a hidden periodic trait within the population, in which case the sampling technique is no longer random and the representative nature of the sample is compromised.

4.2.2.3 Stratified sampling method

Stratified sampling is a probability sampling technique in which the entire population is divided into subgroups, and then the final subjects are randomly selected from the subgroups in a proportional manner. Stratified sampling consists of the following steps:

(a) The populations are coded from 1 to N_p.
(b) The population is divided into Q_s subgroups, called strata.
(c) Each element of the population can be assigned to one, and only one, stratum.
(d) The number of observations within each stratum is known and $N_p = N_1 + N_2 + N_3 + \ldots + N_{Q-1} + N_Q$.
(e) A probability sample is obtained from each stratum.

The Latin hypercube sampling method proposed by McKay et al. (2000) is a more sophisticated stratified sampling scheme and is widely used in sampling for ANNs in the design of indoor environments (Magnier and Haghighat, 2010; Zhou and Haghighat, 2009a; Zhou and Haghighat, 2009b). Consider a variable Y that is a function of other variables $X_1, X_2, \ldots, X_{EE}$. This function may be very complicated. Latin hypercube sampling selects N_s different values from each of EE variables $X_1, X_2, \ldots, X_{EE}$ in the following manner. The range of each variable is divided into N_s non-overlapping intervals on the basis of equal probability. One value from each interval is selected at random with respect to the probability density in the interval. The N_s values thus obtained for X_1 are paired in a random manner with the N_s values of X_2. These N_s pairs are combined in a random manner with the N_s values of X_3 to form N_s triplets, and so on, until N_s *EE*-triplets are formed. This is the Latin hypercube sample. The program for the Latin hypercube sampling method has been provided by the 'toolbox' in MATLAB.

One advantage of the Latin hypercube sample is evident when the output is dominated by only a few of the components of *X*. This method ensures that each of these components is represented in a fully stratified manner, no matter which components might turn out to be important (McKay et al., 2000).

In the inverse design of enclosed environment, every sample will cost large computation time. The Latin hypercube sampling method can generate a small yet representative collection of samples. So the Latin hypercube sampling method is recommended to be used to reduce the computational cost for the inverse design.

4.2.3 Normalization methods for the ANN

Data preprocessing also has a significant effect on ANN performance. The most common preprocessing routine is normalization. This technique accommodates different ranges and units of values in different dimensions of a multi-dimensional data set by means of translation and scaling.

In a multilayer ANN, sigmoid transfer functions are generally used in the hidden layers. These functions essentially become saturated when the net input is greater than three. If this happens at the beginning of the training process, the gradients will be very small, and the network training process will be very slow. In the first layer of the network, the net input is a product of the input times the weight plus the bias. If the input is very large, then the weight must be very small in order to prevent saturation of the transfer function. It is standard practice to normalize the inputs before applying them to the network. Generally, the normalization step is applied to both the input and output vectors in the data set. In this way, the network output always falls into a normalized range. The network output can then be reverse transformed back into the units of the original target data when the network is put to use in the field (Demuth et al., 2008).

In order to ensure that all variables receive equal attention during the training process, they should also be normalized. When different design variables and objectives are involved in the inverse design, they should be normalized to the same range. For example, if the ANN inputs are inlet velocity and temperature, their ranges are usually different. After normalization, the velocity and temperature will have the same range, usually 0–1. The typical normalization method can be expressed as follows:

$$NOR = \frac{OR - OR_{\min}}{OR_{\max} - OR_{\min}} \tag{4.2}$$

where *NOR* is the normalization value, *OR* is the original value, OR_{max} is the maximum original value, and OR_{min} is the minimum original value. This method is known as the linear normalization method. The original value is normalized to the range of 0 to 1. In the inverse design of a building environment, the ranges of design variables are known, while the ranges of design objectives are usually unknown, and the linear normalization method can be revised as follows:

$$NOR = \beta + (1-\beta)\frac{OR - OR_{\min}}{OR_{\max} - OR_{\min}} \tag{4.3}$$

Where β is a coefficient that ranges from 0 to 1. The original value is normalized to the range of β to 1-β. Values beyond the range of the training samples can be predicted by the ANN.

The linear normalization method is appropriate when the prediction accuracy at different locations in the design space is equally important. However, a number of threshold values may exist in the design space, and the prediction accuracy near these values is more important than at locations. For example, the design standard for PMV in a building should be −0.5 to 0.5. If the real value is 0.49 and the value predicted by the ANN is 0.51, the prediction error is small. However, the predicted result indicates that the environment will not satisfy the design requirement, which is not true. If the real value is 2.0 and the value predicted by the ANN is 1.0, then the error is large. In this case, the predicted result and real value both indicate that the environment does not satisfy the design requirement. Therefore, local prediction accuracy is more important than global prediction accuracy when threshold values are present in the design requirements. To enhance the local prediction accuracy of the ANN, a logarithmic normalization method is proposed (Zhang and You, 2015):

$$NOR = \begin{cases} 0.5 + 0.5\log_{10}\left(9\left(\dfrac{OR - OR_t}{OR_{\max} - OR_t}\right)+1\right) & \text{if } OR \geq OR_t \\ 0.5 - 0.5\log_{10}\left(9\left(\dfrac{OR_t - OR}{OR_t - OR_{\min}}\right)+1\right) & \text{if } OR < OR_t \end{cases} \tag{4.4}$$

A specific threshold value (OR_t) is considered, and the discriminability near the threshold value is increased when this logarithmic normalization is used instead of linear normalization. Fig. 4.4 compares the linear and logarithmic normalization methods when the original value ranges from 0 to 10 and the threshold value is 5.

4.2.4 Training algorithms for the ANN

After the training samples have been generated and the dataset has been normalized, the ANN training process is conducted. The training algorithm is one of the most essential ANN theories. There are two different styles of training: incremental training and batch training (Perez-Sanchez et al., 2013). In incremental training, the weights and biases of the network are initially chosen randomly and are then updated each time an input is

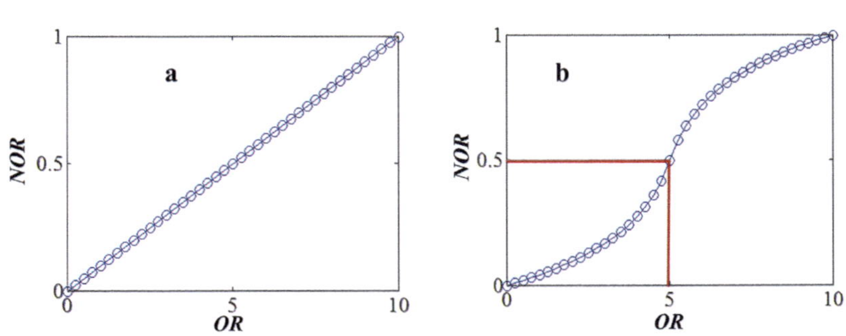

Figure 4.4 Comparison of linear (a) and logarithmic (b) normalization methods

presented to the network. In batch training, the weights and biases are updated only after all the inputs have been presented. Batch training methods are generally more efficient in the MATLAB environment (Demuth et al., 2008).

In the training process, the ANN is fed a set of input–output pairs. An optimization algorithm is used to adjust the weights of neurons in order to minimize the quadratic error between observed data and computed outputs. The training samples are used to accomplish the network learning and fit the network weights by minimizing an appropriate error function. The "backpropagation" technique is typically used for this purpose. A well-trained ANN represents a mathematical transformation between the multidimensional domain spaces of the input and output (Mjalli et al., 2007).

The backpropagation algorithm is a common method of training an ANN and is used in conjunction with an optimization method such as gradient descent. The algorithm calculates the gradient of an error function with respect to all the weights in the network. The gradient is fed to the optimization method, which uses the gradient to update the weights in an attempt to minimize the error function. The backpropagation algorithm requires a real output for each input value in order to calculate the error function gradient. The training algorithm can be divided into two phases: propagation and weight update.

PHASE 1: PROPAGATION

Each propagation involves the following steps:

(a) Forward propagation of a training pattern's input through the neural network in order to generate the propagation's output activations.
(b) Backward propagation of the propagation's output activations through the neural network using the training pattern target in order to generate the deltas (the differences between the input and output values) of all output and hidden neurons.

PHASE 2: WEIGHT UPDATE

For each weight synapse, these steps are followed:

(a) Multiplication of its output delta and input activation to obtain the gradient of the weight.
(b) Subtraction of a ratio (percentage) of the gradient from the weight.

This ratio influences the speed and quality of learning; it is called the learning rate. The greater the ratio, the faster the neuron is trained; the lower the ratio, the more accurate the training. The sign of the gradient of a weight indicates where the error is increasing. This is why the weight must be updated in the opposite direction. Phases 1 and 2 are repeated until the performance of the network is satisfactory. Various training algorithms based on backpropagation have been used to train ANNs, such as the steepest descent algorithm, Newton's method, the Gauss–Newton algorithm, and the Levenberg–Marquardt and Bayesian regularization algorithms. Most of these training algorithms are available in MATLAB. The sum square error (E_s) is defined to evaluate the training process. For all training patterns and network outputs, it is calculated by

$$E_s\left(IN,\varpi\right)=\frac{1}{2}\sum_{pp=1}^{PP}\sum_{cc=1}^{CC}tr_{pp,cc}^{2} \tag{4.5}$$

where IN is the input vector, ϖ is the weight vector, pp is the index of patterns, PP is the number of patterns, cc is the index of outputs, and CC is the number of outputs. The $tr_{pp,cc}$ is the training error at output cc when pattern pp is applied; it is defined as

$$tr_{pp,cc} = DA_{pp,cc} - OUT_{pp,cc} \tag{4.6}$$

where DA is the desired output vector and OUT is the actual output vector.

The relationship among the steepest descent algorithm, the Newton's method, the Gauss–Newton algorithm, and the Levenberg–Marquardt algorithms has been provided by Yu and Wilamowski (2011) as follows.

4.2.3.1 Steepest descent algorithm

The steepest descent algorithm is one of the most significant breakthroughs for ANN training. Normally, the gradient of the sum square error is defined as the first-order derivative function (4.5):

$$gs = \frac{\partial E_s(IN,\varpi)}{\partial \varpi} = \left[\frac{\partial E_s}{\partial \varpi_1} \quad \frac{\partial E_s}{\partial \varpi_2} \quad \cdots \quad \frac{\partial E_s}{\partial \varpi_Z}\right]^T \tag{4.7}$$

where Z is the number of weights. The update rule for the steepest descent algorithm can be expressed as

$$\varpi_{ii+1} = \varpi_{ii} - R_l g_{ii} \tag{4.8}$$

where R_l is the learning rate (step size) and *ii* is the iteration index.

The steepest descent algorithm is usually converges slowly. Small step size should be used to catch the required minima. The result is a slow training process in places where the gradient is gentle. The gradient descent might be increasingly 'zigzags' as the gradients point nearly orthogonally to the shortest direction to the required minima.

4.2.3.2 Newton's method

The convergence speed of the steepest descent method can be improved by increased Newton's method. It assumes that all the gradient components $gs_1, gs_2, \ldots, gs_Z$ are functions of weights and that all weights are linearly independent:

$$\begin{cases} gs_1 = F_1(\varpi_1, \varpi_2, \cdots \varpi_Z) \\ gs_2 = F_2(\varpi_1, \varpi_2, \cdots \varpi_Z) \\ \quad \bullet\bullet\bullet \\ gs_Z = F_N(\varpi_1, \varpi_2, \cdots \varpi_Z) \end{cases} \tag{4.9}$$

where $F_1, F_2, \ldots, F_Z$ are nonlinear relationships between weights and related gradient components. Unfolding each gs_{zz} (zz=1, 2,…, Z) in Eq. (4.9) by the Taylor series and taking the first-order approximation:

$$\begin{cases} gs_1 \approx gs_{1,0} + \frac{\partial gs_1}{\partial \varpi_1}\Delta\varpi_1 + \frac{\partial gs_1}{\partial \varpi_2}\Delta\varpi_2 + \cdots + \frac{\partial gs_1}{\partial \varpi_Z}\Delta\varpi_Z \\ gs_2 \approx gs_{2,0} + \frac{\partial gs_2}{\partial \varpi_1}\Delta\varpi_1 + \frac{\partial gs_2}{\partial \varpi_2}\Delta\varpi_2 + \cdots + \frac{\partial gs_2}{\partial \varpi_Z}\Delta\varpi_Z \\ \quad \bullet\bullet\bullet \\ gs_Z \approx gs_{Z,0} + \frac{\partial gs_Z}{\partial \varpi_1}\Delta\varpi_1 + \frac{\partial gs_Z}{\partial \varpi_2}\Delta\varpi_2 + \cdots + \frac{\partial gs_Z}{\partial \varpi_Z}\Delta\varpi_Z \end{cases} \tag{4.10}$$

The gradient vector $\frac{\partial gs_{zi}}{\partial \varpi_{zj}}$ can be determined as:

$$\frac{\partial gs_{zi}}{\partial \varpi_{zj}} = \frac{\partial\left(\frac{\partial E_s}{\partial \varpi_{zi}}\right)}{\partial \varpi_{zj}} = \frac{\partial^2 E_s}{\partial \varpi_{zi} \partial \varpi_{zj}} \tag{4.11}$$

where zi and zj are the weight indices from 1 to Z. Combining Eqs. (4.10) and (4.11) yields:

$$\begin{cases} gs_1 \approx gs_{1,0} + \frac{\partial^2 E_s}{\partial \varpi_1^2}\Delta\varpi_1 + \frac{\partial^2 E_s}{\partial \varpi_1 \partial \varpi_2}\Delta\varpi_2 + \cdots + \frac{\partial^2 E_s}{\partial w_1 \partial w_Z}\Delta\varpi_Z \\ gs_2 \approx gs_{2,0} + \frac{\partial^2 E_s}{\partial \varpi_1 \partial \varpi_2}\Delta\varpi_1 + \frac{\partial^2 E_s}{\partial \varpi_2^2}\Delta\varpi_2 + \cdots + \frac{\partial^2 E_s}{\partial \varpi_2 \partial \varpi_Z}\Delta\varpi_Z \\ \cdots \\ gs_Z \approx gs_{Z,0} + \frac{\partial^2 E_s}{\partial \varpi_Z \partial \varpi_1}\Delta\varpi_1 + \frac{\partial^2 E_s}{\partial \varpi_Z \partial \varpi_2}\Delta\varpi_2 + \cdots + \frac{\partial^2 E_s}{\partial \varpi_Z^2}\Delta\varpi_Z \end{cases} \tag{4.12}$$

In order to obtain the minima of total error function E, each individual of the gradient vector must be set to zero. Therefore, the left sides of Eq. (4.12) are all zero, and

$$\begin{cases} 0 \approx gs_{1,0} + \frac{\partial^2 E_s}{\partial \varpi_1^2}\Delta\varpi_1 + \frac{\partial^2 E_s}{\partial \varpi_1 \partial \varpi_2}\Delta\varpi_2 + \cdots + \frac{\partial^2 E_s}{\partial \varpi_1 \partial \varpi_Z}\Delta\varpi_Z \\ 0 \approx gs_{2,0} + \frac{\partial^2 E_s}{\partial \varpi_1 \partial \varpi_2}\Delta\varpi_1 + \frac{\partial^2 E_s}{\partial \varpi_2^2}\Delta\varpi_2 + \cdots + \frac{\partial^2 E_s}{\partial \varpi_2 \partial \varpi_Z}\Delta\varpi_Z \\ \cdots \\ 0 \approx gs_{Z,0} + \frac{\partial^2 E_s}{\partial \varpi_Z \partial \varpi_1}\Delta\varpi_1 + \frac{\partial^2 E_s}{\partial \varpi_Z \partial \varpi_2}\Delta\varpi_2 + \cdots + \frac{\partial^2 E_s}{\partial \varpi_Z^2}\Delta\varpi_Z \end{cases} \tag{4.13}$$

Combining Eq. (4.7) with (4.13) yields:

$$\begin{cases} -\frac{\partial E_s}{\partial \varpi_1} = -gs_{1,0} \approx + \frac{\partial^2 E_s}{\partial \varpi_1^2}\Delta\varpi_1 + \frac{\partial^2 E_s}{\partial \varpi_1 \partial \varpi_2}\Delta\varpi_2 + \cdots + \frac{\partial^2 E_s}{\partial \varpi_1 \partial \varpi_Z}\Delta\varpi_Z \\ -\frac{\partial E_s}{\partial \varpi_2} = -gs_{2,0} \approx + \frac{\partial^2 E_s}{\partial \varpi_1 \partial \varpi_2}\Delta\varpi_1 + \frac{\partial^2 E_s}{\partial \varpi_2^2}\Delta\varpi_2 + \cdots + \frac{\partial^2 E_s}{\partial \varpi_2 \partial \varpi_Z}\Delta\varpi_Z \\ \cdots \\ -\frac{\partial E_s}{\partial \varpi_Z} = -gs_{Z,0} \approx + \frac{\partial^2 E_s}{\partial \varpi_Z \partial \varpi_1}\Delta\varpi_1 + \frac{\partial^2 E_s}{\partial \varpi_Z \partial \varpi_2}\Delta\varpi_2 + \cdots + \frac{\partial^2 E_s}{\partial \varpi_Z^2}\Delta\varpi_Z \end{cases} \tag{4.14}$$

There are Z equations for Z parameters, so that all $\Delta\varpi$ can be calculated. With the solutions, the weight space can be updated iteratively. Eq. (4.14) can also be written in matrix form:

$$\begin{bmatrix} -gs_1 \\ -gs_2 \\ \cdots \\ -gs_Z \end{bmatrix} = \begin{bmatrix} -\dfrac{\partial E_s}{\partial \varpi_1} \\ -\dfrac{\partial E_s}{\partial \varpi_2} \\ \cdots \\ -\dfrac{\partial E_s}{\partial \varpi_Z} \end{bmatrix} \begin{bmatrix} \dfrac{\partial^2 E_s}{\partial \varpi_1^2} & \dfrac{\partial^2 E_s}{\partial \varpi_1 \partial \varpi_2} & \cdots & \dfrac{\partial^2 E_s}{\partial \varpi_1 \partial \varpi_Z} \\ \dfrac{\partial^2 E_s}{\partial \varpi_1 \partial \varpi_2} & \dfrac{\partial^2 E_s}{\partial \varpi_2^2} & \cdots & \dfrac{\partial^2 E_s}{\partial \varpi_2 \partial \varpi_Z} \\ & \cdots & & \\ \dfrac{\partial^2 E_s}{\partial \varpi_Z \partial \varpi_1} & \dfrac{\partial^2 E_s}{\partial \varpi_Z \partial \varpi_2} & \cdots & \dfrac{\partial^2 E_s}{\partial \varpi_Z^2} \end{bmatrix} \times \begin{bmatrix} \Delta\varpi_1 \\ \Delta\varpi_2 \\ \cdots \\ \Delta\varpi_Z \end{bmatrix} \tag{4.15}$$

where the square matrix is a Hessian matrix:

$$H = \begin{bmatrix} \dfrac{\partial^2 E_s}{\partial \varpi_1^2} & \dfrac{\partial^2 E_s}{\partial \varpi_1 \partial \varpi_2} & \cdots & \dfrac{\partial^2 E_s}{\partial \varpi_1 \partial \varpi_Z} \\ \dfrac{\partial^2 E_s}{\partial \varpi_1 \partial \varpi_2} & \dfrac{\partial^2 E_s}{\partial \varpi_2^2} & \cdots & \dfrac{\partial^2 E_s}{\partial \varpi_2 \partial \varpi_Z} \\ & \cdots & & \\ \dfrac{\partial^2 E_s}{\partial \varpi_Z \partial \varpi_1} & \dfrac{\partial^2 E_s}{\partial \varpi_Z \partial \varpi_2} & \cdots & \dfrac{\partial^2 E_s}{\partial \varpi_Z^2} \end{bmatrix} \tag{4.16}$$

Combining Eqs. (4.7) and (4.16) with Eq. (4.15) produces the following expressions:

$$-gs = H\Delta\varpi \tag{4.17}$$

$$\Delta\varpi = -H^{-1}gs \tag{4.18}$$

Therefore, the update rule for Newton's method is

$$\varpi_{ii+1} = \varpi_{ii} - H_{ii}^{-1} gs_{ii} \tag{4.19}$$

As the second-order derivatives of the total error function, Hessian matrix H provides the proper evaluation of the change in the gradient vector. A comparison of Eqs. (4.8) and (4.19) shows that well-matched step sizes are provided by the inverted Hessian matrix.

4.2.3.3 Gauss–Newton algorithm

If Newton's method is used to update the weights, then in order to obtain Hessian matrix H, the second-order derivatives of the total error function must be calculated. This calculation can be very complicated. To simplify the process, Jacobian matrix J is introduced as follows:

$$J=\begin{bmatrix} \frac{\partial tr_{1,1}}{\partial \varpi_1} & \frac{\partial tr_{1,1}}{\partial \varpi_2} & \cdots & \frac{\partial tr_{1,1}}{\partial \varpi_Z} \\ \frac{\partial tr_{1,2}}{\partial \varpi_1} & \frac{\partial tr_{1,2}}{\partial \varpi_2} & \cdots & \frac{\partial tr_{1,2}}{\partial \varpi_Z} \\ \cdots & \cdots & \cdots & \cdots \\ \frac{\partial tr_{1,M}}{\partial \varpi_1} & \frac{\partial tr_{1,M}}{\partial \varpi_2} & \cdots & \frac{\partial tr_{1,M}}{\partial \varpi_Z} \\ \cdots & \cdots & \cdots & \cdots \\ \frac{\partial tr_{P,1}}{\partial \varpi_1} & \frac{\partial tr_{P,1}}{\partial \varpi_2} & \cdots & \frac{\partial tr_{P,1}}{\partial \varpi_Z} \\ \frac{\partial tr_{P,2}}{\partial \varpi_1} & \frac{\partial tr_{P,2}}{\partial \varpi_2} & \cdots & \frac{\partial tr_{P,2}}{\partial \varpi_Z} \\ \cdots & \cdots & \cdots & \cdots \\ \frac{\partial tr_{P,M}}{\partial \varpi_1} & \frac{\partial tr_{P,M}}{\partial \varpi_2} & \cdots & \frac{\partial tr_{P,M}}{\partial \varpi_Z} \end{bmatrix} \tag{4.20}$$

Integration of Eqs. (4.5) and (4.7) allows the individuals of the gradient vector to be calculated:

$$gs_{zz}=\frac{\partial E_s}{\partial \varpi_{zz}}=\frac{\partial\left(\frac{1}{2}\sum_{pp=1}^{PP}\sum_{cc=1}^{CC} tr_{pp,cc}^2\right)}{\partial \varpi_{zz}}=\sum_{pp=1}^{PP}\sum_{cc=1}^{CC}\left(\frac{\partial tr_{pp,cc}}{\partial \varpi_{zz}} tr_{pp,cc}\right) \tag{4.21}$$

Combining Eqs. (4.20) and (4.21) yields the following relationship between Jacobian matrix J and gradient vector g:

$$gs = J \cdot tr \tag{4.22}$$

$$tr=\begin{bmatrix} tr_{1,1} \\ tr_{1,1} \\ \cdots \\ tr_{1,CC} \\ tr_{PP,1} \\ tr_{PP,2} \\ \cdots \\ tr_{PP,CC} \end{bmatrix} \tag{4.23}$$

When Eq. (4.5) is inserted into Eq. (4.16), the individual at the zi^{th} row and zj^{th} column of Hessian matrix can be calculated as

$$h_{zi,zj}=\frac{\partial^2 E_s}{\partial \varpi_{zi}\partial \varpi_{zj}}=\frac{\partial^2\left(\frac{1}{2}\sum_{pp=1}^{PP}\sum_{cc=1}^{CC} tr_{pp,cc}^2\right)}{\partial \varpi_{zi}\partial \varpi_{zj}}=\sum_{pp=1}^{PP}\sum_{cc=1}^{CC}\frac{\partial tr_{pp,cc}}{\partial \varpi_{zi}}\frac{\partial tr_{pp,cc}}{\partial \varpi_{zj}}+S_{zi,zj} \tag{4.24}$$

where $S_{zi,zj}$ is equal to

$$S_{zi,zj} = \sum_{pp=1}^{PP} \sum_{cc=1}^{CC} \frac{\partial^2 tr_{pp,cc}}{\partial \varpi_{zi} \partial \varpi_{zj}} tr_{pp,cc} \tag{4.25}$$

As the basic assumption of Newton's method is that $S_{zi,zj}$ is close to zero, the relationship between Hessian matrix H and Jacobian matrix J can be rewritten as

$$H \approx J^T J \tag{4.26}$$

Thus, the update rule of the Gauss–Newton algorithm can be expressed as

$$\varpi_{ii+1} = \varpi_{ii} - \left(J_{ii}^T J_{ii}\right)^{-1} J_{ii} tr_{ii} \tag{4.27}$$

Obviously, the advantage of the Gauss–Newton algorithm over the standard Newton's method is that the former does not require the calculation of second-order derivatives of the total error function, because the Jacobian matrix J is used instead. However, the Gauss–Newton algorithm still faces the same convergence problem as the Newton algorithm for complex error space optimization. Mathematically, it can be said that the matrix $J^T J$ may not be invertible.

4.2.3.4 Levenberg–Marquardt algorithm

The Levenberg–Marquardt algorithm blends the steepest descent method and the Gauss–Newton algorithm. Fortunately, it inherits the speed advantage of the former and the stability of the latter. At the same time, it is more robust than the Gauss–Newton algorithm because in many cases it converges well even if the error surface is much more complex than in the quadratic situation. Although the Levenberg–Marquardt algorithm tends to be slightly slower than the Gauss–Newton algorithm (in convergent situations), it converges much faster than the steepest descent method. Essentially, the Levenberg–Marquardt algorithm performs a combined training process: around areas with complex curvature, the Levenberg–Marquardt algorithm switches to the steepest descent algorithm until the local curvature is suitable for making a quadratic approximation; it then becomes approximately the Gauss–Newton algorithm, which can speed up the convergence significantly.

In the Levenberg–Marquardt algorithm, in order to ensure that the approximated Hessian matrix $J^T J$ is invertible, another approximation to the Hessian matrix is introduced:

$$H \approx J^T J + \mu I \tag{4.28}$$

where μ is called the combination coefficient and is always positive, and I is the identity matrix. In Eq. (4.28) it can be seen that the individuals on the main diagonal of the approximated Hessian matrix will be larger than zero. Therefore, the use of this approximation ensures that matrix H is always invertible. When Eqs. (4.27) and (4.28) are combined, the update rule of the Levenberg–Marquardt algorithm can be expressed as

$$\varpi_{ii+1} = \varpi_{ii} - \left(J_{ii}^T J_{ii} + \mu I\right)^{-1} J_{ii} tr_{ii} \tag{4.29}$$

As a combination of the steepest descent and Gauss–Newton algorithms, the Levenberg–Marquardt algorithm switches between the two algorithms during the training process. When the combination coefficient μ is very small (nearly zero), Eq. (4.29) approaches Eq. (4.27), and the Gauss–Newton algorithm is used. When the combination coefficient is very large, Eq. (4.29) approximates Eq. (4.8), and the steepest descent method is used.

If the combination coefficient μ in Eq. (4.29) is very large, it can be interpreted as the learning coefficient in the steepest descent method:

$$R_l = \frac{1}{\mu} \tag{4.30}$$

The Levenberg-Marquardt algorithm can converge quickly and provide calculations with a high level of precision. It can be used to accelerate the convergence speed of each of the ANN training processes (Du et al., 2014).

4.2.3.5 Bayesian regularization algorithm

Over-fitting is a considerable problem for ANNs and results in poor generalization. In such cases, the network has failed to respond well when tested and simulated with an unseen data set. The network has actually memorized the samples it is trained with, instead of learning to generalize the process to respond to unknown conditions. A data training sample that is small in comparison to the total number of network parameters is one reason for poor generalization. One way of improving the generalization ability of an ANN is to ensure that the sampling data is large enough to provide an adequate fit. However, this measure would increase the computing cost for sampling. Regularization is another statistical approach to the problem of over-fitting (Choudhury et al., 2012).

As it turns out, the Bayesian regularization algorithm has perfect generalization capability (Deng and Hwang, 2006; Titterington, 2004). For this algorithm, a regularization term is added to the sum of squared errors, and the target error for training (E_{train}) is modified as:

$$E_{train}(\varpi) = \zeta_1 E_s + \zeta_2 E_w \tag{4.31}$$

$$E_s = \Sigma_{cc}\left(OUT_{cc} - DA_{cc}\right)^2 \tag{4.32}$$

$$E_w = \Sigma_{zz}\varpi_{zz}^2 \tag{4.33}$$

where E_w is the regularization term, and ζ_1 and ζ_2 are the regularization coefficients. The weight value is regarded as a random variable. The prior probabilities of the training data set and the weight collection set are assumed to obey the Gaussian distribution. The Levenberg-Marquadt algorithm is used to obtain the minimum $E_{train}(w)$ to update the regularization coefficients. The weight is updated according to

$$\Delta\varpi_{ii} = -R_l \frac{\partial E_{train}^{ii}}{\partial \varpi_{ii}} + R_m \Delta\varpi^{(ii-1)} \tag{4.34}$$

where R_m is the momentum rate. The learning rate, momentum rate, and target error are closely related to the convergence of the ANN. Their values are always determined empirically. The learning rate is directly proportional to the size of the steps taken in weight space. However, it is worthwhile to reiterate that learning rate is only one of the parameters that affect the size of the steps taken in weight space. Traditionally, learning rates remain axed during training, and optimal learning rates are determined by trial and error. Guidelines for appropriate learning rates have been proposed for single-layer networks, but it is difficult to extend these guidelines to the multilayer case. Many heuristics have been proposed which adapt the learning rate, and hence the size of the steps taken in weight space, as training progresses according to the shape of the error surface. The momentum term may be considered to increase the effective step size in shallow regions of the error surface and can speed up the training process by several orders of magnitude. It should be noted that the momentum term must be less than 1.0 for convergence (Maier and Dandy, 2000).

4.2.5 Integration of the ANN with CFD

4.2.5.1 Design variables and design objectives

When an ANN is combined with CFD, the design variables and objectives should first be confirmed, and then the structure of the ANN can be determined. The numbers of nodes in the input and output layers are the same as the numbers of design variables and objectives. In the sampling stage, the values of design variables and objectives are obtained by the sampling method and the CFD process, respectively. The sampling method is first used to generate values of the design variables of the samples. The design variables can be the geometric parameters or boundary conditions of a building, such as floor shape (Wang et al., 2006), supplied air velocity and temperature, or other HVAC operating parameters (Nassif, 2014; Wang and Ma, 2008). Most of the design variables involved in the design of buildings are HVAC parameters. Next, values of the design objectives, such as velocity, temperature, thermal comfort, and the cost of energy, can be obtained by use of CFD software such as Contam, FLUENT, EnergyPlus, or Airpack (Zhou and Haghighat, 2009b). In the training stage, normalized values of the design variables and objectives of the samples are used to train the ANN, and ANN will be considered well-trained when the training process has converged. In the design stage, values of the design variables are generated by a search process, and the design objectives are obtained by the well-trained ANN or from the design requirements. The design target is usually to obtain the values of the design variables that would lead to optimal design objectives.

Detailed data for the indoor environment, such as the velocity and temperature at any location, can be obtained by CFD. Thus, a large database can be created from the CFD results. Because an ANN with a large structure is difficult to train, only a few parameters should be selected from the CFD results for the design stage, such as the average velocity and temperature, and the velocity and temperature at some specific positions. Generally, several design variables and objectives, or perhaps a few dozens, are sufficient for the practical design of buildings. The design variables and design objectives are usually selected by the designer or based on the standards for the building environment. Furthermore, several principles should be observed in the selection of design variables and objectives:

(1) The values of the design variables can be changed during the CFD process, and the design objectives can be obtained from the CFD results. Although some design objectives cannot be obtained directly by CFD, they can be calculated by other programs on the basis of the CFD results.
(2) The relationship between the design variables and objectives should be stable. A given value of a design variable corresponds to a specific value of a design objective.
(3) A relationship between the design variables and objectives should exist. If a design variable has no impact on the design objective, it cannot be involved in the design stage.

Either a linear or nonlinear relationship between the design variables and objectives can be represented by an ANN. Mapping of a nonlinear relationship is usually more complicated; more training samples are needed; and the relationship is more difficult to predict. A given set of design variables corresponds to only one set of design objectives. However, a set of design objectives may correspond to multiple sets of design variables. For example, the PMV is relative to the velocity, temperature, relative humidity, and many other parameters of the supplied air. A specific PMV value can be obtained from different combinations of supplied-air parameters. Thus, a multi-solution problem is usually present in the inverse design of a building environment.

The independence of design variables is not necessary, but the design variables must be well identified. If some of the impact factors are not incorporated, the mapping relationship between the design variables and objectives will be unstable. The number of potential values of the design variables must be finite in order for the sampling process to be conducted. Therefore, the range and interval of each design variable should be known before the sampling stage. The interval of a design variable is usually determined by the precision of the design. For example, if the minimum unit of inlet velocity is 0.1 m/s, the interval of the inlet velocity should be 0.1 m/s. A large design space exists when the interval of the design variable is small, which will decrease the performance of the ANN with the same number of training samples. The expression of design variables is flexible. For example, when velocity and temperature are involved, the real values of these parameters can be set as the input to the ANN. If the shape of the building is involved, the potential shapes can be encoded as 1, 2 . . . n, which is the input to the ANN. If the state of a door or window is involved, then "0" can be used to represent "open," "1" can be used to represent "closed," and 0 and 1 are the potential values of the design variable.

The independence of design objectives is also unnecessary. The mapping relationship of each design objective to the design variables is independent of other mapping relationships. Therefore, either one multiple-output ANN or multiple single-output ANNs can be adopted. Multiple single-output ANNs would require multiple training processes, which in turn would increase the computational cost. The range of design objectives is determined by the range of design variables. It is unnecessary to unify the expression of design objective to that of design variable. For example, the design objective can be expressed as a value of temperature (a real number), and the design variable can be expressed as a condition of window, which can be as "open" or "close" (a symbol). The real values of environmental parameters are usually obtained by means of CFD software, and thus the real value is the most common output of an ANN. Most of the design objectives involved in the design of buildings are parameters related to thermal comfort or energy cost. Generally, ANNs are more accurate in predicting system parameters than detailed parameters. Detailed parameters are usually very sensitive to changes in the design variables, which might induce inflection points. It is difficult for an ANN to predict inflection points if they are not included in the training samples.

The CFD databases consist of the design variables and objectives of the training samples. After normalization, the databases are used to train the ANN, and a training algorithm is used to obtain the optimal values of weights and bias for the ANN. The number of training samples and structural parameters of the ANN (the number of hidden layers and the number of neurons in each hidden layer) are empirically determined. The well-trained ANN can then be used in the design of a building environment.

According to the roles of the design variables and objectives in the ANN, integration methods for building design applications can be divided into two types: direct inverse design and indirect inverse design.

4.2.5.2 Direct inverse design method

When the design objectives are set as the input to the ANN and the design variables are set as the output, the method is called direct inverse design. The values of the design variables are obtained from arbitrary values of the design objectives. For example, suppose that the design objectives are the velocity and temperature in specific locations, and the design variable is the size of the inlet. The design result (inlet size) can then be immediately obtained when the required values of velocity and temperature are assigned as the input to the ANN. Four stages are involved in this method:

(a) Sampling stage: The sampling method is used to generate values of the design variables for the training samples, and CFD is used to obtain values of the design objectives for each training sample. The CFD database consists of all the design variables and objectives.
(b) Training stage: The structure and training parameters are set. The database is normalized and used to train the ANN. The inputs are the design objectives, and the outputs are the design variables. The ANN is well trained when the training process converges.
(c) Design stage: The required values of the design objectives are set as the input to the well-trained ANN, and the predicted values from the ANN (design variables) are the design results.
(d) Verification stage: The required design variables or the CFD results obtained from the design results are used to evaluate the design accuracy. In this stage, the CFD process is conducted once, as only one solution is obtained by the ANN.

A flow chart of the direct inverse design process can be seen in Fig. 4.5. This method is widely used to identify parameter in buildings, such as contaminant source, heat transfer coefficient, and boundary conditions.

The essence of direct inverse design is the establishment of an inverse mapping relationship by the ANN. The original mapping relationship might be difficult to invert because of the highly nonlinear nature of the Navier-Stokes equation or other control equations in buildings (Zhang and Chen 2007a; Zhang and Chen 2007b). The ANN is a surrogate for the original model, and it is expressed by an equation consisting of weight, bias, and transfer functions. Differential equations, integral equations, and other complicated equations are not involved in the establishing of an ANN model. Thus, the model is much simpler than

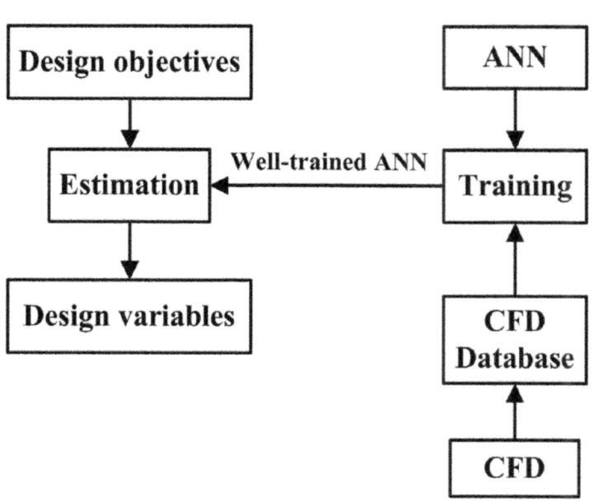

Figure 4.5 Flow chart of ANN-CFD integration by the direct inverse design method

the original CFD model. The inverse mapping relationship between the design variables and design objectives can be established easily by the ANN.

The advantage of direct inverse design is high efficiency. Evaluation of a well-trained ANN from input to output is very fast (Kabir et al., 2008). The ANN is used only once, and other simulation processes are not involved in the design process. Most of the computational costs are related to the CFD process in the sampling stage. The simulation times for CFD processes are similar, and therefore the number of CFD cases can be used to project the computational cost. The cost is proportional to the number of training samples, which is usually determined by designer.

There are several obstacles in the application of direct inverse design:

(1) The design objective must be expressed as an equation. For example, the objective can be PMV = 0.5, but it cannot be PMV < 0.5. Because the ANN input and output must be real numbers, the design objectives must be specific values. If the design requirement is an interval range, the input value is difficult to identify, and the interval solution is also difficult to obtain at the output of the ANN.
(2) It is difficult to solve a multi-solution problem by this method. In such a problem, the inverse mapping relationship is one-to-many. The training process does not converge easily, and only one solution can be obtained in the design stage.
(3) The target value of a design objective is usually difficult to confirm. No-solution and multi-solution problems are likely to occur with an improper target value for the design objective. In some studies, researchers have known the required design results before using an ANN. They have tried to verify or improve the performance of the ANN, and no-solution and multi-solution problems have been artificially avoided. However, in practical applications, the design results are usually unknown, and these problems cannot be avoided.
(4) Deviation exists in the design results. Because an ANN is a surrogate for the original model, there is some deviation between the two, and it affects the accuracy of the results. This deviation can be determined in the verification stage. It cannot be decreased in the design or verification stage. Generally, increasing the number of training samples will decrease the deviation, but the computing cost will increase. What is worse is that deviation is difficult to avoid because of the differences between the ANN model and the original model.

In an inverse design process that is based on an ANN and CFD, design deviation has two parts: the deviation between CFD results and real values, and the deviation between CFD results and ANN results. In previous studies of the inverse design method, real values have usually not been involved, and the deviation between CFD results and real values has been treated as systematic error. Therefore, the CFD results are used to evaluate the accuracy of the inverse design process.

In the direct inverse design method, if the target value of a design variable is known, the design error can be evaluated by the following equation:

$$DE(\xi) = \left| \frac{F(\xi)_{ANN} - F(\xi)}{F(\xi)} \right| \times 100\% \tag{4.35}$$

where $DE(\xi)$ is the design error for the design variable, $F(\xi)_{ANN}$ is the ANN result for the design variable, and $F(\xi)$ is the target value of the design variable.

If the target value for a design variable is unknown, the CFD results can be used to evaluate the design accuracy of the ANN. The design results from the ANN are used as the initial or boundary conditions for CFD, and the real design objectives (CFD results) corresponding to the design results can be obtained by the simulation process. Then the design error can be evaluated by the following equation:

$$DE(\xi) = \left| \frac{F(\xi)_{ANN} - F(\xi)_{CFD}}{F(\xi)_{CFD}} \right| \times 100\% \tag{4.36}$$

where $DE(\xi)$ is the design error for the design objective, $F(\xi)_{ANN}$ is the ANN result for the design objective, and $F(\xi)_{CFD}$ is the CFD result for the design objective. The design error is the same as the prediction error of the ANN. A larger number of training samples might lead to higher design accuracy, but the computational cost would increase.

4.2.5.3 Indirect inverse design method

When the design variables and objectives are set as the input and output of an ANN in the training and designing stages, the method is called indirect inverse design. A well-trained ANN is used to predict the values of design objectives from arbitrary values of design variables. The design requirements for a building environment are usually represented as intervals, such as $-0.5 < \text{PMV} < 0.5$. These requirements can be satisfied by many different combinations of design variables. It is difficult to solve a multi-solution problem by direct inverse design, but the indirect inverse design method is effective. Therefore, it is a more popular method for multi-solution problems. Four stages are also involved in the indirect inverse design method:

(a) Sampling stage: A sampling method is used to generate values of the design variables for the training samples, and CFD is used to obtain values of the design objectives for each sample. The CFD database consists of all the design variables and objectives.
(b) Training stage: The structure and training parameters are set. The database is normalized and used to train the ANN. The inputs are the design variables, and the outputs are the design objectives. If the training process converges, a well-trained ANN is obtained.
(c) Design stage: A search method is used to generate the design variables of potential solutions. The ANN is used to predict the values of the design objectives for these potential solutions. If the predicted values satisfy the design requirement, the current individual is a solution for the design.
(d) Verification stage: The CFD results obtained from the design results are used to evaluate the accuracy of the design. In this stage, every solution should be verified, and the number of CFD processes is the same as the number of solutions. In some studies, the prediction error of an ANN has been considered to be acceptable, and this stage has been ignored because of the large computational cost that would be entailed by verification.

A flow chart of the indirect inverse design method is shown in Fig. 4.6. The well-trained ANN is used repeatedly until the convergence criterion is reached. The criterion is usually the obtaining of a set of solutions. This method has been widely used in building environment studies, and it can be integrated with many optimization algorithms or control methods to reduce computational costs in building design (Ferreira et al., 2012; Jahedi and Ardehali, 2012; Marvuglia et al., 2014).

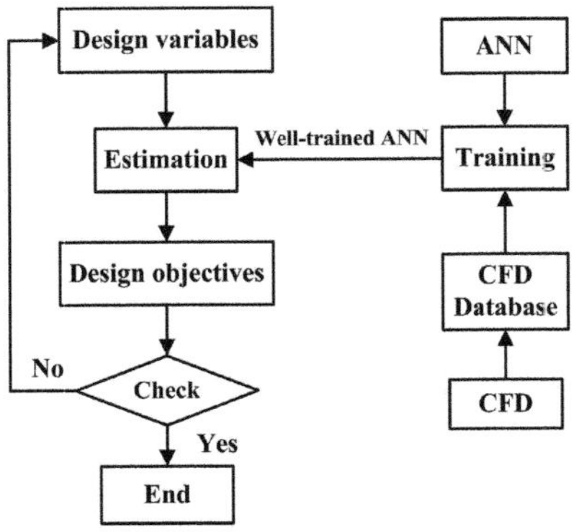

Figure 4.6 Flow chart of ANN-CFD integration by the indirect design method

The well-trained ANN replaces CFD in the design process and is used to predict the design objectives. The role of the ANN in the design process is the same as that of CFD. The ANN can be used to predict any design objectives that can be obtained by CFD. Thus, a multi-solution problem can be solved by this method. Most of the computational cost is related to the CFD process. The computational cost is proportional to the number of CFD cases.

There are also some disadvantages to this method:

(1) The design process is usually arbitrary. When an ANN serves as the only engine for a design, the search process is arbitrary because multiple random sampling methods are used in solving a multi-solution problem. Furthermore, the entire design process would be dedicated to finding only one optimal solution. However, because the prediction process can be completed immediately without the need for any CFD processes, the computational cost is not an obstacle in this method. In addition, the number of CFD cases used in the design process can be used to estimate the cost.
(2) The design result is affected by the prediction accuracy of the ANN. The use of an ANN instead of a CFD model increases the uncertainty and the risk of accumulative error with the existence of prediction error (Nguyen et al. 2014). To achieve high prediction accuracy, a large number of training samples is needed, which would entail an increase in the computational cost. In addition, deviation between the ANN model and the original model is difficult to avoid.

Several measures can be taken to overcome the disadvantages of the indirect inverse design method. When optimization algorithms are integrated with an ANN, the search process is determined by these algorithms, which increases the efficiency of the design process. The integration of a genetic algorithm with an ANN will be presented in Chapter 6.

A correction process can be used to avoid the effect of prediction error on the design result. A flow chart of the indirect inverse design method with correction can be seen in Fig. 4.7. In this method, both the ANN and CFD are used to obtain values of the design objectives.

In the design stage, the ANN is used to predict the design objectives from the design variables. If the prediction value of the ANN satisfies the design requirements, the CFD process will be conducted to obtain the real values (CFD results) of the design objectives.

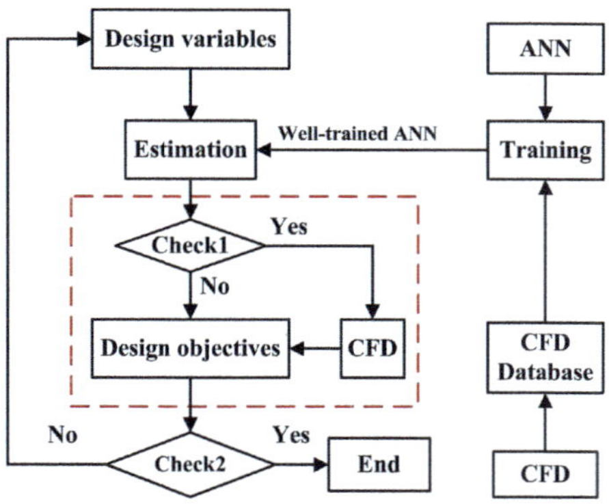

Figure 4.7 Flow chart of ANN-CFD integration by the indirect inverse design method with correction

If the CFD results satisfy the design requirements, the current individual is a solution for the design, and values of the design objectives for this individual are obtained by CFD. In this way, the design result is not influenced by the prediction error of the ANN. If the CFD results do not satisfy the design requirements, the current individual is not a solution for the design, and the search process continues. A threshold value can also be used to start the correction process. For example, if the requirement for temperature is less than 20°C, the threshold value can be set as 19 or 21°C. If the predicted value is less than 19 or 21°C, the correction process is initiated. When the threshold value is stricter than the design requirement, better solutions can be obtained, but the number of solutions will be reduced, and a no-solution problem may exist. When the threshold value is less strict than the design requirement, more CFD processes will be required. The verification stage is not involved in this method, as all the design results are obtained by CFD.

The computational cost is generated in the sampling and design stages. The number of samples is specified by the designer. The number of CFD cases in the design process is uncertain, and it is related to the prediction accuracy of the ANN and the convergence criterion for the design. Improving the prediction accuracy can reduce the unnecessary CFD processes. If the convergence criterion is easy to reach, few individuals will be involved in the design stage, and the number of CFD processes will be small. Although the correction process increases the computational cost, convergence and design accuracy are maintained.

Two indices can be used to evaluate ANN performance in the design of a building environment: average relative error (ARE) and success rate (SR).

ARE is used to evaluate the prediction accuracy of the ANN in the whole design space, and it is expressed as follows:

$$ARE = \sum_{hh=1}^{HH} \frac{\left| F(O)_{CFD}^{hh} - F(O)_{ANN}^{hh} \right|}{F(O)_{CFD}^{hh} \times HH} \times 100\% \tag{4.37}$$

where $F(O)_{CFD}^{hh}$ is the CFD result for the hh^{th} individual, $F(O)_{ANN}^{hh}$ is the ANN result for the hh^{th} individual, and HH is the number of individuals. For comparison of the relative errors for different design objectives, normalized values should be used to calculate the errors.

SR can be used to evaluate prediction accuracy of the ANN in the local design space, and it is defined as:

$$SR = 100\% \times \frac{n(E_{CFD})}{n(E_{ANN})} \tag{4.38}$$

where $n(E_{CFD})$ is the number of excellent individuals obtained by CFD, and $n(E_{ANN})$ is the number of excellent individuals obtained by the ANN.

The choice of ARE or SR is based on the purpose for the ANN. If the ANN is used to obtain the real values of environmental parameters, then ARE is more suitable for evaluating ANN performance. If the ANN is used to determine whether or not the current environment parameters satisfy the design requirements then SR is a more suitable index. There is no positive correlation between these two indices. Even if the ARE was very low, the prediction accuracy near some positions might be unsatisfied, leading to low SR.

The principles of the ANN and the integration of the ANN with CFD have been detailed in this section. In the application of ANN in the inverse design, the Latin hypercube sampling method is recommended to generate the training sample, and the Bayesian regularization algorithm is recommended to train the ANN. When the threshold values exist in the design objective, the logarithmic normalization method is a better choice compared to the linear normalization method. The direct inverse design method based on the ANN can be used when there is only one solution for the inverse design. The indirect inverse design method based on the ANN can be used to obtain multiple solutions.

4.3 Application of the CFD-based ANN method

4.3.1 Inverse identification of a built environment

This study validated the ANN method by applying it to two ventilation cases, one with isothermal flow and the other with non-isothermal flow.

4.3.1.1 Two-dimensional isothermal case

1. MODEL DESCRIPTION

A two-dimensional case from Nielsen et al. (1978) was used to evaluate the performance of the inverse design method. Fig. 4.8 shows the geometry of this case. The inlet velocity is $U_x = 0.455$ m/s and $U_y = 0$ with a turbulence intensity of 4%.

The design variables for the Nielsen model are the inlet velocity components (U_x and U_y). The ranges for the design variables are specified, where U_x is 0–1m/s (the actual value is 0.455 m/s) and U_y is 0–1m/s (the actual value is 0 m/s). The design objective for the model is the air velocity profile at the lower part of line x = 6 m (Fig. 4.9), and the objective function is:

$$O(\mathbf{U}_{inlet}) = \int_{\Omega} \left|\mathbf{U} - \mathbf{U}_0\right|^2 d\Omega \tag{4.39}$$

where $\mathbf{U}_0$ is the air velocity obtained by CFD simulation when $U_x = 0.455$ m/s and $U_y = 0$ m/s.

A smiller value of $O(\mathbf{U}_{inlet})$ represents a higher fitness value, and the goal of the inverse design process is to find the solution with minimum $O(\mathbf{U}_{inlet})$. The minimum value of $O(\mathbf{U}_{inlet})$ is 0, which in this case can be obtained when $U_x = 0.455$ m/s and $U_y = 0$ m/s. The goal of the design process is to identify the combination of U_x and U_y at which $O(\mathbf{U}_{inlet}) = 0$.

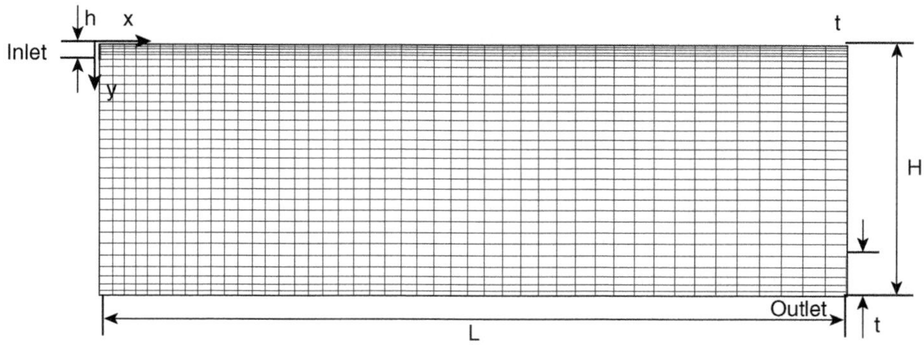

Figure 4.8 Structure of the Nielsen case.

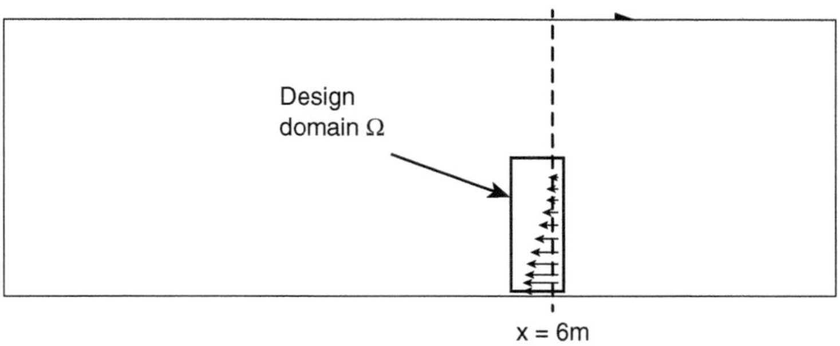

Figure 4.9 Design domain for the Nielsen model

2. SAMPLING

The software platform in the present study was built by MATLAB and FLUENT. The sampling, training, and design programs were compiled in MATLAB, and the CFD process was conducted with FLUENT software. The Latin hypercube sampling method was used to generate the design variables for the samples, and FLUENT was used to generate the design objectives. Forty samples were generated in the sampling stage to train the ANN, and another 10 samples were generated to test the ANN. The intervals of both U_x and U_y were 0.005 m/s, and the numbers of potential values for U_x and U_y were both 201.

The FLUENT software was used to obtain the simulated velocity at the design domain. The RNG k-ε turbulence model was adopted to determine the Reynolds stress, and the SIMPLE algorithm was adopted to solve the discrete equations. The value of $O(\mathbf{U}_{inlet})$ varied from 0.014 to 0.319 in the training samples. The time requirement for each CFD process was less than 5 minutes and was similar for different samples.

3. TRAINING

(1) Direct inverse design method A three-layer ANN was used for training and design of the direct inverse design method. The inputs were the design objectives of the samples, and the outputs were the design variables of the samples. If the value of cost function $O(\mathbf{U}_{inlet})$ had been set as the input, then a multi-solution problem might have existed. A particular value of $O(\mathbf{U}_{inlet})$ might have been obtained from several different combinations

of U_x and U_y. Thus, the simulated velocities at the design domain were instead set as the inputs to ANN, and the outputs were U_x and U_y. The linear normalization method was used to normalize the input and output parameters. Because the real input ranges were unknown, the inputs were normalized to 0.2–0.8 by the linear normalization method, where the bounds of the training samples were used. The outputs were normalized to 0–1 because the real output ranges were known.

There were 15 nodes in the input layer, ten nodes in the hidden layer, and two nodes in the output layer. The Bayesian regularization algorithm was used to train the ANN, and the convergence criterion for training was a mean squared error (MSE) of less than 0.001 with a maximum iteration epoch of 3,000. The training process can be seen in Fig. 4.10. In this case, the process converged at epoch 97. The time requirement for training was several seconds. Thus, the computing cost for ANN training was negligible in comparison with that for the CFD process.

(2) Indirect inverse design method A three-layer ANN was also used for training and design of the indirect inverse design method. The inputs were the design variables of the samples, and the output was the value of the $O(\mathbf{U}_{inlet})$ value for the samples. The inputs were normalized to 0~1, and the output was normalized to 0.2–0.8. There were two nodes in the input layer, eight in the hidden layer and one in the output layer. The Bayesian regularization algorithm was used to train the ANN, and the convergence criterion for training was a MSE of less than 0.001 with a maximum iteration epoch of 3,000. The training process is shown in Fig. 4.11. The process converged at epoch 131. The time requirement for training was also only a few seconds. Thus, the computing cost for ANN training could be neglected in comparison with that for the CFD process.

4. TESTING

To test the training result of the ANN, another 10 samples were generated. They had not been included in the training samples. The relative error (*RE*) of an ANN can be used to evaluate the ANN's prediction accuracy and is expressed as follows:

$$RE = \frac{\left|F(O)_{CFD} - F(O)_{ANN}\right|}{F(O)_{CFD}} \times 100\% \tag{4.40}$$

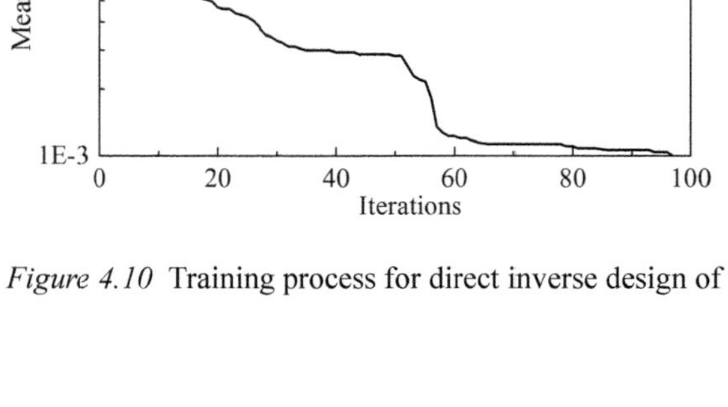

Figure 4.10 Training process for direct inverse design of the Nielsen model

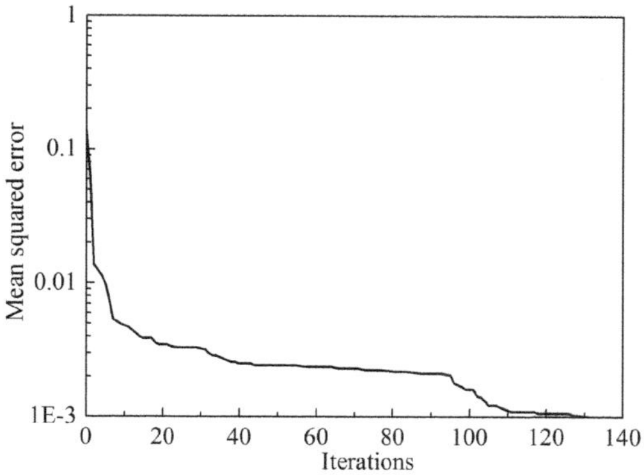

Figure 4.11 Training process for indirect inverse design of the Nielsen model

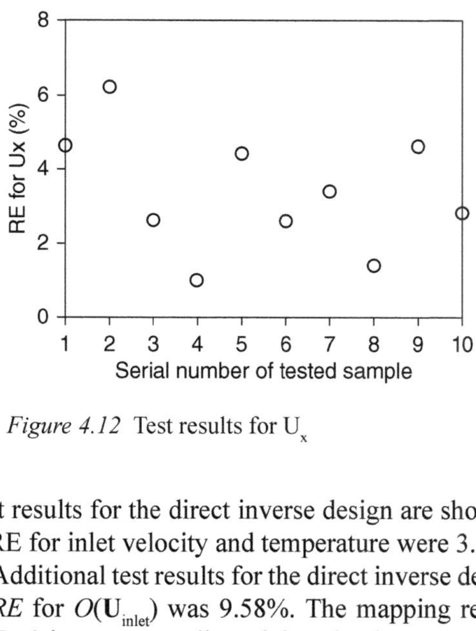

Figure 4.12 Test results for U_x

Test results for the direct inverse design are shown in Figs. 4.12 and 4.13. The mean values of RE for inlet velocity and temperature were 3.36% and 3.76%, respectively.

Additional test results for the direct inverse design are shown in Fig. 4.14. The mean value of *RE* for $O(\mathbf{U}_{inlet})$ was 9.58%. The mapping relationship between the inlet parameter and $O(\mathbf{U}_{inlet})$ is more complicated than that between the inlet parameter and simulation results for velocity, and therefore the prediction error increased.

5. DESIGN RESULT

(1) Direct inverse design method The well-trained ANN was used to design the inlet velocity for the Nielsen model. The ANN inputs were the simulation results obtained when $U_x = 0.455$ m/s and $U_y = 0$ m/s. The output was the inlet velocity, which is the design result shown in Table 4.1. The targets for U_x and U_y were achieved. The deviations were caused by the prediction error of the ANN.

(2) Indirect inverse design method When the indirect inverse design method was used to design the inlet velocity for the Nielsen model, the inputs of the ANN were arbitrary values

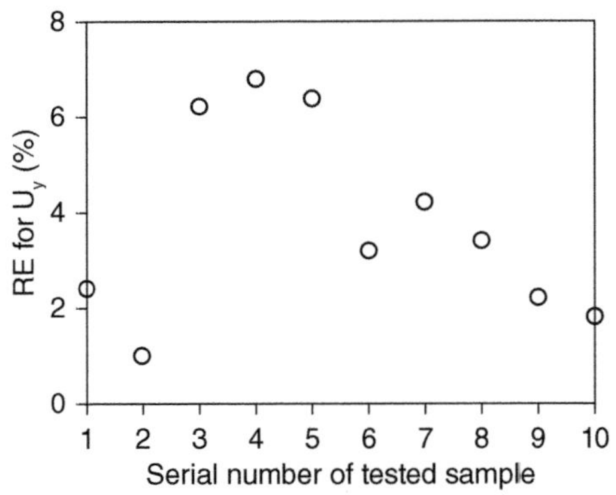

Figure 4.13 Test results for U_y

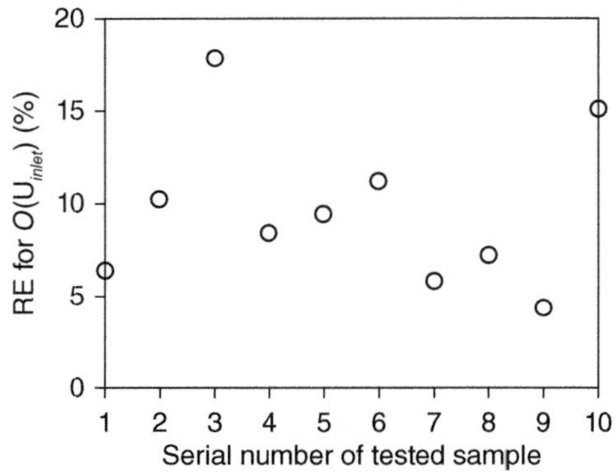

Figure 4.14 Test results for $O(\mathbf{U}_{inlet})$

Table 4.1 Design results obtained by the direct inverse design method.

	Real value	*Design result*
U_x (m/s)	0.455	0.465
U_y (m/s)	0	0.045

of U_x and U_y. Thousands of prediction values could be obtained within one second, and thus the entire design space was involved in the design process. Initially, the design method was used without correction, and the results are shown in Fig. 4.15. Although the value of $O(\mathbf{U}_{inlet})$ was less than 0.1 for 11.1% of the individuals, it never reached 0. The best value was 0.0015, and the corresponding inlet parameters were $U_x = 0.445$ m/s and $U_y = 0.05$ m/s. Therefore, it was not possible to obtain the real inlet parameters by this design method.

Next, the design method with correction was used. The design process was as follows:

(a) Set the threshold value A = 0.001.
(b) If the prediction value of $F(O)_{ANN}$ is less than A, conduct the CFD process to obtain the real value of $F(O)_{CFD}$.
(c) If $F_{CFD} = 0$, the design process ends. Otherwise, A = A+0.001; return to step (b).

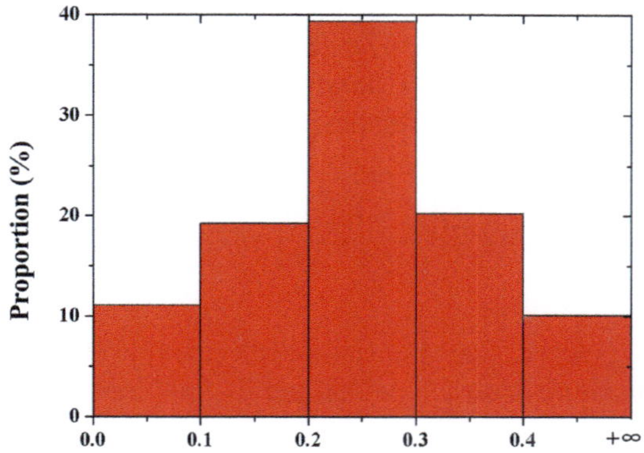

Figure 4.15 Design results without correction

The design process converged at A = 0.031. Although 171 new CFD cases were required, the real inlet parameters were obtained.

For this inverse design, only one optimal solution existed, and the prediction error had a significant effect on the design result. It was not possible to obtain the solution by the direct inverse design method, or by the indirect inverse design method without correction. Use of the indirect method with correction ensured convergence of the design.

4.3.1.2 Blay case

1. MODEL DESCRIPTION

The Blay case (Blay et al., 1992) was used in the present study to evaluate the performance of the inverse design method. It is a widely-used representative model for floor heating (natural convection) coupled with a mechanical ventilation system (forced convection) in a confined space. This case has clearly denned geometries and boundary conditions, and thus it is ideal for assessing new methods and models for indoor environments (Xue et al., 2013). The structure and boundary conditions of the Blay case are shown in Fig. 4.16.

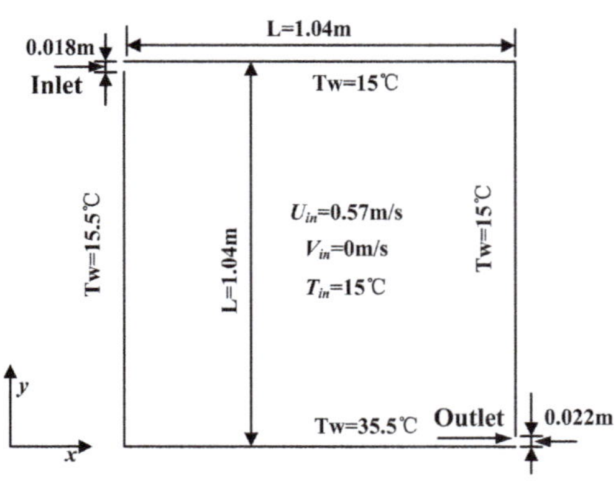

Figure 4.16 Structure and boundary conditions of the Blay case

The design variables for the Blay model are inlet velocity ($|U|_{in}$) and inlet temperature (T_{in}). The ranges for the design variables are specified as 0–1 m/s for $|U|_{in}$ (the actual value is 0.57 m/s) and 0–20°C for T_{in} (the actual value is 15°C). The design variables are simulated values of velocity and temperature at ten measurement points. A cost function has been established for the simulated and measured variables at the ten points (Xue et al. 2013):

$$F_{Blay} = \frac{1}{3}\sqrt{\frac{\Sigma\left(U_{xi}^{'}-U_{xi}\right)^2}{\Sigma U_{xi}^2} + \frac{\Sigma\left(U_{yi}^{'}-U_{yi}\right)^2}{\Sigma U_{yi}^{'2}} + \frac{\Sigma\left(T_i^{'}-T_i\right)^2}{\Sigma T_i^2}} \quad (i = 1,2,...10) \tag{4.41}$$

where F_{Blay} is the cost function value, $U_{xi}^{'}$ and $U_{yi}^{'}$ are the simulated velocity components at the ten points, $T_i^{'}$ is the simulated temperature at the ten points, and U_{xi}, U_{yi}, and T_i are the measured results. A smaller value of F_{Blay} represents a higher fitness value, and the goal of the inverse design process was to find the solution with the minimum F_{Blay}. The minimal value of F_{Blay} is 0, which can be obtained by CFD when $|U|_{in} = 0.57$ m/s and $T_{in} = 15$°C.

In essence, this inverse design process was an identification problem. The goal was to identify the combination of inlet velocity and temperature at which $F_{Blay} = 0$.

2. SAMPLING

Forty samples were generated in the sampling stage to train the ANN, and another 10 samples were generated to test the ANN. The intervals of inlet velocity and temperature were 0.01 m/s and 0.1°C, respectively. The numbers of potential values for the inlet velocity and temperature were 101 and 201, respectively.

The FLUENT software was used to obtain the simulated velocity and temperature at the ten points. The RNG k-ε turbulence model was adopted to determine the Reynolds stress, and the SIMPLE algorithm was adopted to solve the discrete equations. The value of F_{Blay} varied from 0.011 to 0.677 in the training samples. The time requirement for each CFD process was about 10 minutes and was similar for different samples.

3. TRAINING

(1) Direct inverse design method A three-layer ANN was used for training and design of the direct inverse design method. The inputs were the design objectives of the samples, and the outputs were the design objectives of the samples. If the value of cost function had been set as the input, a multi-solution problem might have existed. A particular value of *F* might have been obtained by several different combinations of inlet velocity and temperature. Therefore, the simulated velocity and temperature at the ten points were instead set as the inputs to the ANN, and the outputs were the inlet velocity and temperature. The linear normalization method was used to normalize the input and output parameters. Because the real input ranges were unknown, the inputs were normalized to 0.2–0.8 by the linear normalization method, where the bounds of the training samples were used. The outputs were normalized to 0–1 because the real output ranges were known.

There were 30 nodes in the input layer, ten nodes in the output layer, and two nodes in the output layer. The Bayesian regularization algorithm was used to train the ANN, and the convergence criterion for training was a MSE of less than 0.001 with a maximum iteration epoch of 3,000. The training process can be seen in Fig. 4.17. The process converged at epoch 61. The time requirement for training was less than one minute. Thus, the computing cost for ANN training was negligible in comparison with that for the CFD process.

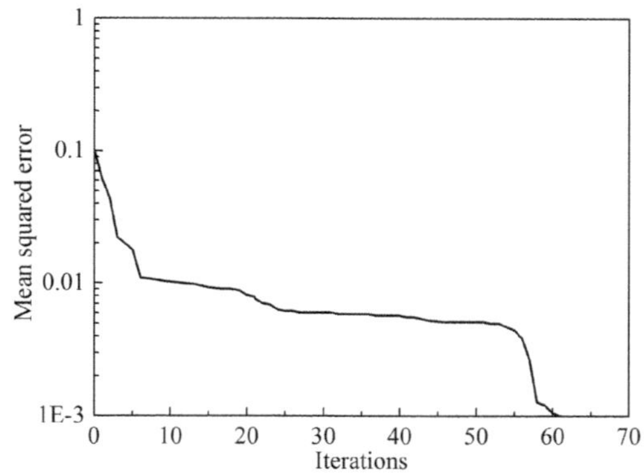

Figure 4.17 Training process for direct inverse design of the Blay model

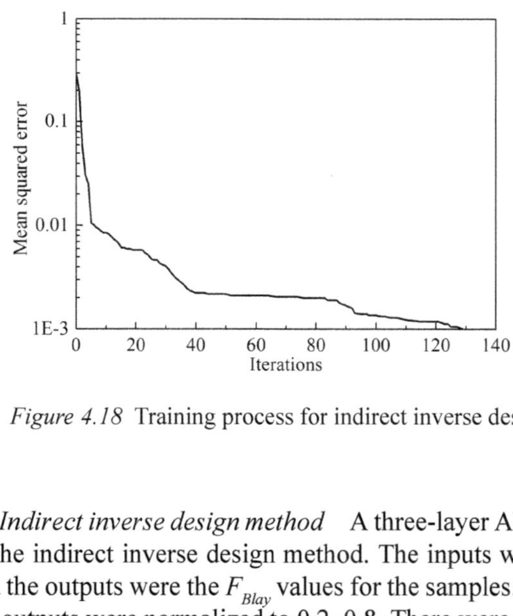

Figure 4.18 Training process for indirect inverse design of the Blay model

(2) Indirect inverse design method A three-layer ANN was also used for training and design of the indirect inverse design method. The inputs were the design variables of the samples, and the outputs were the F_{Blay} values for the samples. The inputs were normalized to 0–1, and the outputs were normalized to 0.2–0.8. There were two nodes in the input layer, eight in the hidden layer, and one in the output layer. The Bayesian regularization algorithm was used to train the ANN, and the convergence criterion for training was a MSE of less than 0.001 with a maximum iteration epoch of 3,000. The training process, shown in Fig. 4.18, converged at epoch 129. The time required for training was only a few seconds. The time requirement for the Blay model was less than that for the Nielsen model because the ANN structure was much simpler in the Blay case. The computing cost for ANN training was again negligible in comparison with that for the CFD process.

4. TESTING

Test results for the direct inverse design method are shown in Figs. 4.19 and 4.20. The mean values of RE for inlet velocity and temperature were 1.86% and 2.17%, respectively.

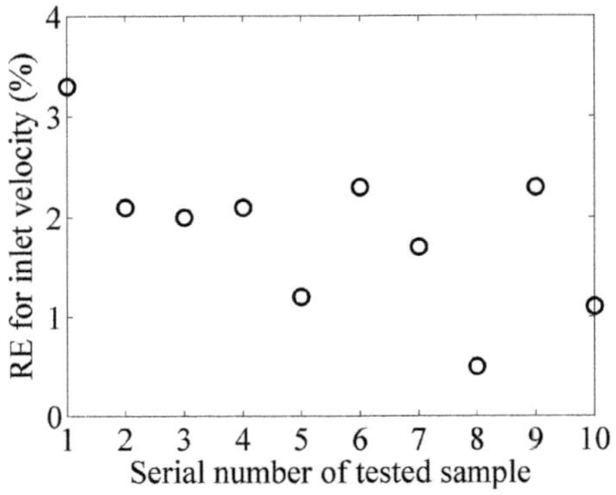

Figure 4.19 Test results for inlet velocity

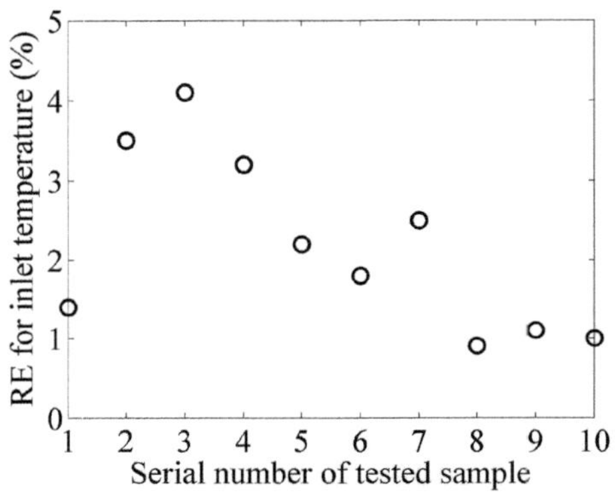

Figure 4.20 Test results for inlet temperature

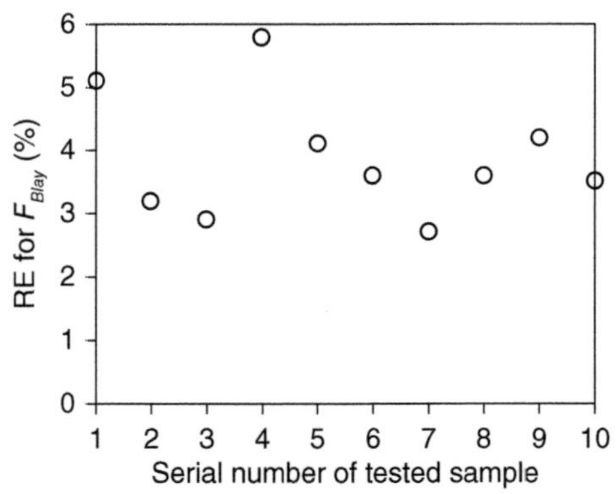

Figure 4.21 Test results for F_{Blay}

Additional test results for the direct inverse design method are shown in Fig. 4.21. The mean value of *RE* for F_{Blay} was 3.87%. The mapping relationship between the inlet parameter and F_{Blay} is more complicated than that between the inlet parameter and the simulation results for velocity and temperature, and therefore the prediction error increased.

5. DESIGN RESULTS

(1) Direct inverse design method The well-trained ANN was used to design the inlet velocity and temperature for the Blay model. The inputs to the ANN were the simulation results obtained when the inlet velocity and temperature were 0.57 m/s and 15°C, and the outputs were the inlet velocity and temperature, which were the design results. These results are shown in Table 4.2. The target inlet velocity was successfully designed, but the target inlet temperature was not obtained. The deviation in inlet temperature was 2.7% and was caused by prediction error.

(2) Indirect inverse design method When the indirect inverse design method was used to design the inlet velocity and temperature for the Blay model, the ANN inputs were arbitrary values of the inlet velocity and temperature. It was possible to obtain thousands of prediction values within one second, and thus the entire design space was involved in the design process. Initially, the design method was used without correction, and the results are shown in Fig. 4.22. Although the F value was less than 0.1 for 4.2% of the individuals, it never reached 0. The best value was 0.002, and the corresponding inlet parameters were 0.56 m/s and 15.1°C. Therefore, it was not possible to obtain the real inlet parameters by this method.

Next, the design method with correction was used. The design process was as follows:

(a) Set the threshold value A = 0.001;
(b) If the prediction value of F_{ANN} is less than A, conduct the CFD process to obtain the real value of $F(O)_{\text{CFD}}$.
(c) If $F(O)_{\text{CFD}} = 0$, the design process ends. Otherwise, A = A + 0.001; return to step (b).

The design process converged at A = 0.024. Although 126 new CFD cases were required, the real inlet parameters (0.57 m/s, 15.0 °C) were obtained. The design method with correction ensured convergence of the design.

Table 4.2 Design results obtained by the direct inverse design method

	Real value	*Design result*	*Deviation*
Inlet velocity	0.57 m/s	0.57 m/s	0%
Inlet temperature	15.0°C	15.4°C	2.7%

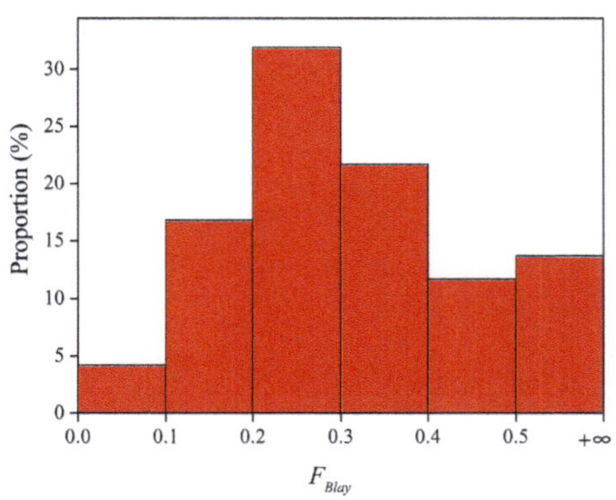

Figure 4.22 Design results without correction

For this inverse design, only one optimal solution existed, and the prediction error had a significant effect on the design result. It was not possible to obtain the solution by the direct inverse design method, or by the indirect inverse design method without correction.

4.3.2 Optimal design of a built environment

The ANN-based inverse design process was evaluated for an office model and an aircraft cabin model. Because multiple solutions existed, the indirect inverse design method was adopted in each case

4.3.2.2 Yuan case

1. MODEL DESCRIPTION

First, the Yuan case (Yuan et al., 1999) was used to study the inverse design method. The structure of this case is shown in Fig. 4.23.

The design variable for the Yuan model is the inlet angle. A_x is the angle between the velocity direction and x direction, and A_y is the angle between the velocity direction and y direction. The magnitude and direction of the inlet velocity were specified in the FLUENT software. The velocity magnitude was set as 0.09 m/s, and the x-component of the flow direction was set as −1. The x- and z-components varied within the range of −1~1.

The design objective was thermal comfort. The PMV (predicted mean vote) was used to evaluate the thermal comfort in this space. PMV expresses warm and cold discomfort for the body as a whole. Six variables are involved in the calculation of PMV: the user's activity level and clothing insulation, indoor air temperature, mean radiant temperature, relative air velocity, and humidity. Furthermore, the seven-point psycho-physical ASHRAE scale was used to measure the thermal comfort level as follows: −3 (cold), −2 (cool), −1 (slightly cool), 0 (neutral), +1 (slightly warm), +2 (warm), and +3 (hot). According to International Organization for Standardization (2005) and ASHRAE Standard 55-2004 (2004), the recommended value for PMV is $-0.5 < PMV < 0.5$. To provide a more comfortable environment, a stricter requirement is $|PMV| < 0.2$. The design domain for PMV (Fig. 4.24) was the surface 10 cm away from the human body. The PMV value was obtained by means of a user-defined function based on the FLUENT results.

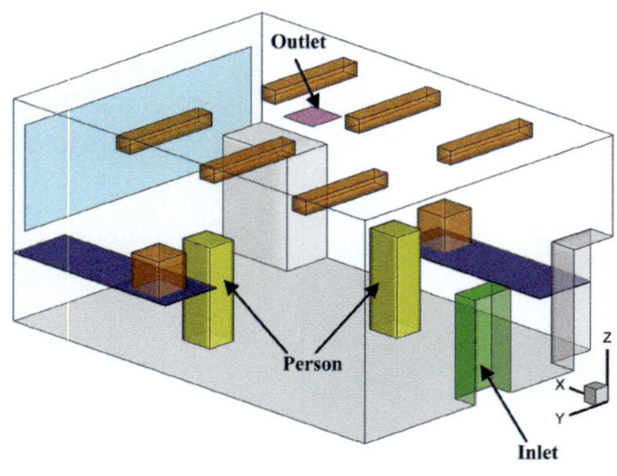

Figure 4.23 Structure of the Yuan model

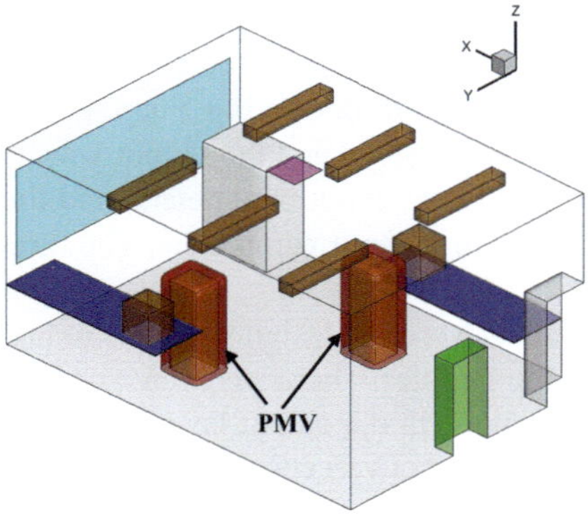

Figure 4.24 Design domains for the Yuan model

2. SAMPLING

Forty samples were generated in the sampling stage to train the ANN, and another 10 samples were generated to test the ANN. The intervals of inlet velocity and temperature were each 0.01. The number of potential values for the inlet angle was 201 × 201.

The RNG k-ε turbulence model was adopted to determine the Reynolds stress. The PRESTO algorithm was adopted to solve the discrete equations for pressure, and a second-order upwind algorithm was used to solve the discrete equations for other variables. The time required for each CFD process was about 2 hours and was similar for different samples. The values of the design objectives are shown in Table 4.3. The proportion of samples with |PMV| less than 0.2 was 32.5%.

The minimum value of |PMV| was 0.096, and the corresponding design variables were $A_x = -11.3°$, $A_y = 42.6°$, and T = 23.0°C. The optimal solution is shown in Fig. 4.25.

Table 4.3 The range of design objectives in the training samples.

Design objectives	*Maximum value*	*Minimum value*	*Mean value*	*Proportion*
|PMV|	0.715	0.096	0.400	32.5%

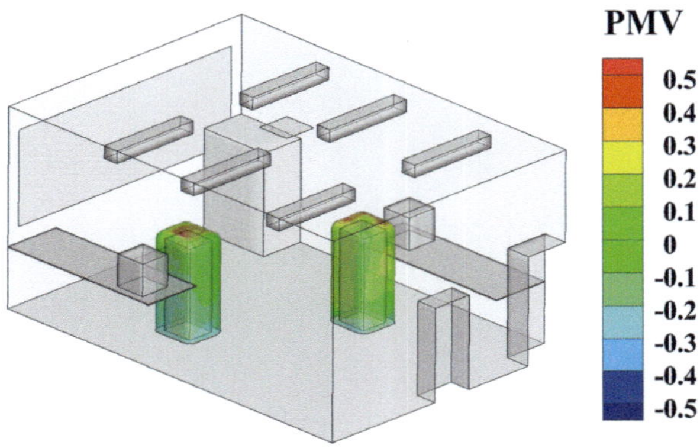

Figure 4.25 Design objective of the optimal solution obtained by sampling

3. TRAINING

The design objective of the Yuan model was expressed as an inequality, and it was possible for multiple solutions to exist. Therefore, the direct inverse design method could not be used here. Instead, the indirect inverse design method was used to design the Yuan model. The inputs to the ANN were arbitrary values of the inlet parameters, and the output was |PMV|. The linear normalization method was used to normalize the input parameters, and the logarithmic normalization method was used to normalize the output parameters. The threshold value for |PMV| was 0.2.

A three-layer ANN was used for training. The inputs were the design variables of the samples, and the outputs were the design objectives of the samples. There were three nodes in the input layer, eight nodes in the hidden layer, and one node in the output layer. The Bayesian regularization algorithm was used to train the ANN. The convergence criterion for training was a MSE of less than 0.001. The training process can be seen in Fig. 4.26, and it converged at epoch 759. The time requirement for training was less than 1 minute, which was negligible in comparison to that for the CFD process.

4. TESTING

To test the ANN training results, another ten samples were generated. They had not been included in the training samples. Four of them were predicted to satisfy the design objective, and four of these samples actually satisfied the objective. Thus, the SR was 50%. The REs of the ANN were also used to evaluate the prediction accuracy. The mean RE value for |PMV| was 22.1%. The REs here were larger than those for the Blay and Nielsen models because the relationships between the design variables and objectives for the Yuan model were more difficult to predict than those for the other models. Increasing the number of training samples might improve the prediction accuracy, but the computational cost would also increase. The maximum *RE* values for these design objectives were greater than 50%. Therefore, the correction process should be incorporated into the design process to eliminate deviations in the results.

5. DESIGN RESULTS

Because there may have been multiple solutions for this design of the cabin environment, it was not necessary to involve the entire design space. Moreover, the computational cost

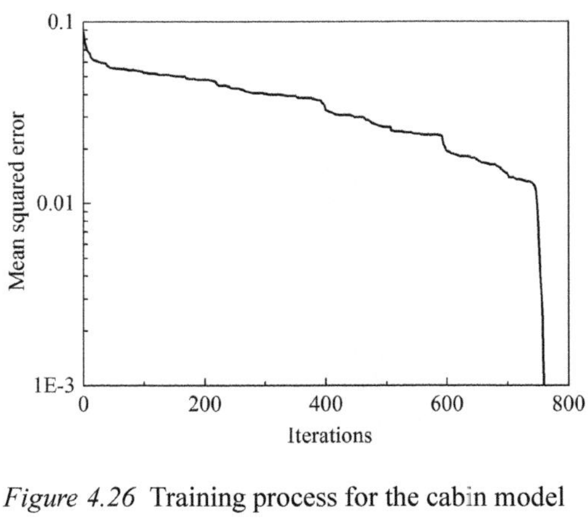

Figure 4.26 Training process for the cabin model

would have increased if more individuals had been involved in the design process. The convergence criterion is usually the obtaining of a set of solutions. In the present research, the criterion was a total of 20 solutions. In the sampling stage, 40 individuals were used to obtain 13 solutions. Therefore, it was necessary for at least 62 individuals to be involved in the design stage. To ensure convergence, 200 individuals were used (numbered from 1 to 200). The values of the design variables were obtained by the Latin hypercube sampling method, and the values of the design objectives were obtained initially by the ANN. The range of predicted design objectives is shown in Table 4.4. The proportions satisfying the requirement for the design objectives were similar to those obtained in the sampling stage. The time requirement for the prediction was only a few seconds, which was negligible. Ninety-two of the 200 individuals were predicted to simultaneously satisfy all the design requirements. They can be called the "predicted solutions."

The CFD process was conducted to obtain the real values of the design objectives for the 92 predicted solutions. They were calculated one by one until 13 solutions had been obtained. The design process converged when 28 predicted solutions had been calculated by CFD. Thus, 13 solutions were obtained by a total of 31 CFD cases. The efficiency was significantly better than in the sampling stage (13/31 versus 13/40). The range of design objectives for the solutions is shown in Table 4.5.

The minimum value of |PMV| was found to be 0.094. The design variables were $A_x = 38.7^{\circ}$, $A_y = -18.8^{\circ}$, and $T = 21^{\circ}C$. The distributions of |PMV| are shown in Fig. 4.27.

Table 4.4 Range of predicted design objectives among the training samples.

Design objectives	*Maximum value*	*Minimum value*	*Mean value*	*Proportion*
|PMV|	0.835	0.094	0.425	46.0%

Table 4.5 Range of design objectives for the solutions.

Design objectives	*Maximum value*	*Minimum value*	*Mean value*
|PMV|	0.195	0.094	0.125

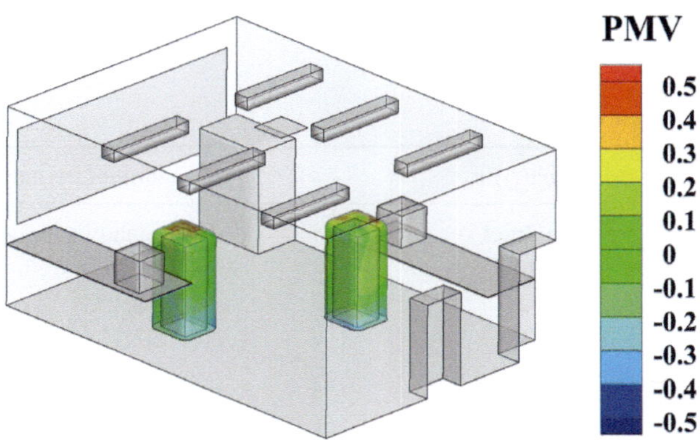

Figure 4.27 Design objective of the optimal solution obtained by inverse design

4.3.2.2 Aircraft cabin case

1. MODEL DESCRIPTION

The environment on board a commercial aircraft is different from that traditionally found in buildings. Occupant density is typically high, and the aircraft must be regarded as both a public place (for passengers) and a workplace (for crew members). The controlled atmosphere inside the aircraft during flight has a lower pressure and relative humidity than those found in many other environments. The environment outside the aircraft during flight is hostile to human life, and, while aircraft are operated with the comfort of passengers and crew in mind, their safety and health must always be paramount. Therefore, the cabin environment is more important and more difficult to design than the environment in a traditional building. An aircraft cabin model was used for the present study of the inverse design method. As a cabin is usually symmetrical, only half of a cabin was considered, with 15 passengers in five rows. The structure of the cabin is shown in Fig. 4.28.

The boundary conditions are listed in Table 4.6. All boundaries except the inlet and outlet were set as no-slip walls. The temperatures on these boundaries were set as constants, as has been done in other studies (Zhang and Chen 2007c; Zhang et al. 2009). The relative pressure at the outlet was 0 Pa.

Four design variables (inlet temperature, angle, size, and location) were considered because they significantly influence the indoor environment in an aircraft cabin. The flow

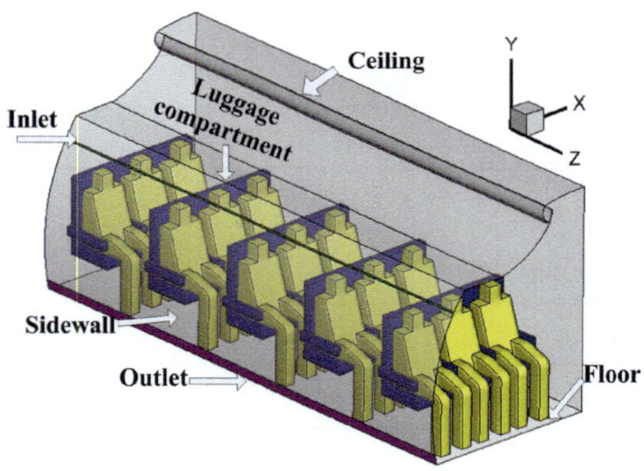

Figure 4.28 Structure of the cabin used to study the inverse design method

Table 4.6 Boundary conditions for the cabin model

Items	*Boundary type*	*Boundary variables*
Air supply	Velocity Inlet	Steady supply
Air outlet	Pressure out	0
Passengers	No-slip wall	30.3°C
Ceiling	No-slip wall	22°C
Luggage	No-slip wall	21°C
Side wall	No-slip wall	22°C
Floor	No-slip wall	23°C
Others	No-slip wall	Adiabatic

rate was 9.4 L/(s·per), as recommended by ASHRAE Standard 161-2007: Air Quality within Commercial Aircraft, published by the American Society of Heating, Refrigerating and Air-Conditioning Engineers. Therefore, the total flow rate at the inlet was fixed, and the inlet velocity was determined by the inlet size. The inlet angle is the one between the velocity direction and x direction. The ranges of the design variables are shown in Table 4.7. The T_{in} is the inlet temperature, A_x is the inlet angle, S_{in} is the size of the inlet, and L_{in} is the location of the inlet (the distance between the center of the inlet and the luggage compartment). The values of S_{in} and L_{in} were set when establishing the geometric model of the cabin, and the values of T_{in} and A_x were set at the "boundary conditions" in the FLUENT software.

The design objectives were thermal comfort, local velocity, and air quality. A revised PMV was proposed, according to the characteristics of the cabin environment:

$$PMV_c = -0.0758PMV^2 + 0.6757PMV - 0.1262 \tag{4.42}$$

To provide a more comfortable environment, a stricter requirement for PMV was set: $|PMVc| < 0.3$. The design domain for $|PMVc|$ (Fig. 4.29) was the surface 10 cm away from the human body.

The PMV index expresses warm and cold discomfort for the body as a whole. However, thermal dissatisfaction can also be caused by unwanted cooling or heating of one particular part of the body. This is known as local discomfort. The most common cause of local discomfort is draft. The local velocity can be used to evaluate the draft. According to ASHRAE Standard 161-2007, the local velocity for draft-sensitive bare body areas should be less than 0.2 m/s, and at head level in the absence of a personal airflow outlet, it should be greater than 0.1 m/s. Therefore, the velocities at head level (U_{head}) and near the feet (U_{feat}) were included in the design objectives. The design requirements for local velocity were 0.1 m/s < U_{head} < 0.2 m/s and U_{feat} < 0.2 m/s. The design domains for local velocity are also shown in Fig. 4.29.

Table 4.7 Ranges of design variables

Variable	*Lower bound*	*Upper bound*	*Unit*
T_{in}	18	24	°C
A_x	−45	45	°C
S_{in}	0.01	0.03	m
L_{in}	0.05	0.35	m

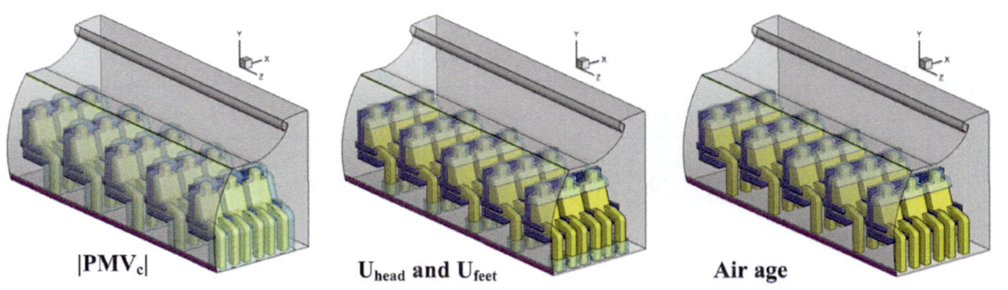

Figure 4.29 Design domains for different design objectives

The age of air can be used to evaluate the air quality in an indoor environment. The mean age of air in a building zone is defined as the average time that air has spent in a zone of the building accumulating contaminants, and it is an important factor in assessing the quality of ventilation (Buratti et al., 2011). It is a statistical value that represents the time required for all the particles of fresh air to travel from the supply point to the point under investigation. Air age is useful in the analysis of a room because fresh air distribution, air velocity, (drafts), and temperature maps can be deployed. The average age of air for a room is a number that characterizes the treatment of the inlet air and its performance within the enclosure, as the room's average age of air depends on both the volumetric flow rate and the direction in which the air moves through the space. A room's air age can be used to obtain the air change efficiency for the room (Meiss et al., 2013). In this study, the mean age of air was used to evaluate the air quality in the cabin. The air age at head level was one of the design objectives. The design domains for the age of air are shown in Fig. 4.29. The design requirement for air age was less than 120 s which can be achieved under typical operating conditions.

In conclusion, the design requirements for the different design objectives were $|PMVc| < 0.3$, $0.1\ m/s < |\mathbf{U}|_{head} < 0.2\ m/s$, $|\mathbf{U}|_{feat} < 0.2\ m/s$, and air age < 120 s. The local velocity was obtained directly by the FLUENT software. The PMVc was obtained by means of a user-defined function that was based on the FLUENT results, and air age was obtained by means of a user-defined scalar in FLUENT.

2. SAMPLING

Forty samples were generated in the sampling stage in order to train the ANN, and another 10 samples were generated to test the ANN. The intervals of inlet velocity and temperature were 0.01 m/s and 0.1°C, respectively. The numbers of potential values for the inlet velocity and temperature were 101 and 201, respectively.

The RNG $k-\varepsilon$ turbulence model was adopted to determine Reynolds stress. The PRESTO algorithm was adopted to solve the discrete equations for pressure, and the second-order upwind algorithm was used to solve the discrete equations for other variables. The time requirement for each CFD process was about 10 hours and was similar for different samples. The values of the design objectives are shown in Table 4.8. Most of the samples satisfied the design requirement for U_{feat}, whereas few of them satisfied the design requirement for the age of air.

Only 2.5% of the samples, i.e., only one sample, satisfied all the design requirements simultaneously. The design variables were T_{in} = 23.8°C, A_x = −36°, S_{in} = 0.017 m, and L_{in} = 0.23 m. The distributions of the different design objectives are shown in Fig. 4.30.

Table 4.8 Ranges of design objectives for the training samples

Design objective	*Maximum value*	*Minimum value*	*Mean value*	*Proportion*
|PMVc|	0.748	0.121	0.377	37.5%
$\|\mathbf{U}\|_{head}$	0.080 m/s	0.240 m/s	0.118 m/s	52.5%
$\|\mathbf{U}\|_{feat}$	0.087 m/s	0.224 m/s	0.133 m/s	82.5%
Age of air	90.4 s	280.4 s	164.2 s	17.5%

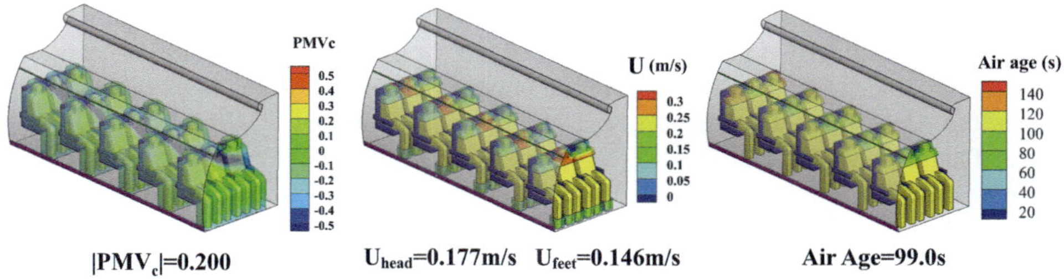

Figure 4.30 Design objectives for the solution obtained by sampling

3. TRAINING

The design objectives for the cabin environment were expressed as an inequality, and it was possible for multiple solutions to exist. Therefore, the direct inverse design method could not be used here. Instead, the indirect inverse design method was used to design the cabin environment. The ANN inputs were arbitrary values of the inlet parameters, and the outputs were |PMVc|, $\mathbf{U}_{head}$, U_{feat}, and the age of air. The linear normalization method was used to normalize the input parameters, and the logarithmic normalization method was used to normalize the output parameters. The threshold values were 0.3 for |PMVc|, 0.1 m/s and 0.2 m/s for $\mathbf{U}_{head}$, 0.2 m/s for U_{feet}, and 120 s for the age of air.

A three-layer ANN was used for training. The inputs were the design variables of the samples, and the outputs were the design objectives of the samples. There were four nodes in the input layer, ten nodes in the hidden layer, and four nodes in the output layer. The Bayesian regularization algorithm was used to train the ANN. Because the relationship between the design variables and objectives of the cabin model was complicated, it was difficult for the training process to converge. Therefore, the convergence criterion for training was a MSE of less than 0.01, which was larger than that for the other models. The training process, shown in Fig. 4.31, converged at epoch 109. The time requirement for training was less than 1 minute, which was negligible in comparison to that for the CFD process.

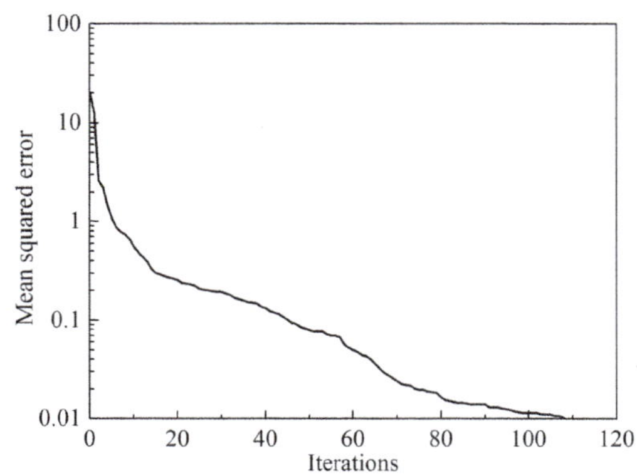

Figure 4.31 Training process for the cabin model

4. TESTING

To test the ANN training results, another ten samples were generated. They had not been included in the training samples. Because none of these ten samples simultaneously satisfied all the design objectives, the SR could not be obtained. The REs of the ANN were used to evaluate the ANN's prediction accuracy. The mean RE values for |PMVc|, U_{head}, U_{feat}, and the age of air age were 12.4%, 11.4%, 11.1%, and 9.5%, respectively. Increasing the number of training samples might have improved the prediction accuracy, but the computational cost would also have increased. The maximum RE values for these design objectives were greater than 30%. Therefore, the correction process should be incorporated into the design process to eliminate deviations in the results.

5. DESIGN RESULTS

Because there might have been multiple solutions for the design of this cabin environment, it was not necessary to involve the entire design space. Moreover, the computational cost would have increased if more individuals had been involved. The typical convergence criterion is the obtaining of a set of solutions. In the present research, the criterion was a total of five solutions. In the sampling stage, 40 individuals were used to obtain 1 solution. Therefore, it was necessary to involve at least 200 individuals in the design stage. To ensure convergence, 1,000 individuals were used (numbered from 1 to 1,000). The values of the design variables were obtained by the Latin hypercube sampling method, and the values of the design objectives were initially obtained by the ANN. The ranges of predicted design objectives are shown in Table 4.9. The proportions that satisfied the requirements for the different objectives were similar to those obtained in the sampling stage. The time requirement for the prediction was only a few seconds, which was negligible. Forty-eight of the 1,000 individuals were predicted to simultaneously satisfy all the design requirements. They were again called "predicted solutions."

The CFD process was conducted in order to obtain the real values of the design objectives for the 48 predicted solutions. They were calculated one by one until five solutions had been obtained. The design process converged when 23 predicted solutions had been calculated by CFD. Thus, five solutions were obtained by a total of 23 CFD cases. The efficiency was significantly higher than that for the sampling stage (5/23 versus 1/40). The ranges of design objectives for the solutions are shown in Table 4.10.

Table 4.9 Ranges of predicted design objectives for the training samples

Design objective	*Maximum value*	*Minimum value*	*Mean value*	*Proportion*
|PMVc|	0.858	0.078	0.412	34.5%
$\|\mathbf{U}\|_{head}$	0.061 m/s	0.261 m/s	0.125 m/s	59.0%
$\|\mathbf{U}\|_{feat}$	0.047 m/s	0.264 m/s	0.106 m/s	83.3%
Age of air	78.2 s	311.5 s	178.4 s	20.2%

Table 4.10 Ranges of design objectives for the solutions

Design objective	*|PMVc|*	$\|U\|_{head}$ *(m/s)*	$\|U\|_{feat}$ *(m/s)*	*Age of air (s)*
Maximum value	0.285	0.187	0.175	125.6
Minimum value	0.191	0.103	0.125	98.2
Mean value	0.245	0.145	0.142	114.2

There is no agreement as to wether a large or small U_{head} is better. Only |PMVc|, $\mathbf{U}_{feat}$, and the age of air were used to evaluate the solutions when U_{head} satisfied the preset requirement. None of the solutions was better than the other solutions for all the design objectives. The optimal solution for all the design objectives can be obtained by aggregating and weighting different design objectives into one assessment index, F_{agg} (Zhang and You, 2014b):

$$F_{agg} = \varpi_{PMV} \frac{|PMVc| - Min(|PMV_c|)}{Max(|PMVc|) - Min(|PMV_c|)} + \varpi_U \frac{U_{feet} - Min(U_{feet})}{Max(U_{feet}) - Min(U_{feet})} + \varpi_{AGE} \frac{Age - Min(Age)}{Max(Age) - Min(Age)} \tag{4.43}$$

where the weights are ϖ_{PMV}, ϖ_U, and ϖ_{AGE} for the design objectives of |PMVc|, U_{feat}, and the age of air. The *Max* (|PMVc|), *Max* (U_{feat}), and *Max* (Age) are the maximum values of |PMVc|, U_{feat}, and air age, respectively; and the *Min* (|PMVc|), *Min* (U_{feat}), and *Min* (Age) are the minimum values of the same parameters. The minimum value of F_{agg} was found to be 0.687 when the design objectives were equally weighted. The corresponding optimal design variables were T_{in} = 23.2°C, A_x = -39°, S_{in} = 0.02 m, and L_{in} = 0.17 m. The optimal solution is shown in Fig. 4.32.

4.4 Discussion

4.4.1 Design accuracy

In the CFD-based inverse design method, CFD results are typically used to evaluate the accuracy of the design results. The design accuracy can be expressed as the deviation between the CFD results and ANN results, and this deviation is determined by the prediction accuracy of the ANN. The prediction accuracy is more important in a single-solution problem. When the accuracy is very low, there is a high risk of missing the solution. In a multi-solution problem, there is some tolerance of prediction error. Prediction accuracy is affected by the training samples and training parameters. If the target results are included in the training samples, high prediction accuracy will be achieved for the design result. A correction process can be used to eliminate the impact of prediction error on the design result, but doing so will increase the computational cost. The threshold value for starting the CFD process should be determined carefully.

4.4.2 Design efficiency

The calculation time for the ANN is negligible in comparison to that for CFD. In the direct inverse design method, all the CFD cases are generated in the sampling stage. In the indirect

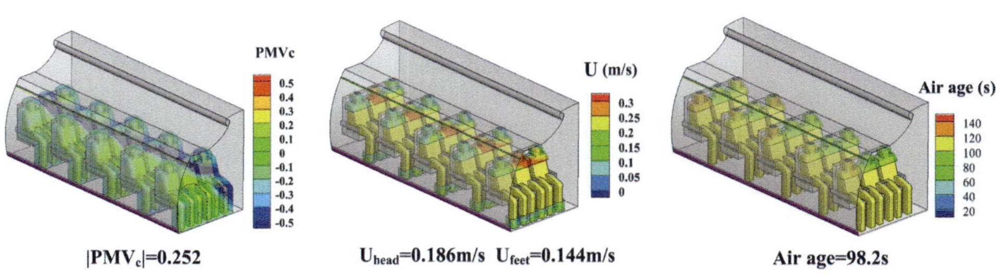

Figure 4.32 Optimal solution when the design objectives were equally weighted

method, CFD cases can be generated in both the sampling and design stages. Increasing the number of training samples might improve the ANN's prediction accuracy, which would in turn reduce the number of CFD cases in the design stage. The number of training samples is usually specified by designer. Therefore, the method of determining the appropriate number of samples should be studied in order to improve the efficiency of the ANN-based inverse design process.

If the solutions can be easily obtained, especially in a multi-solution problem, and if only a few solutions are needed, then an ANN is not an appropriate design tool. The training samples must be generated before the start of the design process, and the computational cost for sampling might be larger than that for the trial-and-error method.

4.5 Conclusions

In this chapter, the use of a CFD-based ANN for the inverse design of built environment was examined. The ANN served as a surrogate for the CFD model in the inverse design process. It was able to determine the mapping relationship between the design variables and objectives. Prediction accuracy was influenced by the ANN structure, the method used to select the samples, the data processing method, and the algorithm used to train the ANN.

In the direct inverse design method, the inverse relationship between the design variables and objectives was established. High efficiency was achieved, but it was difficult to solve multi-solution problems, and deviations were present in the design results. In the indirect inverse design method, the ANN was used in place of CFD to predict the design objectives. Deviations in the design results were eliminated by use of a correction process, but doing so increased the computational costs. The design results demonstrated that the CFD-based ANN method was more efficient than the traditional CFD-based design method.

Notes

1 This chapter is partially reproduced from the paper entitled Comparing the linear and logarithm normalized artificial neural networks in inverse design of aircraft cabin environment, *Building Simulation*, 9: 729–734. The reprint is with permission from Springer.

2 This chapter is partially reproduced from the book entitled *The Industrial Electronics Handbook*, Second Edition – Five Volume Set, CRC Press, 2011. The reprint is with permission from Taylor and Francis Group LLC Books.

References

ASHRAE. Air quality within commercial aircraft. Atlanta, GA, 2007.

Ayata T, Arcaklioglu E, Yildiz O. Application of ANN to explore the potential use of natural ventilation in buildings in Turkey. *Applied Thermal Engineering*. 2007;27(1):12–20.

Balabin RM, Lomakina EI, Safieva RZ. Neural network (ANN) approach to biodiesel analysis: Analysis of biodiesel density, kinematic viscosity, methanol and water contents using near infrared (NIR) spectroscopy. *Fuel*, 2011;90(5):2007–2015.

Bastani A, Haghighat F, Kozinski JA. Contaminant source identification within a building: toward design of immune buildings. *Building and Environment*, 2012;51(2012):320–329.

Blay D, Mergui S, Niculae C. Confined turbulent mixed convection in the presence of a horizontal buoyant wall jet. *Fundamentals of Mixed Convection, ASME HTD*, 1992;213:65–72.

Buratti C, Mariani R, Moretti E. Mean age of air in a naturally ventilated office: Experimental data and simulations. *Energy and Buildings*, 2011;43(8):2021–2027.

Choudhury TA, Hosseinzadeh N, Berndt CC. Improving the generalization ability of an artificial neural network in predicting in-flight particle characteristics of an atmospheric plasma spray process. *Journal of Thermal Spray Technology*, 2012;21(5):935–949.

Demuth H, Beale M, Hagan M. *Neural network toolbox™ 6. User's guide*. MathWorks Inc, 2008. Available from https://filer.case.edu/pjt9/b378s10/nnet.pdf

Deng S, Hwang Y. Applying neural networks to the solution of forward and inverse heat conduction problems. *International Journal of Heat and Mass Transfer*, 2006;49(25–26):4732–4750.

Du Z, Fan B, Jin X, Chi J. Fault detection and diagnosis for buildings and HVAC systems using combined neural networks and subtractive clustering analysis. *Building and Environment*, 2014;73(2014):1–11.

Ferreira PM, Ruano AE, Silva S, Conceicao EZE. Neural networks based predictive control for thermal comfort and energy savings in public buildings. *Energy and Buildings*, 2012;55(2012):238–251.

Jahedi G, Ardehali MM. Wavelet based artificial neural network applied for energy efficiency enhancement of decoupled HVAC system. *Energy Conversion and Management*, 2012;54(1): 47–56.

Kabir H, Wang Y, Yu M, Zhang Q. Neural network inverse modeling and applications to microwave filter design. *IEEE Transactions on Microwave Theory and Techniques*, 2008;56(4):867–879.

Khashman A. Neural networks for credit risk evaluation: Investigation of different neural models and learning schemes. *Expert Systems with Applications*, 2010;37(9):6233–6239.

Liang J, Du R. Thermal comfort control based on neural network for HVAC application. *Proceedings of the 2005 IEEE Conference on Control Applications*. 2005;819–824.

Liang J, Du R. Design of intelligent comfort control system with human learning and minimum power control strategies. *Energy Conversion and Management*, 2008;49(4):517–528.

Lu T, Viljanen M. Prediction of indoor temperature and relative humidity using neural network models: Model comparison. *Neural Computing & Applications*, 2009;18(4):345–357.

Magnier L, Haghighat F. Multiobjective optimization of building design using TRNSYS simulations, genetic algorithm, and artificial neural network. *Building and Environment*, 2010;45(3): 739–746.

Maier HR, Dandy GC. Neural networks for the prediction and forecasting of water resources variables: a review of modelling issues and applications. *Environmental Modelling & Software*, 2000;15(1): 101–124.

Marvuglia A, Messineo A, Nicolosi G. Coupling a neural network temperature predictor and a fuzzy logic controller to perform thermal comfort regulation in an office building. *Building and Environment*, 2014;72(2014):287–299.

Mazhar F, Khan AM, Chaudhry IA, Ahsan M. On using neural networks in UAV structural design for CFD data fitting and classification. *Aerospace Science and Technology*, 2013;30(1):210–225.

McKay MD, Beckman RJ, Conover WJ. A comparison of three methods for selecting values of input variables in the analysis of output from a computer code. *Technometrics*, 2000;42(1);42:55–61.

Meiss A, Feijo-Munoz J, Garcia-Fuentes MA. Age-of-the-air in rooms according to the environmental condition of temperature: a case study. *Energy and Buildings*, 2013;67(2013);67:88–96.

Mjalli FS, Al-Asheh S, Alfadala HE. Use of artificial neural network black-box modeling for the prediction of wastewater treatment plants performance. *Journal of Environmental Management*, 2007;83(3):329–338.

Nassif N. Modeling and optimization of HVAC systems using artificial neural network and genetic algorithm. *Building Simulation*, 2014;7(3):237–245.

Neto AH, Fiorelli FAS. Comparison between detailed model simulation and artificial neural network for forecasting building energy consumption. *Energy and Buildings*, 2008;40(12):2169–2176.

Nguyen A, Reiter S, Rigo P. A review on simulation-based optimization methods applied to building performance analysis. *Applied Energy*, 2014;113(2014):1043–1058.

Nielsen PV, Restivo A, Whitelaw JH. The velocity characteristics of ventilated rooms. *Journal of Fluids Engineering*, 1978;100:291–298.

Ozbalta TG, Sezer A, Yildiz Y. Models for prediction of daily mean indoor temperature and relative humidity: Education building in Izmir, Turkey. *Indoor and Built Environment*, 2012;21(6):772–781.

Perez-Sanchez B, Fontenla-Romero O, Guijarro-Berdinas B, Martinez-Rego D. An online learning algorithm for adaptable topologies of neural networks. *Expert Systems with Applications*, 2013;40(18):7294–7304.

Qin R, Yan D, Zhou X, Jiang Y. Research on a dynamic simulation method of atrium thermal environment based on neural network. *Building and Environment*, 2012;50(2012):214–220.

Saijal KK, Ganguli R, Viswamurthy SR. Optimization of helicopter rotor using polynomial and neural network metamodels. *Journal of Aircraft*, 2011;48(2):553–566.

Singh RM, Datta B, Jain A. Identification of unknown groundwater pollution sources using artificial neural networks. *Journal of Water Resources Planning and Management-ASCE*, 2004; 130(6):506–514.

Thomas B, Soleimani-Mohseni M. Artificial neural network models for indoor temperature prediction: Investigations in two buildings. *Neural Computing & Applications*, 2007;16(1):81–89.

Titterington DM. Bayesian methods for neural networks and related models. *Statistical Science*, 2004;19(1):128–139.

Vukovic V, Tabares-Velasco PC, Srebric J. Real-time identification of indoor pollutant source positions based on neural network locator of contaminant sources and optimized sensor networks. *Journal of the Air& Waster Management Association*, 2010;60(9):1034–1048.

Wang SW, Ma ZJ. Supervisory and optimal control of building HVAC systems: a review. *HVAC&R Research*. 2008;14(1):3–32.

Wang W, Rivard H, Zmeureanu R. Floor shape optimization for green building design. *Advanced Engineering Informatics*, 2006;20(4):363–78.

Xue Y, Zhai ZJ, Chen Q. Inverse prediction and optimization of flow control conditions for confined spaces using a CFD-Based genetic algorithm. *Building and Environment*, 2013;64(2013):77–84.

Yi YK, Malkawi AM. Integrating neural network models with computational fluid dynamics (CFD) for site-specific wind condition. *Building Simulation*, 2011;4(3):245–254.

Yu H, Wilamowski BM. Levenberg–Marquardt training. The Industrial Electronics Handbook, Second Edition – Five Volume Set. *CRC Press*, Boca Raton. 2011.

Yuan XX, Chen Q, Leon RG. Measurements and computations of room airflow with displacement ventilation. *ASHRAE Transactions*, 1999;105(1):340–52.

Zhang T, You X. Applying neural networks to solve the inverse problem of indoor environment. *Indoor and Built Environment*, 2014a;23(8):1187–1195.

Zhang T, You X. A simulation-based inverse design of preset aircraft cabin environment. *Building and Environment*, 2014b;82(2014):20–26.

Zhang T, You X. Improvement of the training and normalization method of artificial neural network in the prediction of indoor environment. *Procedia Engineering*, 2015;121(2015):1245–1251.

Zhang TF, Chen QY. Novel air distribution systems for commercial aircraft cabins. *Building and Environment*, 2007a;42(4):1675–1684.

Zhang TF, Chen Q. Identification of contaminant sources in enclosed spaces by a single sensor. *Indoor Air*, 2007b;17(6):439–449.

Zhang TF, Chen Q. Identification of contaminant sources in enclosed environments by inverse CFD modeling. *Indoor Air*, 2007c;17(3):167–177.

Zhang Z, Chen X, Mazumdar S, Zhang T, Chen Q. Experimental and numerical investigation of airflow and contaminant transport in an airliner cabin mockup. *Building and Environment*, 2009;44(1):85–94.

Zhou L, Haghighat F. Optimization of ventilation systems in office environment, Part II: Results and discussions. *Building and Environment*, 2009a;44(4):657–665.

Zhou L, Haghighat F. Optimization of ventilation system design and operation in office environment, Part I: Methodology. *Building and Environment*, 2009b;44(4):651–656.

5 CFD-based proper orthogonal decomposition method[1]

To speed up an inverse design, an efficient solution to the Navier-Stokes equations for indoor environmental performance is needed. Proper orthogonal decomposition (POD) of the thermo-flows is a way to meet this need. POD extracted the coherent structures of the thermo-flows into POD modes, providing a lower-order description of the thermo-flow fields by linear combinations of a small number of POD modes. This chapter presented the basic principles of the POD-based approach for proper design of indoor environments. The method first sampled representative thermo-flows with full CFD simulations, and then the POD modes and their coefficients were extracted. To construct a much larger amount of thermo-flow data, a cubic spline scheme was adopted to interpolate the coefficients of the POD modes. Once the cause-effect maps from all the design variables to the resulting environmental performance were established, the design parameters that led to satisfactory performance were circumscribed. The above method has been tested in an empty two-dimensional air cavity, an idealized three-dimensional office, and a three-dimensional aircraft cabin. The number of design variables was varied, and the adopted design target ranged from one to multiple objectives. The results showed that the proposed method was able to reproduce the thermo-flow fields, and they were in good agreement with the experimental data. The POD-based approach could provide not only the ranges of solution sets that simultaneously satisfy multiple design targets but also a specific solution corresponding to the best performance of each design target.

5.1 Introduction

When optimizing an enclosed environment, the CFD-based genetic algorithm method, CFD-based adjoint method, and CFD-based artificial neural network method repetitiously solved the Navier-Stokes (N-S) equations, which slowed down the design process. To speed up the process, an efficient solution to the nonlinear N-S equations is needed, and numerous methods have been attempted. The essential component is a quick mapping of the indoor environmental performance results from various design parameters. If such cause-effect mapping could be instantly established, then one could choose the design parameters according to the performance results. In such a situation, optimization is not necessary because all the cause-effect maps are available, and the best design corresponding to a given performance criterion is explicit. This constitutes the basic idea of inverse design via the CFD-based proper orthogonal decomposition (POD) method.

The POD procedure provides a basis for the modal decomposition of an ensemble of data, such as the thermo-flow fields obtained from numerical simulations or experiments. The coherent structures of these data that contain the basic features and trends of the fields are stored in the basis, which is also called the POD mode. The most prominent advantage of POD analysis is its efficiency in reproducing the data with only a few modes by means of linear combinations (Holmes et al., 2012). Hence, POD offers a lower-order description of the thermo-flow fields. The POD method was proposed by Lumley (1967) nearly a half century ago to analyze inhomogeneous turbulent flows. Early applications of the POD method were limited to a small number of near-wall and jet-induced mixing flows (Bakewell and Lumley, 1967; Herzog, 1986; Glauser and George, 1987). Later, the snapshot method (Sirovich, 1987) was proposed to minimize the dimensions of the POD eigenvalue system. Snapshots are the data samples used for extracting the POD modes, through which the dimensions of the eigenvalue system are reduced from the number of grid points to the number of snapshots. Generally, the number of snapshots is far less than the number of spatial grid points, and thus the computing costs are significantly reduced.

In enclosed environments, POD has been used primarily for rapid prediction of indoor thermo-flow and spatial concentrations of pollutants, optimization of air-supply parameters, and development of controllers for dynamic ventilation. Elhadidi and Khalifa (2005) employed POD analysis to efficiently predict the velocity and temperature distributions inside an empty office. Ding et al. (2008) proposed a POD-based data interpolation to predict natural and forced convection flows. Sempey et al. (2009) performed a POD-based prediction of the temperature distribution in air-conditioned rooms in a fixed-flow context. Allery et al. (2005) tracked particle motion in a two-dimensional ventilated cavity with airflow provided by a POD construction. Li et al. (2013a) integrated a genetic algorithm with the POD method to efficiently optimize the air-supply velocity and temperature in a room. Ahuja et al. (2012) developed POD-based controllers to eliminate the heat disturbance in an indoor environment. Tallet et al. (2012) used POD analysis to dynamically control a window opening and improve indoor comfort and air quality. Li et al. (2013b) proposed a POD-based prediction model for dynamic control of room temperature.

The above discussion shows an efficient inverse design can be realized by POD. Only a small number of full CFD simulations are required for extraction of the POD modes, and then much larger amounts of thermo-flow data can be constructed. Once the cause-effect maps from all the design variables to the resulting environmental performance are established, the design parameter that leads to satisfactory performance can be easily determined. In this chapter, the basic principles of POD were introduced, and a number of formulas were derived. Next, the method was used to optimize the air-supply parameters for an empty two-dimensional cavity, a three-dimensional office, and an aircraft cabin with complex geometry. Finally, the solution accuracy and computing costs of the POD-based design were evaluated.

After completing this chapter, the reader will be able to:

- Extract the POD modes and inversely solve for the coefficients of the modes.
- Interpolate the coefficients of the POD modes for constructing the thermo-flow fields that were not used to extract the modes.
- Determine the necessary air-supply parameters to achieve specific indoor environmental performance targets.

5.2 Principles of inverse design with the CFD-based POD method

In this section, the fundamentals of POD are briefly introduced, followed by the method for preparing full CFD-simulated data and the criteria for data saturation. The solution procedure for POD-based prompt design is then presented.

5.2.1 Fundamentals of POD

Proper orthogonal decomposition is the most efficient linear decomposition method and is also known as Karhunen-Loeve decomposition, principal component analysis, and singular system analysis (Holmes et al., 2012). The POD method extracts the coherent structures of the analyzed data, and it is optimal for data modeling and reconstruction (Holmes et al., 2012). The method originated from the decomposition of turbulent flows, but it can be extended to decompose any data ensemble. The basic principles are addressed below.

Let **f** represent a dummy vector for velocity components, temperature, or some other variables that describe indoor environmental performance at all grid points and which constitute a vector. Suppose that we have a number of **f**s in a data ensemble, each expressed as a function of the spatial coordinates, i.e., $\mathbf{f} = \mathbf{f}(\mathbf{x})$, where **x** is a position vector. Apparently, different **f**s can result if the air-supply parameters vary. If a sufficient number of **f**s are provided within the span of all possible air-supply parameters, then the POD modes extracted from the data ensemble will be representative for later data construction.

Next, let us define an orthogonal basis (or spatial mode, or POD mode) $\boldsymbol{\varphi}$ that captures more information on a vector than any other basis. The averaged projection of each **f** onto $\boldsymbol{\varphi}$ will be maximized (Holmes et al., 2012) as:

$$\max \frac{\left\langle |(\mathbf{f}, \boldsymbol{\varphi})|^2 \right\rangle}{\|\boldsymbol{\varphi}\|^2} \tag{5.1}$$

where $(\cdot,\cdot)$ denotes the inner product, $|\cdot|$ is the modulus, $\langle\cdot\rangle$ is an averaging operation, and $\|\cdot\|$ is the second-norm operation, which obtains the root square sum of all elements of a vector. The modulus operation can be omitted in real Euclidean space. To maximize $\left\langle |(\mathbf{f}, \boldsymbol{\varphi})|^2 \right\rangle$ subject to the constraint $\|\boldsymbol{\varphi}\|^2 = 1$, the function for this constrained variational problem can be written as (Holmes et al., 2012):

$$C_f[\boldsymbol{\varphi}] = \left\langle |(\mathbf{f}, \boldsymbol{\varphi})|^2 \right\rangle - \lambda_L \left(\|\boldsymbol{\varphi}\|^2 - 1 \right) \tag{5.2}$$

where λ_L is a Lagrangian multiplier. A necessary condition for extrema is that the derivative is zero for all variations $\boldsymbol{\varphi} + \delta_E \Xi$ (Holmes et al., 2012), where $\boldsymbol{\Xi}$ represents another basis and δ_E is an arbitrary small number in real Euclidean space. The mathematical formula is:

$$\left. \frac{d}{d\delta_E} C_f[\boldsymbol{\varphi} + \delta_E \Xi] \right|_{\delta_E \to 0} = 0 \tag{5.3}$$

According to Eq. (5.2), $C_f[\boldsymbol{\varphi} + \delta_E \Xi]$ can be written as:

$$C_f[\boldsymbol{\varphi} + \delta_E \Xi] = \left\langle |(\mathbf{f}, \boldsymbol{\varphi} + \delta_E \Xi)|^2 \right\rangle - \lambda_L \left(\|\boldsymbol{\varphi} + \delta_E \Xi\|^2 - 1 \right) \tag{5.4}$$

Substituting Eq. (5.4) into Eq. (5.3) and noting that $\|\mathbf{b}\|^2 = (\mathbf{b}, \mathbf{b})$, we obtain:

$$\begin{aligned} \left. \frac{d}{d\delta_E} C_f[\boldsymbol{\varphi} + \delta_E \Xi] \right|_{\delta_E \to 0} &= \left. \frac{d}{d\delta_E} \left[\left\langle |(\mathbf{f}, \boldsymbol{\varphi} + \delta_E \Xi)|^2 \right\rangle - \lambda_L \left(\|\boldsymbol{\varphi} + \delta_E \Xi\|^2 - 1 \right) \right] \right|_{\delta_E \to 0} \\ &= \left. \frac{d}{d\delta_E} \left\langle (\mathbf{f}, \boldsymbol{\varphi} + \delta_E \Xi)^2 \right\rangle \right|_{\delta_E \to 0} - \lambda_L \left. \frac{d}{d\delta_E} (\boldsymbol{\varphi} + \delta_E \Xi, \boldsymbol{\varphi} + \delta_E \Xi) \right|_{\delta_E \to 0} \\ &= 0 \end{aligned} \tag{5.5}$$

If all the operations are in real space, and using the property in real space that (**b**, R**d**) = R (**b**, **d**) = R (**d**, **b**), where R is a real number, and (**b** + **e**, **d**) = (**b**, **d**) + (**e**, **d**), Eq. (5.5) can be expanded as:

$$\begin{aligned} \left.\frac{d}{d\delta_E} C_f[\boldsymbol{\varphi}+\delta_E\boldsymbol{\Xi}]\right|_{\delta_E\to 0} &= \left.\left\{ \begin{array}{l} 2\left\langle (\mathbf{f},\boldsymbol{\varphi}+\delta_E\boldsymbol{\Xi})\dfrac{d}{d\delta_E}(\mathbf{f},\boldsymbol{\varphi}+\delta_E\boldsymbol{\Xi})\right\rangle \\ -\lambda_L \dfrac{d}{d\delta_E}\left[(\boldsymbol{\varphi},\boldsymbol{\varphi})+2\delta_E(\boldsymbol{\varphi},\boldsymbol{\Xi})+\delta_E^{\,2}(\boldsymbol{\Xi},\boldsymbol{\Xi})\right] \end{array} \right\}\right|_{\delta_E\to 0} \\ &= \left[2\left\langle(\mathbf{f},\boldsymbol{\varphi}+\delta_E\boldsymbol{\Xi})(\mathbf{f},\boldsymbol{\Xi})\right\rangle - 2\lambda_L(\boldsymbol{\varphi},\boldsymbol{\Xi}) - 2\lambda_L\delta_E(\boldsymbol{\Xi},\boldsymbol{\Xi})\right]\Big|_{\delta_E\to 0} \\ &= 2\left\langle(\mathbf{f},\boldsymbol{\varphi})(\mathbf{f},\boldsymbol{\Xi})\right\rangle - 2\lambda_L(\boldsymbol{\varphi},\boldsymbol{\Xi}) \\ &= 2\left[\left\langle(\mathbf{f},\boldsymbol{\Xi})(\boldsymbol{\varphi},\mathbf{f})\right\rangle - \lambda_L(\boldsymbol{\varphi},\boldsymbol{\Xi})\right] = 0 \end{aligned} \tag{5.6}$$

Since the inner product of two vectors is a scalar, Eq. (5.6) can be further written as:

$$\begin{aligned} \left\langle(\mathbf{f},\boldsymbol{\Xi})(\boldsymbol{\varphi},\mathbf{f})\right\rangle - \lambda_L(\boldsymbol{\varphi},\boldsymbol{\Xi}) &= \left\langle((\boldsymbol{\varphi},\mathbf{f})\mathbf{f},\boldsymbol{\Xi})\right\rangle - \lambda_L(\boldsymbol{\varphi},\boldsymbol{\Xi}) \\ &= \left(\left\langle(\boldsymbol{\varphi},\mathbf{f})\mathbf{f} - \lambda_L\boldsymbol{\varphi}\right\rangle,\boldsymbol{\Xi}\right) = 0 \end{aligned} \tag{5.7}$$

where we have rearranged the inner product and brought the average "inside" using the commutativity of $\langle\cdot\rangle$ and $(\cdot,\cdot)$. Since $\boldsymbol{\Xi}$ is an arbitrary vector, the projection of $\left\langle(\boldsymbol{\varphi},\mathbf{f})\mathbf{f} - \lambda_L\boldsymbol{\varphi}\right\rangle$ onto $\boldsymbol{\Xi}$ is zero, which means that $\left\langle(\boldsymbol{\varphi},\mathbf{f})\mathbf{f} - \lambda_L\boldsymbol{\varphi}\right\rangle$ must also be zero. Hence, Eq. (5.7) can be expressed as:

$$\left\langle(\boldsymbol{\varphi},\mathbf{f})\mathbf{f} - \lambda_L\boldsymbol{\varphi}\right\rangle = \left\langle\mathbf{f}(\boldsymbol{\varphi},\mathbf{f})\right\rangle - \lambda_L\boldsymbol{\varphi} = 0 \tag{5.8a}$$

or

$$\left\langle\mathbf{f}(\boldsymbol{\varphi},\mathbf{f})\right\rangle = \lambda_L\boldsymbol{\varphi} \tag{5.8b}$$

The ensemble average $\left\langle\mathbf{f}(\boldsymbol{\varphi},\mathbf{f})\right\rangle$ can be written as an average of all available vectors:

$$\left\langle\mathbf{f}(\boldsymbol{\varphi},\mathbf{f})\right\rangle = \frac{1}{N_v}\sum_{j=1}^{N_v}\mathbf{f}^j(\boldsymbol{\varphi},\mathbf{f}^j) \tag{5.9}$$

where N_v is the number of vectors in a data ensemble, and the superscript *j* represents the jth data sample or vector. Substituting Eq. (5.9) into Eq. (5.8b) yields:

$$\frac{1}{N_v}\sum_{j=1}^{N_v}\mathbf{f}^j(\boldsymbol{\varphi},\mathbf{f}^j) = \lambda_L\boldsymbol{\varphi} \tag{5.10}$$

Eq. (5.10) forms an $N_g \times N_g$ eigenvalue system, where N_g is the number of spatial grid points in the numerical solution.

To minimize computing costs, the snapshot method (Sirovich, 1987) can be used to reduce the dimensions of the system to $N_v \times N_v$, where N_v is typically significantly less than N_g. Snapshots are the representative thermo-flow fields used for extracting the POD modes. Let us define $a_j = \dfrac{1}{\lambda_L N_v}\left(\boldsymbol{\varphi},\mathbf{f}^j\right) = \dfrac{1}{\lambda_L N_v}\left(\mathbf{f}^j,\boldsymbol{\varphi}\right)$ (where the subscript *j* represents the jth element in the vector **a**). Then Eq. (5.10) can be simplified to:

$$\sum_{j=1}^{N_v} \mathbf{f}^j a_j = \varphi(\mathbf{x}). \tag{5.11}$$

Projecting $\mathbf{f}^i(\mathbf{x})$ onto both sides yields the following equation:

$$\left(\mathbf{f}^i, \sum_{j=1}^{N_v} \mathbf{f}^j a_j\right) = \left(\mathbf{f}^i, \varphi\right) \text{ and } i = 1,\ldots, N_v \tag{5.12}$$

where the superscript i represents the ith snapshot vector in a data ensemble. According to the above definition of a_j, the right-hand side of Eq. (5.12) can be written as $\lambda_L N_v a_i$. Because a_j on the left side of Eq. (5.12) is a number, it can be taken outside of the inner product. Then, interchanging the order of the sum operation and the inner product, Eq. (5.12) can be rewritten as:

$$\frac{1}{N_v} \sum_{j=1}^{N_v} \left(\mathbf{f}^i, \mathbf{f}^j\right) a_j = \lambda_L a_i. \tag{5.13}$$

If we define tensor $s_{ij} = \frac{1}{N_v}\left(\mathbf{f}^i, \mathbf{f}^j\right)$ and both i and $j = 1, \cdots, N_v$, then Eq. (5.13) becomes:

$$\begin{bmatrix} s_{11} & \cdots & s_{1N_v} \\ \cdots & \ddots & \vdots \\ s_{N_v 1} & \cdots & s_{N_v N_v} \end{bmatrix} \begin{bmatrix} a_1 \\ \vdots \\ a_{N_v} \end{bmatrix} = \lambda_L \begin{bmatrix} a_1 \\ \vdots \\ a_{N_v} \end{bmatrix} \tag{5.14a}$$

or

$$\mathbf{s} \cdot \mathbf{a} = \lambda_L \mathbf{a} \tag{5.14b}$$

Eq. (5.14) becomes an eigenvalue system of $N_v \times N_v$, where vector $\mathbf{a}$ can be treated as an eigenvector, and λ_L is the corresponding eigenvalue. The eigenvalue represents the energy captured by the POD mode. The total energy is defined as the sum of all of the eigenvalues. Commonly, 95% of the total energy is enough for most applications, and such a criterion can be adjusted in accordance with the purpose and the desired accuracy level (Elhadidi and Khalifa, 2005). Once the above equations have been solved for the eigenvector $\mathbf{a}$ and the eigenvalue λ_L, Eq. (5.11) can be used to obtain the POD mode φ from the snapshots. Other derivations of the POD modes can be found in previous textbooks (Holmes et al., 2012; Smith et al., 2005).

5.2.2 Interpolation of coefficients of POD modes

Suppose that a total of N_e numbers of POD modes are required to obtain the desired accuracy. Then each data field can be expressed in linear combinations of a series of POD modes:

$$\mathbf{f}_{\mathrm{R}} = \sum_{k=1}^{N_e} c_k \varphi^k \tag{5.15}$$

where $\mathbf{f}_{\mathrm{R}}$ is a reproduced field vector, c_k is the kth element in the coefficient vector $\mathbf{c}$, and φ^k is the kth POD mode. Eq. (5.15) provided a lower-order description of a data field. To construct a data field, not only the POD modes but also the coefficients of the POD modes must be known.

Recall that the snapshot fields can also be reconstructed by Eq. (5.15). According to the orthogonality principle, the coefficient of a POD mode for a snapshot can be expressed as:

$$c_k = (\mathbf{f}, \varphi^k) \tag{5.16}$$

If the snapshots are sufficiently representative, the extracted POD modes will contain the coherent structures of the data in the whole span. Hence, the POD modes are nearly fixed. However, the coefficients of the POD modes must be varied to reproduce a specific data field. Although the coefficients of the POD modes for a non-snapshot field are unknown, their coefficients can be estimated from an interpolation to the coefficients of the snapshot fields. This provides a quick way to construct the non-snapshot fields for inverse design.

To preserve the monotonicity of interpolated data and their smooth profiles, a piecewise cubic Hermite interpolation (a shape-preserving interpolation) can be used. Let us illustrate the basic principle with a case of one-dimensional interpolation. Suppose that $g = x_{p,1} < x_{p,2} < \ldots < x_{p,n} = l$ is a partition of the interval $I_t = [g, l]$, and $\{q_{p,i}: i = 1, 2, \cdots, n\}$ is a given set of data values at the partition points $x_{p,i}$ ($i = 1, 2, \cdots, n$). Our goal is to construct a piecewise cubic Hermite interpolation $H_p(x_{p,i})$ such that:

$$H_p\left(x_{p,i}\right) = q_{p,i},\ i = 1, 2, \cdots, n. \tag{5.17}$$

In each subinterval $I_{t,i} = [x_{p,i}, x_{p,i+1}]$, $H_p(x_p)$ is a cubic polynomial expressed as (Fritsch and Carlson, 1980):

$$H_p\left(x_p\right) = q_{p,i}H_{c1}\left(x_p\right) + q_{p,i+1}H_{c2}\left(x_p\right) + d_iH_{c3}\left(x_p\right) + d_{i+1}H_{c4}\left(x_p\right) \tag{5.18}$$

where x_p represents an arbitrary variable between $x_{p,i}$ and $x_{p,i+1}$; $d_i = H_p'(x_{p,i})$ and $d_{i+1} = H_p'(x_{p,i+1})$ are the derivatives with respect to x_p, where the superscript $'$ represents the derivative operation; and $H_{c1}(x_p)$, $H_{c2}(x_p)$, $H_{c3}(x_p)$, and $H_{c4}(x_p)$ are the cubic Hermite basis functions for the interval $I_{t,i}$, which are defined as (Fritsch and Carlson, 1980):

$$H_{c1}\left(x_p\right) = \phi_1\left(\left(x_{p,i+1} - x_p\right)/h_{s,i}\right) \tag{5.19}$$

$$H_{c2}\left(x_p\right) = \phi_1\left(\left(x_p - x_{p,i}\right)/h_{s,i}\right) \tag{5.20}$$

$$H_{c3}\left(x_p\right) = -h_{s,i}\phi_2\left(\left(x_{p,i+1} - x_p\right)/h_{s,i}\right) \tag{5.21}$$

$$H_{c4}\left(x_p\right) = h_{s,i}\phi_2\left(\left(x_p - x_{p,i}\right)/h_{s,i}\right) \tag{5.22}$$

where $h_{s,i} = x_{p,i+1} - x_{p,i}$, $\phi_1\left(x_p\right) = 3x_p^{\ 2} - 2x_p^{\ 3}$, and $\phi_2\left(x_p\right) = x_p^{\ 3} - x_p^{\ 2}$

Arranging Eqs. (5.19) through (5.22) and substituting into them the expressions for φ_1 and φ_2, we obtained the following:

$$\begin{aligned}
H_{c1}\left(x_p\right) &= \frac{3\left(x_{p,i+1} - x_p\right)^2}{h_{s,i}^{\ 2}} - \frac{2\left(x_{p,i+1} - x_p\right)^3}{h_{s,i}^{\ 3}} \\
&= \frac{3\left(h_{s,i} + x_{p,i} - x_p\right)^2}{h_{s,i}^{\ 2}} - \frac{2\left(h_{s,i} + x_{p,i} - x_p\right)^3}{h_{s,i}^{\ 3}} \\
&= 1 - \frac{3\left(x_p - x_{p,i}\right)^2}{h_{s,i}^{\ 2}} + \frac{2\left(x_p - x_{p,i}\right)^3}{h_{s,i}^{\ 3}}
\end{aligned} \tag{5.23}$$

$$H_{c2}\left(x_p\right)=\frac{3\left(x_p-x_{p,i}\right)^2}{h_{s,i}^{\ 2}}-\frac{2\left(x_p-x_{p,i}\right)^3}{h_{s,i}^{\ 3}} \tag{5.24}$$

$$\begin{aligned} H_{c3}\left(x_p\right)&=-h_{s,i}\frac{\left(x_{p,i+1}-x_p\right)^3}{h_{s,i}^{\ 3}}+h_{s,i}\frac{\left(x_{p,i+1}-x_p\right)^2}{h_{s,i}^{\ 2}}\\ &=-h_{s,i}\frac{\left(h_{s,i}+x_{p,i}-x_p\right)^3}{h_{s,i}^{\ 3}}+h_{s,i}\frac{\left(h_{s,i}+x_{p,i}-x_p\right)^2}{h_{s,i}^{\ 2}}\\ &=(x_p-x_{p,i})-\frac{2\left(x_p-x_{p,i}\right)^2}{h_{s,i}}+\frac{\left(x_p-x_{p,i}\right)^3}{h_{s,i}^{\ 2}} \end{aligned} \tag{5.25}$$

$$\begin{aligned} H_{c4}\left(x_p\right)&=h_{s,i}\frac{\left(x_p-x_{p,i}\right)^3}{h_{s,i}^{\ 3}}-h_{s,i}\frac{\left(x_p-x_{p,i}\right)^2}{h_{s,i}^{\ 2}}\\ &=\frac{\left(x_p-x_{p,i}\right)^3}{h_{s,i}^{\ 2}}-\frac{\left(x_p-x_{p,i}\right)^2}{h_{s,i}} \end{aligned} \tag{5.26}$$

Substituting Eqs. (5.23) through (5.26) into Eq. (5.18) and defining $\Delta_i=(q_{p,i+1}-q_{p,i})/(x_{p,i+1}-x_{p,i})$ $\Delta_i=(q_{p,i+1}-q_{p,i})/(x_{p,i+1}-x_{p,i})$, $H_p(x)$ can then finally be explicitly expressed as:

$$H_p\left(x_p\right)=\left[\frac{d_i+d_{i+1}-2\Delta_i}{h_{s,i}^{\ 2}}\right](x_p-x_{p,i})^3+\left[\frac{-2d_i-d_{i+1}+3\Delta_i}{h_{s,i}}\right](x_p-x_{p,i})^2+d_i(x_p-x_{p,i})+q_{p,i} \tag{5.27}$$

According to Eq. (5.27), a cubic polynomial within each interval between two adjacent points can be obtained. Furthermore, to construct a piecewise cubic interpolation, it is essential to calculate the derivatives d_1, d_2..., d_n. A weighted average of $\Delta_{i\text{-}1}$ and Δ_i is used for interior points to calculate d_i (i = 2,..., (n–1)) when $\Delta_{i\text{-}1}$ and have the same sign (MATLAB, 2010), while for cases in which $\Delta_{i\text{-}1}$ and Δ_i have opposite signs or either is zero, $d_i = 0$ (i = 2,..., (n–1)). In addition, for the two end derivatives, d_1 and d_n, the non-centered three-point difference formula is applied (MATLAB, 2010).

The theoretical error in cubic Hermite interpolation can be assessed by comparing the interpolation $H_p(x)$ with the known function that generated the data. For example, if the data $q_{p,i}$ are values of a function $q_p(x_p)$ and $q_p(x_p)$ has a continuous fourth derivative, then we have (Kahaner et al., 1989):

$$\left|H_p\left(x_p\right)-q_p\left(x_p\right)\right|<\text{constant}\cdot h_{s,m}^{\ 4}\cdot\max\left[\left|q_p^{(4)}(x_p)\right|\right]=O\left(h_{s,m}^{\ 4}\right) \tag{5.28a}$$

or

$$\left|H_p\left(x_p\right)-q_p\left(x_p\right)\right|<O\left(h_{s,m}^{\ 4}\right) \tag{5.28b}$$

where $h_{s,m}$ is the maximum spacing between adjacent knots and $q_p^{(4)}(x_p)$ is the fourth-order derivative. The interpolation improves when the number of knots increases and their spacing decreases, and the rate of improvement is proportional to the fourth power of the knot spacing. Hence, as long as the knot spacing is appropriate, the accuracy of the piecewise cubic Hermite interpolation is ensured.

This type of interpolation is also called a shape-preserving method because on intervals where the data are monotonic, $H_p(x_p)$ is also monotonic; and at points where the data have a local extremum, so does $H_p(x_p)$. A smooth interpolation has one or more continuous derivatives on $[x_{p,1}, x_{p,n}]$ (Kahaner et al., 1989). In the above equations we can observe that the piecewise cubic Hermite interpolation has a continuous first derivative, and therefore the interpolated data have a smooth shape. However, the cubic Hermite interpolation is not the only type that can ensure a smooth shape; other smooth interpolations that preserve the monotonicity may do so as well. Nevertheless, we recommend that whenever possible the interpolation be plotted to confirm that it has provided the desired properties.

The above interpolation method can be extended to two-dimensional or higher-dimensional cases when conducting multi-variable inverse design. In practice, one may use the shape-preserving interpolation package in the MATLAB software to interpolate the coefficients of POD modes.

5.2.3 Original CFD data preparation and criteria for data saturation

To extract POD modes for inverse design, a small number of original data samples must be provided by full CFD simulations as snapshots. Because the expense of a full simulation is extremely high, an efficient design would maintain a minimal load of such simulations while still ensuring accuracy. A criterion for the saturation of the original data provided by full CFD simulations is needed.

To determine the appropriate number of data samples from full CFD simulations, we proposed a trial-and-data-sample-increase method (Wei et al., 2015). This is a conservative way of supplementing the necessary full CFD data to reduce the sharp nonlinearity in the original data samples. Most models, including POD, perform poorly when non-linearity prevails. To illustrate the trial-and-data-sample-increase method, Fig. 5.1 shows dummy errors between linearly interpolated sample points and full CFD sample points for an air-supply parameter. This method can be extended to cases involving multiple air-supply parameters. The method began with two original data sample points provided by full CFD simulations, which corresponded to the smallest and largest air-supply parameters in the constrained range, as indicated by the squares labeled S1 and S2 in Fig. 5.1. The errors at these two points were zero, under the assumption that there was no error from the full CFD simulation. The data sample halfway between S1 and S2, labeled S3 in the figure, was estimated by applying a linear interpolation of the data. A full CFD simulation of the air-supply parameter at S3 should also be conducted, as indicated by the point S3′. Then the error between the linearly interpolated data and the full CFD simulation data can be evaluated, where the error was the distance between points S3 and S3′ in Fig. 5.1. The dotted curves in the figure represent the errors between the data provided by the full CFD simulation and the linearly interpolated fields. The error evaluation equations were defined as (Li et al., 2013a):

$$\mathrm{RE}_{\mathrm{average}} = \frac{\sqrt{\sum_{m=1}^{N_g} (f_{\mathrm{interpolated},m} - f_{\mathrm{CFD},m})^2 / N_g}}{f_{\mathrm{CFD,normalized}}} \tag{5.29}$$

and

$$\mathrm{RE}_{\max} = \frac{\max(f_{\text{interpolated},m} - f_{\text{CFD},m})}{f_{\text{CFD,normalized}}}, \ m = 1, \cdots, N_g \tag{5.30}$$

where $\mathrm{RE}_{\text{average}}$ represents the average relative error of the interpolated fields and $\mathrm{RE}_{\max}$ is the maximum relative error; $f_{\text{interpolated}}$ is the interpolated field data; f_{CFD} is the field data from the full CFD simulation; subscript m represents the mth grid cell; and $f_{\text{CFD,normalized}}$ represents the specific value to normalize the relative error. If the relative errors were greater than the specified threshold value indicated by the dash dot line in Fig. 5.1, the currently available data samples (S1 and S2) were not sufficient. In that case, the original data samples were supplemented with the new data sample from the full CFD simulation (S3′), and the relative error of S3′ was zero in Fig. 5.1. Following a similar procedure, the method gradually increased the number of data samples (S4′ and S5′) until all the errors between the linearly interpolated data fields and the data fields from the full CFD simulation, halfway between two neighboring samples, were less than the threshold value, as shown by S6, S7, S8, and S9 in Fig. 5.1. Hence, the final required data samples in this case were determined to be S1, S2, S3′, S4′, and S5′. In this way, the non-linearity among data samples was weakened, which improved the accuracy of POD mode extraction and the subsequent data interpolation.

5.2.4 POD-based design procedure

Fig. 5.2 illustrates the POD-based design procedure. The design criteria and the original spans of the air-supply parameters should be set first. Next, a number of representative discrete air-supply parameters in the constrained spans were selected for full CFD simulation. On the basis of the trial-and-data-sample-increase scheme proposed in the previous section, the number of CFD data samples was gradually increased until data saturation was achieved. POD analysis can then be applied to extract the spatial modes and their coefficients. In order to construct the field data that were not in the snapshots, an interpolation was implemented for the coefficients of the POD modes for the non-snapshots. Next, cause-effect maps between air-supply parameters and the resulting indoor environmental performance were established. Finally, the air-supply parameters that satisfy the specified design targets were circumscribed.

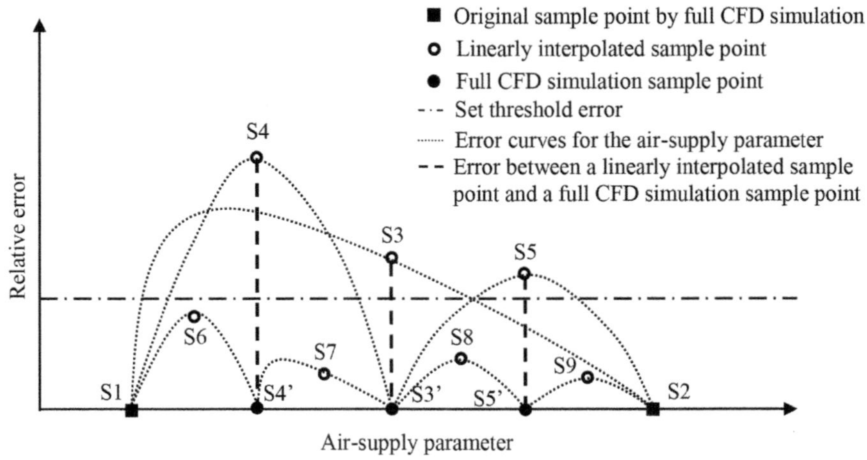

Figure 5.1 Illustration of the trial-and-data-sample-increase method, used to determine whether the CFD-generated data samples are adequate

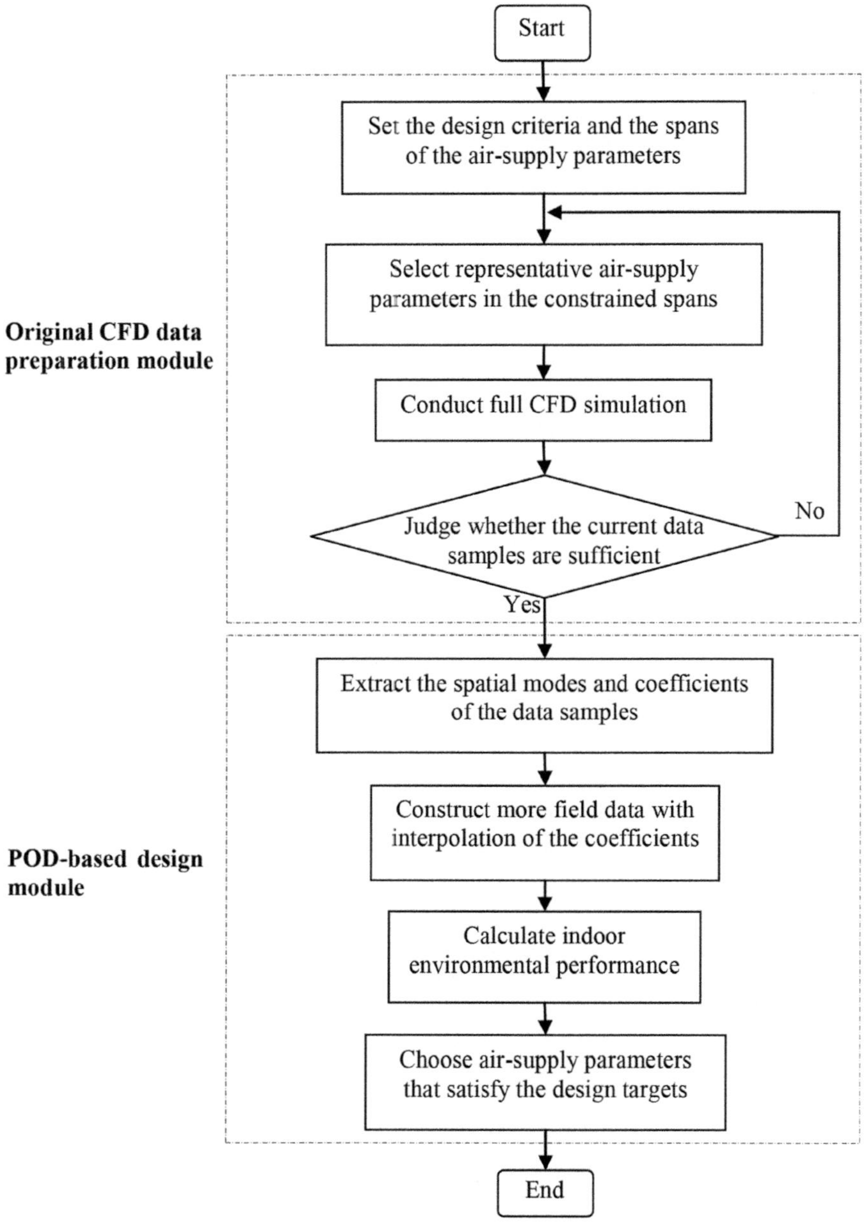

Figure 5.2 Solution flow chart for the POD-based design approach

5.3 Application of the CFD-based POD method

This section describes the use of the proposed POD method to determine the air-supply speed and temperature in an empty two-dimensional air cavity (Blay et al., 1992) for which detailed experimental data are available. The measured air velocity and temperature at several points in the cavity were set as the design targets, and the required air-supply parameters were inversely solved. The air-supply parameters were also inversely determined for a three-dimensional office and an aircraft cabin. In the office, only a single performance parameter was adopted as the design target, while in the aircraft cabin, multiple performance parameters were set as the design targets.

5.3.1 Inverse solution of air-supply parameters in a two-dimensional air cavity

5.3.1.1 Case description

The first case is that of a two-dimensional square air cavity investigated by Blay et al. (1992), as shown in Fig. 5.3. The length of the cavity was 1.04 m. Conditioned air was supplied to the cavity through a slot inlet in the upper left corner with a speed of 0.57 m/s and a temperature of 15°C. The slot inlet had a height of 18 mm. The interior air was exhausted from the cavity through an outlet in the bottom right corner. The height of the outlet was 24 mm. The bottom boundary was heated to 35°C, while the rest of the boundaries were controlled at 15°C. In the present study, the measured horizontal air velocity components and temperatures at discrete points on the central vertical line, together with the vertical air velocity components and air temperatures at discrete points on the central horizontal line, were adopted as the design targets. The design variables, for which we inversely solved, were the air-supply speed and temperature at the slot inlet. These inversely determined parameters would then be compared with the actual values, 0.57 m/s and 15°C, in the original experiment.

5.3.1.2 Solution strategies

According to the solution flow chart in Fig. 5.2, the design criteria and the spans of the air-supply parameters should be determined first. The solution goal was to find the air-supply speed and temperature that led to the closest match with the measured data. The design target was evaluated as:

$$\mathrm{RE} = \frac{1}{2} \times \left(\sqrt{\frac{\Sigma\left(U'_i - U_i\right)^2}{\Sigma U_i^2}} + \sqrt{\frac{\Sigma\left(T'_i - T_i\right)^2}{\Sigma\left(T_{\text{inlet}} - T_i\right)^2}} \right) \tag{5.31}$$

where U'_i and U_i were the velocity components at a measuring point, according to the POD method and the measurements, respectively; T'_i and T_i were the POD-determined and measured air temperatures, respectively, at the measuring point; and T_{inlet} was the air-supply temperature used in the experiment. As mentioned previously, the discrete velocity components and temperatures were measured on the central horizontal line and the central vertical line. The air-supply parameters that led to a minimum value in Eq. (5.31) were our target solution.

During the inverse solution process, the air-supply parameters were assumed to be unknown. The air-supply speed was estimated to vary from 0.1 m/s to 1.1 m/s, and the air-supply temperature from 10°C to 20°C. Hence, spans of air-supply speed and air temperature were set. Discrete air-supply speeds of 0.1 m/s, 0.6 m/s, and 1.1 m/s and temperatures of 10°C, 15°C, and 20°C were sampled for establishment of the original CFD data pool. Thus, there were a total of nine different cases, and these cases were checked for data saturation according to the procedure proposed in Section 5.2.3.

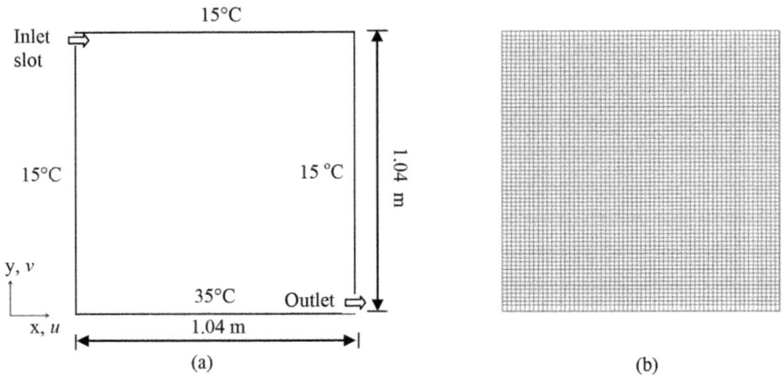

Figure 5.3 Sketch (a) and grid mesh (b) of the cavity used for inverse design of the air-supply speed and temperature

Full CFD simulations were conducted to resolve the original thermo-flow data. Prior to these simulations, the geometry and numerical grid meshes were created. Because the geometry shape of the air cavity was quite simple, the geometry shown in Fig. 5.3(a) was constructed with the GAMBIT software. Next, the rectangular grid meshes were generated. The total grid cell number is 5,460, as shown in Fig. 5.3(b). To better resolve the near-wall boundary effect, the first grid cell size was refined to 0.09 mm, and the growth rate was 1.2 for the next three layers. The grid cell size is approximately 1 cm in the core region. The above grid meshes are sufficient to provide grid-independent results.

The numerical solutions to the N-S equations were carried out in the FLUENT software (version 6.3.26). The turbulent combined convection flows were resolved using the re-normalization group (RNG) k-ε model (Yakhot et al., 1992), because the model has generally good accuracy at a reasonable expense (Chen, 1995; Zhang et al., 2007). The Boussinesq approximation was adopted to represent the varying density with temperature in the momentum equations. In addition, the pressure staggering option (PRESTO) scheme was adopted when discretizing the pressure term to fully account for the buoyancy contribution. The remaining variables were discretized using the second-order upwind scheme. The semi-implicit pressure linked equation (SIMPLE) scheme was used for coupling the pressure and velocity. Convergence was deemed satisfactory when all of the relative residuals were smaller than 10^{-5} and there was no significant change in the velocity and temperature with numerical iterations at a few monitoring points.

5.3.1.3 Solution results

With representative full CFD data, the POD analysis can be carried out. For extraction of the spatial modes, the eigenvalues and the corresponding eigenvectors in Eq. (5.14) should be solved first. Fig. 5.4(a) shows the respective contribution of each eigenvalue to the total energy. In total, nine sets of eigenvalues and eigenvectors can be obtained from the nine different representative cases. It can be seen that the first eigenvector contains nearly all of the energy. The energy captured by each eigenvector decreases dramatically from the first to the second and third. This sharp decrease may be attributed to the fact that the first one or two eigenvectors contain most of the basic features of the data samples. In addition, Fig. 5.4(b)

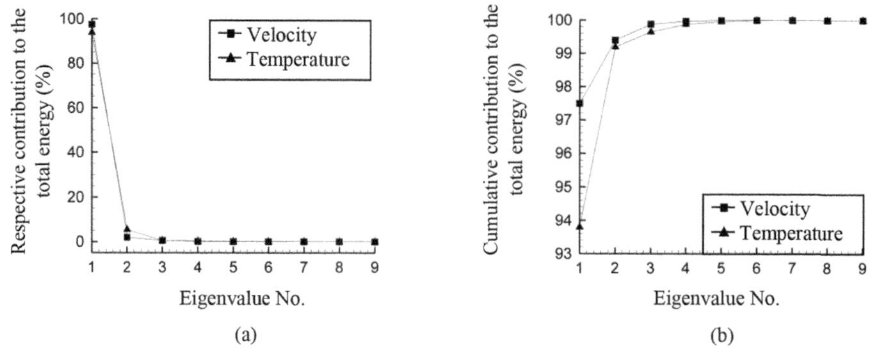

Figure 5.4 Contribution of the eigenvalues to the total energy: (a) respective contribution; (b) cumulative contribution

depicts the cumulative contribution of the eigenvalues. To capture more than 99.99% of the total energy, the first four eigenvectors are sufficient for the velocity, and the first five are sufficient for temperature.

With the resulting eigenvectors, the POD modes can be extracted by use of Eq. (5.11). Fig. 5.5 shows the POD modes that contain the largest and second largest portions of energy, corresponding to the first two eigenvectors discussed above. The spatial modes for the velocity vectors and air temperature are exhibited. Among them, the POD modes that contain the largest portion of energy are shown in Figs. 5.5(a) and (c), while those containing the second largest portion of energy are shown in Figs. 5.5(b) and (d). The most dominant POD modes almost reproduce the rough profiles of the flow and temperature distributions. The POD modes with relatively small portions of energy correspond to much smaller-scale flow structures, which are not captured by the most dominant modes.

Once the POD modes were extracted, the coefficients of the POD modes for each case were calculated by use of Eq. (5.16). The coefficients of the POD modes for the remaining air-supply parameters were then interpolated by the two-dimensional cubic Hermite scheme. Fig. 5.6 shows the coefficients of the POD modes that contain the most energy. A total of 1,111 cases were considered here, with a changing step size of 0.01 m/s for the air-supply speed and 1°C for the air-supply temperature in the specified spans. The coefficients of the nine snapshots obtained from Eq. (5.16) were marked in Fig. 5.6. These original coefficients were fully accepted by the cubic Hermite interpolation. It can be seen in the figure that the interpolation preserved the monotonicity of the coefficients and provided a smooth profile.

Next, the resulting coefficients were used together with Eq. (5.15) to construct the thermo-flow data for the 1,111 cases. Fig. 5.7 shows the target performance defined by Eq. (5.31) for various air-supply parameters in the specified spans. The best match was indicated by the red dot, where the value is 17.3% as defined by Eq. (5.31). The best match corresponds to air-supply parameters of 0.54 m/s and 15°C, which are quite close to the actual values in the original experiment, i.e., 0.57 m/s and 15°C. Next, Fig. 5.8 compares the velocity components and temperature profiles for the best solution by the POD method with the experimental data. It can be seen that the POD-generated profiles agree well with the experimental data. Thus, it can be concluded that the proposed POD method is capable of inversely solving for the air-supply parameters with good accuracy.

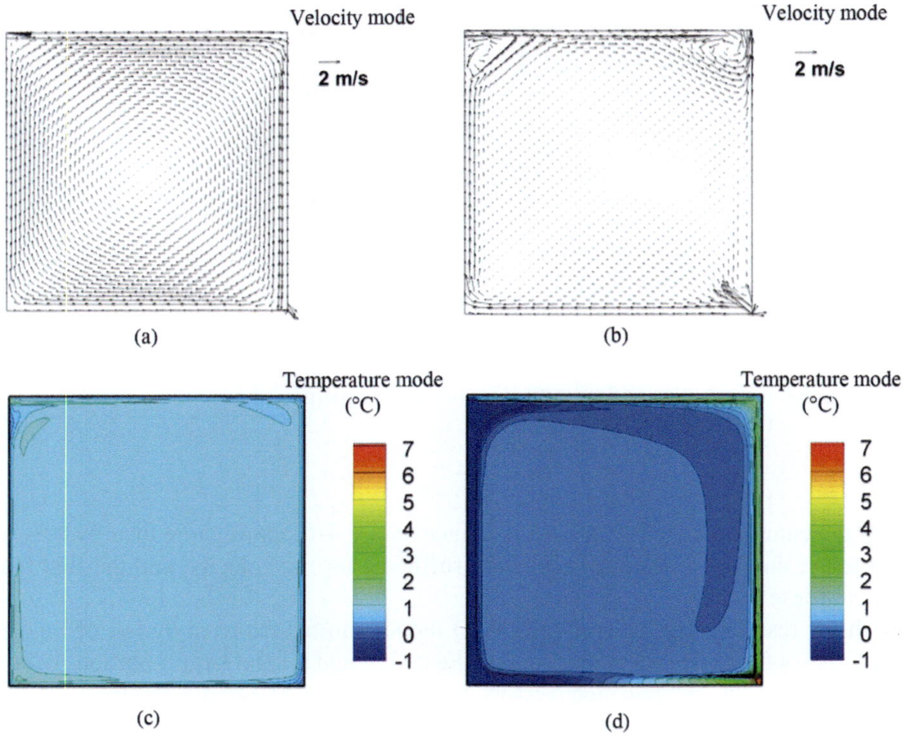

Figure 5.5 Distribution of the POD modes: (a) velocity vectors with the largest portion of energy; (b) velocity vectors with the second largest portion of energy; (c) temperature with the largest portion of energy; (d) temperature with the second largest portion of energy

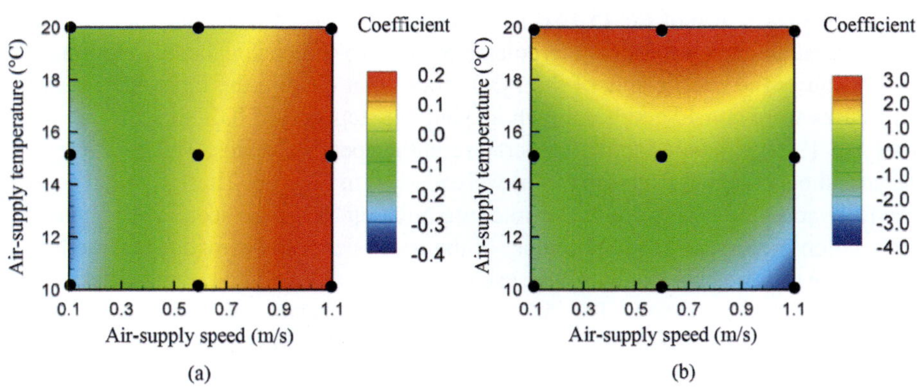

Figure 5.6 Coefficients of the POD modes that contain the most energy for different air-supply speeds and temperatures and the dots represent those for the original snapshot cases: (a) velocity and (b) temperature

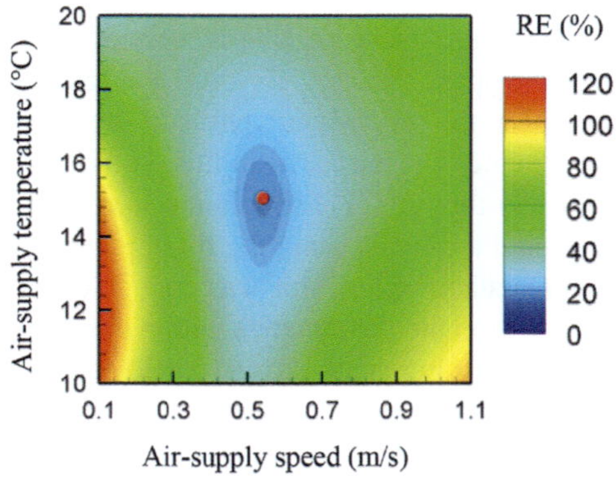

Figure 5.7 Relative error, defined by Eq. (5.31), between the POD method and the experiment, and the solution leading to the closest match with the measured data (red dot)

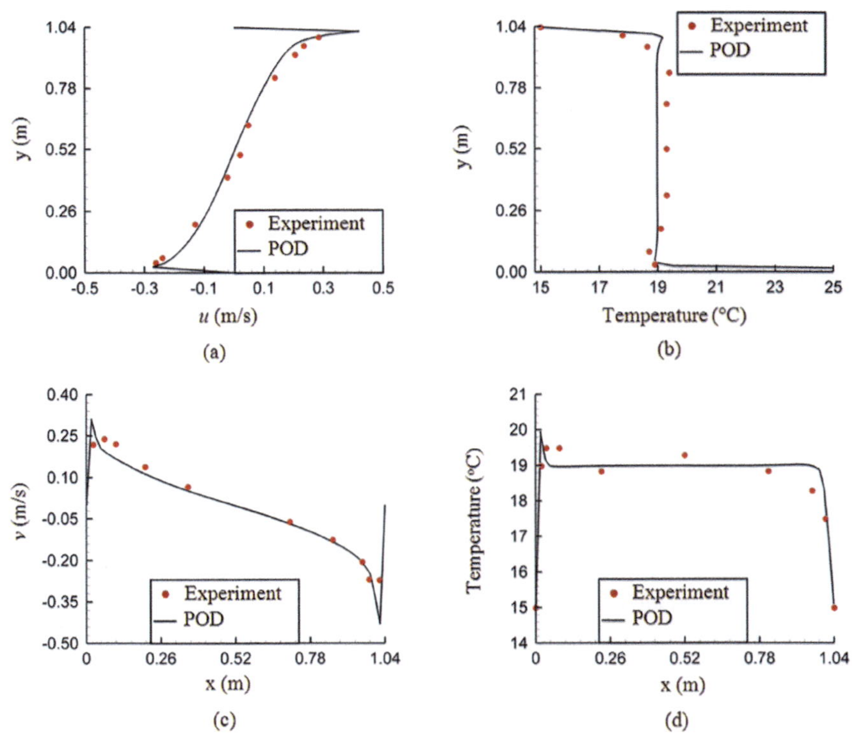

Figure 5.8 Comparison of the best solution by the POD method and the experimental data: (a) horizontal U-velocity profile on the central vertical line; (b) temperature profile on the central vertical line; (c) vertical V-velocity profile on the central horizontal line; (d) temperature profile on the central horizontal line

5.3.1.4 Discussion

The key factors in the accuracy of the POD method are the representativeness of the spatial modes and the accuracy of the coefficients of the spatial modes. Because the modes were extracted from the originally provided data samples, the representativeness of these samples was critical. A trial-and-data-sample-increase scheme was proposed in this study. To determine whether the original data samples provided by the full CFD simulation were sufficient, the gaps between the linearly interpolated fields and the fields using the full CFD simulation were evaluated halfway between two neighboring data samples. This was done to ensure that there was not a severe nonlinear distribution of the evaluated parameters within either sample. A limitation of the method was the possibility of omitting some characteristic data samples when the errors halfway between neighboring air-supply parameters were within the specified threshold error but large errors were present elsewhere. However, the possibility of such an omission was low; to be conservative, one could check a larger number of data samples.

The strategies used to interpolate the coefficients of the spatial modes are important because different strategies assumed different trends in the variation between the two snapshot data samples. An ideal interpolation used a minimum number of snapshots while still providing an excellent approximation of the data. Because the proposed trial-and-data-sample-increase scheme guaranteed that there would be no severe nonlinearity between the two original data samples, a smooth interpolation method that preserved the monotonicity and shape of the samples would work well. For the purpose of demonstration, this textbook adopted the piecewise cubic Hermite interpolation. Future research may develop intelligent coordination schemes between the original data sample preparation and the interpolation of the coefficients of the POD modes. In this way, the number of snapshots could be further reduced, but the data interpolation would compensate for this reduction. Thus, the solution accuracy would be maintained, while the solution efficiency is enhanced.

In regard to computing cost, approximately 4.5 h was required to obtain solutions to the nine snapshot cases by means of full CFD simulation, and less than 1 h was needed for the POD construction of 1,102 cases and inverse solution for the air-supply parameters. The computation was performed on a server equipped with two 2.3 GHz Intel CPUs (four cores per CPU) and total memory of 144 GB. Hence, the POD solution is quite fast, whereas preparation of the snapshots by means of full CFD simulations requires a considerable amount of time.

5.3.1.5 Summary

In this section, the proposed POD approach was used to efficiently determine the air-supply speed and temperature in a two-dimensional air cavity. The goal was to obtain the closest match with the measured air velocity and temperature at discrete points. The following conclusions can be drawn from the reported work:

- The POD modes of the snapshots and the interpolation of the coefficients of the modes can be used to construct the thermo-flow fields with generally good accuracy. The proposed method provided values of the air-supply parameters that were quite close to the actual values used in the experiment. The velocity and temperature profiles reproduced by the POD method agreed well with the measured data.
- The POD-based design required a relatively small amount of computing resources. Most of the computing time for the POD method was devoted to preparation of the original field data by CFD; the POD analysis, interpolation of the spatial mode coefficients, and determination of the air-supply speed and temperature were much more efficient.

5.3.2 Inverse design of air-supply parameters in an office

The POD method was also used to determine the required air-supply parameters for a three-dimensional office, to ensure satisfactory thermal comfort.

5.3.2.1 Case description

The office used in the second inverse-design case is shown in Fig. 5.9(a). The office environment was studied by Yuan et al. (1999), and detailed experimental data are available. The office was 5.16 m long, 3.65 m wide, and 2.43 m high. There was a window in the exterior wall. Two cabinets were located in corners of the office and six fluorescent lamps were mounted on the ceiling. Two seated occupants worked in front of desktop computers. The occupants (0.4 m × 0.35 m × 1.1 m) and computers (0.4 m × 0.4 m × 0.4 m) were all simulated by heated boxes. For the purpose of displacement ventilation, conditioned air was supplied through a wall-mounted diffuser with an opening area of 0.53 m × 1.14 m. The internal air was exhausted through an overhead outlet with dimensions of 0.43 m × 0.43 m.

Our goal was to determine the air-supply parameters that would create a better thermal environment than that reported by Yuan et al. (1999). The design criterion was an absolute value of PMV less than 0.2. The area-averaged PMV was evaluated on the surrounding surfaces at a distance of 0.1 m from the occupants, as illustrated by the pink surfaces in Fig. 5.9(b). Here $|PMV|_{ave} = 0.2$ coincided with the comfort status for the air-supply parameters used in the original experiment. The air-supply temperature and direction were the design variables for which we inversely solved.

5.3.2.2 Solution strategies

According to the solution flow chart in Fig. 5.2, the original spans of the air-supply parameters should be set first. The air-supply temperature was assumed to vary from 15°C to 25°C, and the air-supply direction from −45° to 45° with respect to the positive x coordinate on the horizontal plane (i.e., the xy plane). The air-supply direction along the xz plane was not considered, because in the summer season cooled air was supplied to the office and dropped to the floor because of thermal buoyancy. Hence, the air-supply direction along the xz plane did not significantly impact the results. The air-supply rate was fixed at four air changes per hour (ACH), the same as that used in the experiment. However, the air-supply speed varied when different air-supply directions were used.

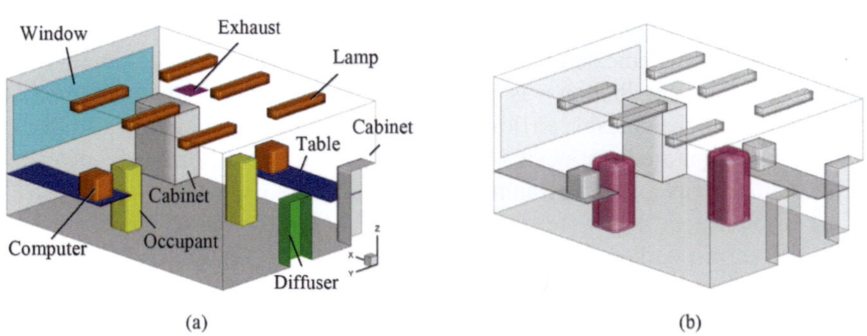

Figure 5.9 An office for inverse design of the air-supply temperature and direction: (a) overview; (b) surrounding surfaces at a distance of 0.1 m from the occupants, where the comfort status was evaluated

To establish the data pool for POD analysis, a total of 15 cases were resolved by full CFD simulations. These cases contain discrete air-supply temperatures of 15°C, 17.5°C, 20°C, 22.5°C, and 25°C, and discrete air-supply angles of −45°, 0°, and 45°, which, in combination, constitute 15 cases. The temperatures were sampled at higher resolution because the comfort status in this case was affected more by the air-supply temperature than the air-supply direction. The 15 cases have been checked for data saturation according to the procedure presented in Section 5.2.3.

For solution of thermo flows with CFD, the office geometry shown in Fig. 5.9(a) was constructed with the GAMBIT software. A total of 112,897 hexahedral grids with an averaged size of 7 cm were generated and were shown to satisfy grid independence. All boundary conditions were the same as those reported by Yuan et al. (1999). The numerical settings were identical to those described in Section 5.3.1.2, except for the wall Prandtl number. This number was adjusted to 0.009 according to the suggestion made by Zhang et al. (2013). The convergence was deemed satisfactory when the unbalanced heat transfer rate for all boundaries was less than 0.5% of the maximum heat gain, and the monitored velocity and temperature at typical points were relatively unchanged with further numerical iterations.

The PMV calculation was based on formulas by Fanger (1970). Some of the parameters involved in the PMV calculation were addressed as follows. The metabolic rate for an adult was 75 W/m^2, and there was assumed to be no mechanical work production. The water vapor pressure was equal to 1586.8 Pa, which corresponds to a relative humidity of 50% at 25°C. The clothing insulation was 0.57 clo, under the assumption that the occupants wore trousers and short-sleeved shirts. The background mean radiation temperature for the clothed body was simplified as the air temperature. The convective heat transfer coefficient and the mean temperature of the outer surface of the clothed body were estimated in accordance with the ASHRAE Handbook (2013). The PMV is an explicit function of velocity and temperature. Therefore, once the velocity and temperature fields are known, the PMV distribution can be calculated instantly.

5.3.2.3 Solution results

With the representative CFD data, the POD bases can be extracted. Because the PMV was evaluated on the surrounding surfaces at a distance of 0.1 m from the occupants, the POD bases of the thermo-flow were analyzed only on these surfaces for the purpose of simplicity. Fig. 5.10 presents the respective and cumulative contributions of the eigenvalues to the total energy when the POD modes were extracted. The first eigenvector of temperature contained more than 90% of the energy, while the first eigenvector of velocity contained only 40% of the energy. These results indicated that the temperature distribution had a dominant mode, while the velocity field distribution was diverse. The lack of a strong dominant mode in the velocity may be due to the unstable airflow near the occupants, which was caused by thermal buoyancy. To capture more than 99.99% of the total energy, a total of 14 eigenvalues were required for both velocity and temperature fields.

With the eigenvectors that were obtained, Eq. (5.11) was used to establish the POD modes, and the corresponding coefficients were extracted by Eq. (5.16). For brevity, the POD modes and their coefficients were not presented here. The coefficients for the non-snapshot cases were interpolated by the two-dimensional cubic Hermite scheme. A total of 384 additional cases were reproduced using Eq. (5.15), with a changing step size of 0.5°C in the air-supply temperature and a changing step size of 5° in the air-supply direction.

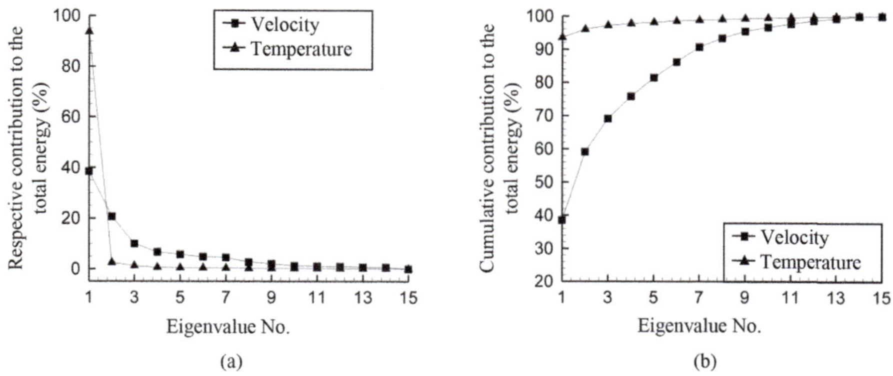

Figure 5.10 Contribution of the eigenvalues to the total energy: (a) respective contribution; (b) cumulative contribution

Fig. 5.11(a) presents the computed comfort status of both occupants as it varied with air-supply temperature and direction. As the temperature increased, the $|PMV|_{ave}$ initially decreased and then increased. Either a lower or a higher air-supply temperature produced discomfort. In contrast, the air-supply direction had little effect on the comfort status. This may be due in part to the fact that both occupants were seated far away from the diffuser. The occupants were predicted to have a lower $|PMV|_{ave}$ when the air-supply temperature was in the range of 19°C to 23.5°C. The air-supply parameters in the shaded area in Fig. 5.11(b) can provide $|PMV|_{ave}$ of less than 0.2. Moreover, the black dot in Fig. 5.11(b), at an air-supply temperature of 23°C and angle of −45°, corresponds to the minimum $|PMV|_{ave}$ of 0.13. When 23°C and −45° were used as the air-supply parameters for a full CFD simulation, the $|PMV|_{ave}$ was found to be 0.12. The similarity in $|PMV|_{ave}$ between the POD method and CFD simulation indicated that the POD method performed well.

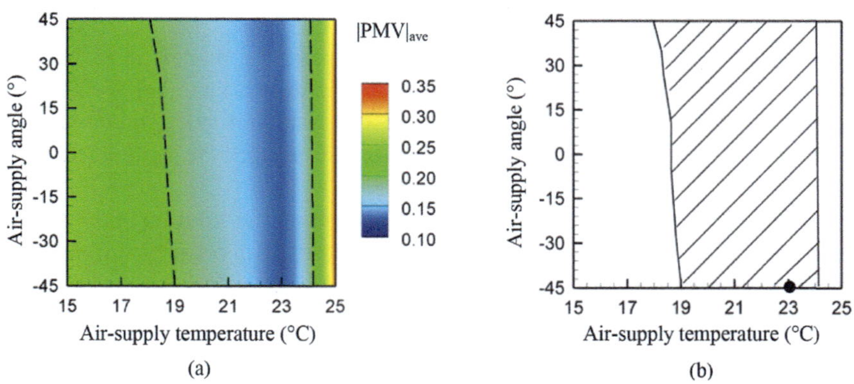

Figure 5.11 Comfort status surrounding the occupants as determined by the POD method: (a) area-averaged absolute value of PMV; (b) air-supply parameters that satisfy the specified design target (shaded zone), and the solution that corresponds to the minimum $|PMV|_{ave}$ (black dot)

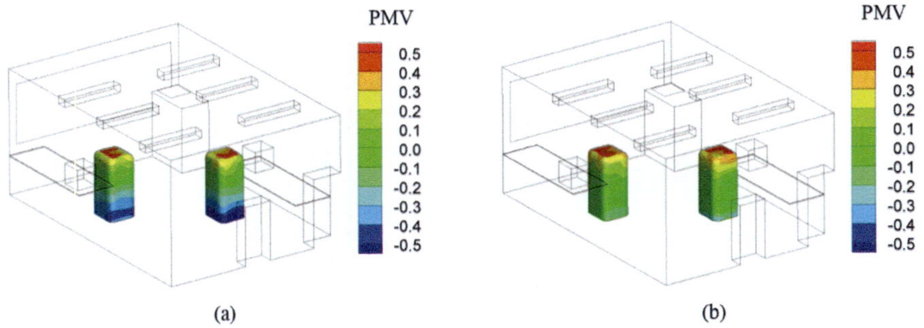

Figure 5.12 Comparison of PMV distributions on surfaces at a distance of 0.1 m from the occupants before and after optimization: (a) the experimental case; (b) the case with optimal air-supply parameters

Next, Fig. 5.12 shows the PMV distribution near both occupants, for the experimental case and the case using the optimal air-supply parameters that were obtained as described above. Both cases were resolved by full CFD simulation to remove the computing errors caused by model reduction. It can be seen that in the experimental case, both occupants might feel that it was cold near their feet but hot at head level. When the optimal air-supply parameters were adopted, the comfort of the occupants greatly increased, as shown in Fig. 5.12(b). The PMV was close to zero in most areas, and the difference in PMV between the head and feet was minimized. Thus, the proposed method provided a better thermal environment than that in the experimental case.

5.3.2.4 Discussion

In the optimized case, POD analysis was conducted on the surrounding surfaces at a distance of 0.1 m from the occupants. Because the comfort status was evaluated near the occupants, the thermo-flow fields were required only on these surfaces. If the thermo-flow fields in the entire office were analyzed, the computing cost would be greater. Theoretically, the POD method can deal with the data in any given domain. Interested readers may compare the results when different domain sizes are specified. It is expected that slightly better results would be obtained for a smaller domain because less information would be involved. With the same computing resources, higher-precision information could be captured in a smaller domain.

The computing time was also counted when implementing the solution in the same computing facility as the air-cavity case. A total of 60 h was required to obtain solutions for the 15 original cases by means of full CFD simulations, and approximately 1.5 h for POD analysis and circumscription of the final results.

5.3.2.5 Summary

In this section, the CFD-based POD approach was used to determine the air-supply temperature and direction that would improve the comfort status in an office. The proposed method was able to circumscribe the air-supply parameters that satisfy the design target. The design that provided the best comfort status was also highlighted. In comparison with the original, non-optimized experimental case, the new design greatly increased the occupants' feeling of comfort. Thus, the proposed method is effective for optimizing the design of an indoor environment.

5.3.3 Inverse design of air-supply parameters in an aircraft cabin

The previous sections described the application of the inverse method to two simple geometries, an empty air cavity and an idealized office. The POD method can be extended to enclosures with more complicated geometric configurations. In this section, the method was used to design the air-supply parameters in a five-row aircraft cabin, in which multiple design targets must be satisfied.

5.3.3.1 Case description

The third case involved a half section of a five-row aircraft cabin, as shown in Fig. 5.13. There were a total of 15 seated passengers. The aircraft cabin used mixing ventilation; conditioned air was supplied to the cabin through a slot inlet on the side wall near the luggage rack. The flow rate for each passenger was fixed at 9.4 liters per second. The internal air was extracted through an outlet on the side wall near the floor. The front and rear boundaries of the aircraft cabin and all the seats were set as adiabatic. The temperature boundary conditions for the rest of the solid surfaces were summarized in Table 5.1. The vertical surface in the aisle was assumed to be symmetrical.

The design goal was to determine the air-supply parameters that minimized the draft risk, ensured thermal comfort, and guaranteed air quality. The design parameters included the air-supply opening size, air-discharge direction, and air temperature. The draft risk was evaluated in both the head and ankle regions, as shown in Fig. 5.13(b). The average air speed was restricted to 0.2 m/s in the head region and 0.35 m/s in the ankle region, as recommended by ASHRAE Standard 161 (2007). Thermal comfort was assessed on the enclosure surfaces surrounding the passengers, at a distance of 0.1 m from their bodies. The design target was an area-averaged absolute value of the cabin predicted mean vote ($|\mathrm{PMVc}|_{\mathrm{ave}}$) of less than 0.3. ASHRAE Standard 55 (2010) recommended a PMV within ±0.5. Because the comfort status surrounding the passengers was analyzed, a stricter threshold of 0.3 was used in this case. Cui and Zhu (2014) proposed a cabin predicted mean vote (PMVc) that has accounted for the impact on thermal comfort of the reduced air pressure during a flight. The PMVc can be easily calculated from the regular PMV proposed by Fanger (1970) as follows:

$$\mathrm{PMV_c(summer)} = -0.0758\mathrm{PMV}^2 + 0.6757\mathrm{PMV} - 0.1262 \tag{5.32}$$

Table 5.1 Boundary conditions for the aircraft cabin

Item	*Value*	*Item*	*Value*
Passenger surface temperature	30.3°C	Side wall temperature	22°C
Ceiling temperature	22°C	Floor temperature	23°C
Luggage-rack temperature	21°C		

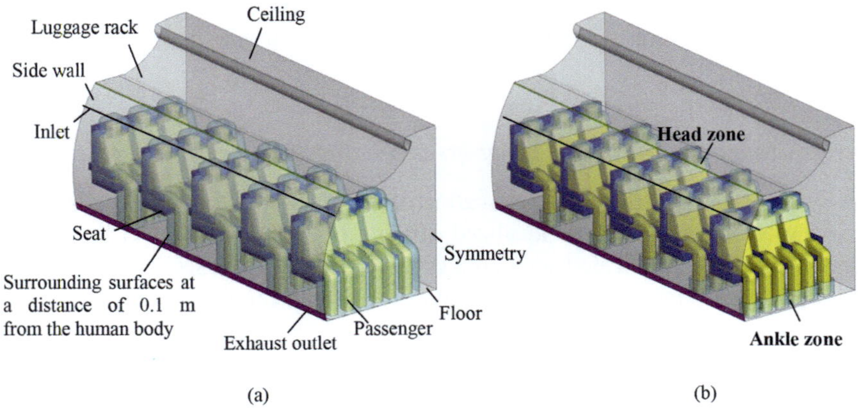

Figure 5.13 Half section of a five-row aircraft cabin for inverse design of the air-supply parameters: (a) overview of the cabin and the surrounding surfaces at a distance of 0.1 m from the bodies of the passengers, used to evaluate the comfort status; (b) head zone, used to evaluate draft risk and air quality, and ankle zone, used to evaluate draft risk

Eq. (5.32) applies only to the summer season, and the calculation of PMV is the same as that widely used in buildings. The cabin air quality was evaluated only in the head zone, and the area-averaged mean age of air (Sandberg, 1981) was restricted to 120 seconds.

5.3.3.2 Solution strategies

As in the previous two cases, the spans of the designed air-supply parameters were estimated first. It was assumed that the air-supply opening size varied between 1 cm and 3 cm, the air was supplied in a downward direction at an angle ranging from 0 to 45° with respect to the horizontal plane, and the air-supply temperature ranged from 22°C to 26°C. Obviously, the air-supply speed varied in accordance with the air-supply opening size and air-discharge direction. To establish the thermo-flow data pool, a total of 45 cases were resolved by full CFD simulations. These cases corresponded to discrete air-supply opening sizes of 1 cm, 1.5 cm, 2 cm, 2.5 cm, and 3 cm; air-supply directions of 0°, 20°, 40°; and air-supply temperatures of 22°C, 24°C, and 26°C. These cases were checked for their representativeness according to the procedure described in Section 5.2.3.

For resolution of the thermo-flows by means of full CFD simulations, the cabin geometry and suitable grid meshes were generated in GAMBIT. A total of 2,648,631 tetrahedral grid cells were created. The grid cell size in the core region was approximately 4 cm, and near the solid walls and occupant surfaces the grids were further refined into approximately 2.5 cm. The grids in proximity to the air-supply inlet were the finest, with the cell size reducing to approximately 0.1 cm. The above grids were checked having achieved grid independence. The numerical solutions to the N-S equations were performed by FLUENT software (version 6.3.26), in which all the settings were the same as those described in the office case. The convergence was deemed satisfactory when the unbalanced heat transfer rate for all the boundaries was less than 0.5% of the maximum heat gain, and there was no meaningful change in the temperature, velocity, and mean age of air with further numerical iterations at a number of typical monitoring points.

Table 5.2 Values of some parameters in PMV calculation

Item	*Value*	*Item*	*Value*
Metabolic rate	75 W/m^2	Clothing insulation	0.57 clo
Mechanical work production	0	Water vapor pressure	228 Pa

It is fortunate that the area-averaged air speed and PMV are explicit functions of the velocity and temperature distributions. Table 5.2 outlines some of the known parameters in the calculation of PMV. Once the velocity and temperature fields are known, the draft risk and comfort status can be evaluated instantly. However, the mean age of air is an implicit function of the velocity field, which must be solved as a passive scalar variable. Hence, the data pool of the mean age of air was also established here for POD analysis.

5.3.3.3 Solution results

This section discussed the implementation of POD analysis for velocity, temperature, and mean age of air in the evaluated regions. By means of a similar process to that in the previous two cases, the POD modes and their coefficients were extracted from the 45 full CFD cases. Next, 1,044 additional cases were constructed from the POD modes and the interpolated coefficients, which were obtained by means of a three-dimensional cubic Hermite interpolation. Among these cases, the changing step size between two consecutive cases was 0.2 cm for the air-supply opening size, 5° for the air-supply direction, and 0.4°C for the air-supply temperature.

Fig. 5.14 shows the computed environmental performance in the cabin for a total of 1,089 cases with different air-supply parameters. Figs. 5.14(a) and (b) present the area-averaged speed in the head and ankle zones, respectively. As the air-supply opening size decreased, the air-supply speed increased, both the speed in the entire cabin, and the averaged speeds in the head and ankle zones. A larger air-supply angle also caused the averaged speed in the head zone to increase slightly, but it did not greatly impact the averaged speeds in the ankle zone. The averaged speed for the whole ankle zone was less than 0.35 m/s, which implied that the design parameters did not cause discomfort to the ankles and feet. Fig. 5.14(c) presents the area-averaged |PMVc|. Because the cabin envelope was assumed to have a fixed surface temperature, which was relatively low as shown in Table 5.1, a warm air supply was needed to maintain a neutrally comfortable cabin. It is thus understandable that the $|PMVc|_{ave}$ decreased with the air-supply temperature in the specified span. In addition, a high air-supply speed combined with a low air temperature and a large angle would undoubtedly cause thermal discomfort. Fig. 5.14(d) shows the area-averaged mean age of air in the head zone. As the air-supply opening size decreased, the mean age of air also decreased. This trend shows that a reduction in the size of the air-supply opening can improve the air circulation in the cabin. Note that the cabin environmental performance parameters presented here have been averaged on the basis of the surface areas, so that a performance parameter might vary with the design variables in a complex manner. Obviously, a sophisticated design cannot be achieved without a large amount of data on the cabin environmental performance, which can be provided by the POD method.

On the basis of the computed environmental performance for the 1,089 cases, the best performance for each specific design target can be provided. The minimum $|PMVc|_{ave}$ was 0.09, attained when the air-supply opening size was 2.2 cm, the air-supply angle was 30°, and the

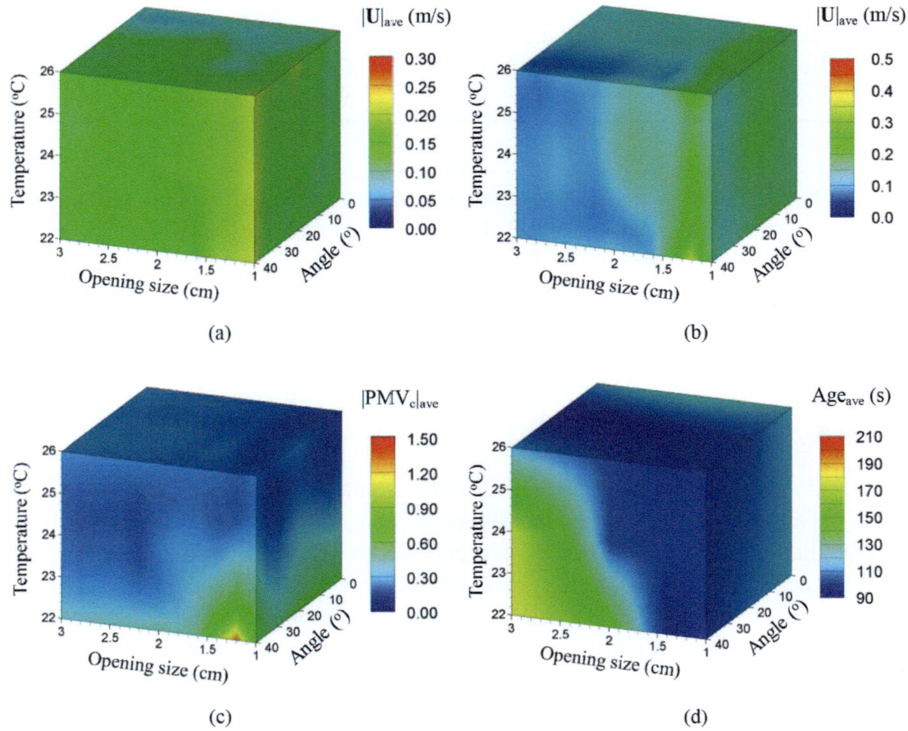

Figure 5.14 Computed cabin environmental performance versus air-supply parameters: (a) area-averaged speed in the head zone; (b) area-averaged speed in the ankle zone; (c) $|PMVc|_{ave}$ on the surrounding surfaces at a distance of 0.1 m from the passengers' bodies; (d) area-averaged mean age of air in the head zone

air-supply temperature was 24°C (Case 1). The minimum Age_{ave} was 93.9 s, attained when the air-supply opening size was 2.6 cm, the air-supply angle was 20°, and the air-supply temperature was 26°C (Case 2). To validate the accuracy of the above design parameters, Cases 1 and 2 were resolved by full CFD simulations. Table 5.3 compares the performance parameters provided by the POD method and CFD simulations. Overall, the results were in good agreement with one another. The POD method provided better agreement with the CFD in

Table 5.3 Comparison of cabin environmental performance parameters obtained by POD and CFD

Design performance parameter	*Case 1 (2.2 cm, 30°, 24°C)*		*Case 2 (2.6 cm, 20°, 26°C)*	
	POD	*CFD*	*POD*	*CFD*
$\|PMVc\|_{ave}$	0.09	0.18	0.27	0.26
$\|\mathbf{U}_{head}\|_{ave}$	0.08 m/s	0.13 m/s	0.13 m/s	0.14 m/s
$\|\mathbf{U}_{foot}\|_{ave}$	0.07 m/s	0.10 m/s	0.12 m/s	0.11 m/s
Age_{ave}	114.7 s	115.6 s	93.9 s	92.6 s

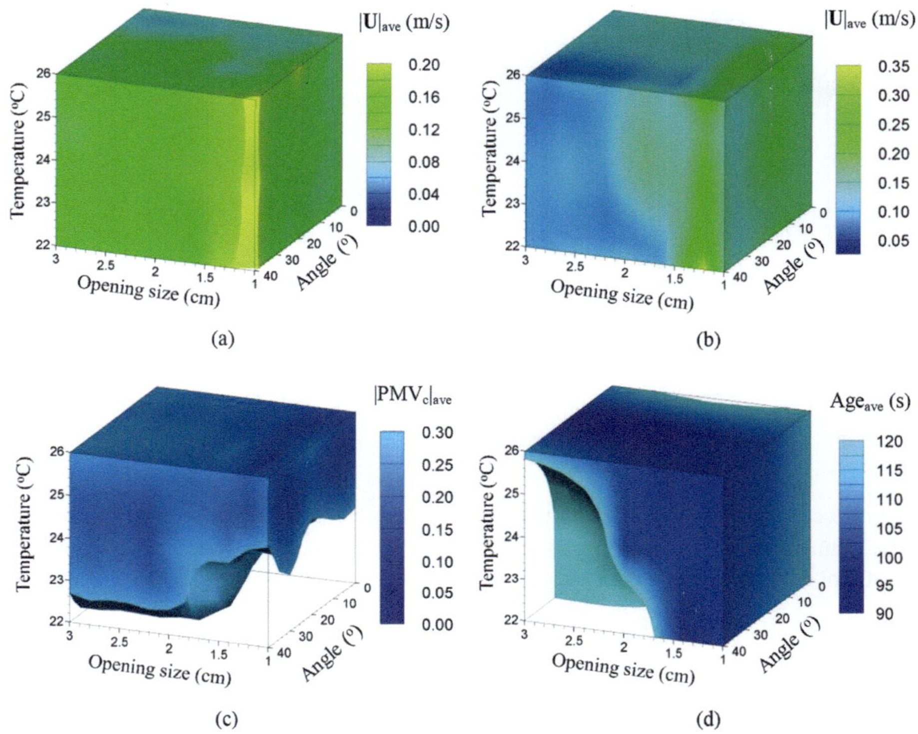

Figure 5.15 Air-supply parameters that satisfy each design target: (a) area-averaged air speed, less than 0.2 m/s in the head zone; (b) area-averaged air speed, less than 0.35 m/s in the ankle zone; (c) area-averaged absolute value of cabin predicted mean vote (|PMVc|), less than 0.3 in the region around the passengers; (d) area-averaged mean age of air, less than 120 s in the head zone

Case 2, which was due to the coincidence of the air-supply parameters closer to the snapshot cases. Nevertheless, the POD method is able to determine the optimal air-supply parameters corresponding to each specific evaluation criterion.

Fig. 5.15 shows the air-supply parameters that satisfy each design target and the corresponding cabin environmental performance. In terms of the air speeds in the head and ankle zones, nearly all the air-supply parameters in the varying range are acceptable. Only with small air-supply opening sizes and large air-supply angles are the area-averaged air speeds greater than 0.2 m/s in the head zone, as shown in Fig. 5.15(a), which has been removed from the solution domain and thus has no color in the figure. For evaluation of performance in terms of $|PMVc|_{ave}$, as shown in Fig. 5.15(c), the bottom right part of the domain should be removed, because the air-supply parameters in this region resulted in discomfort caused by chilling. From the perspective of cabin air quality, large air-supply opening sizes combined with large angles were also excluded, as indicated by Fig. 5.15(d), since weak airflow reduced the air circulation.

The air-supply parameters that satisfy all the design targets can be obtained by combining the results presented in Fig. 5.15 in terms of specific target performance parameters. Fig. 5.16 shows the ranges of air-supply parameters within which any parametric set would

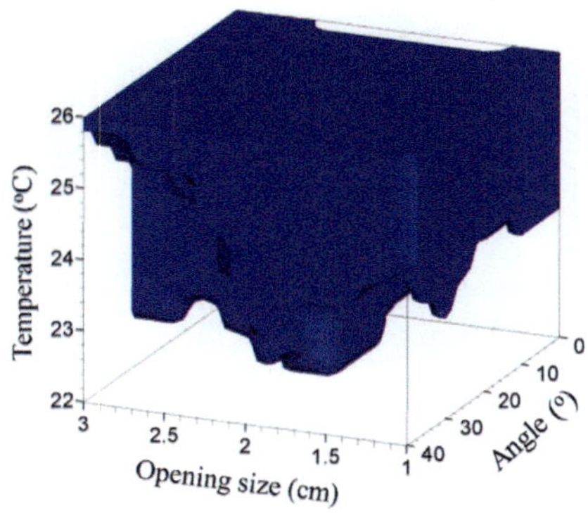

Figure 5.16 Air-supply parameters that satisfy multiple design targets for draft risk, thermal comfort, and cabin air quality

result in acceptable comfort and air quality. The parameters in the lower outer part of the parametric domain of Fig. 5.16 have been removed because they violated one or more of the target performance thresholds that were set in the design.

5.3.3.4 Discussion

In Table 5.3, it can be clearly seen that a minimum value of $|PMVc|_{ave}$ might not correspond to a minimum value of Age_{ave}. For example, when the minimum $|PMVc|_{ave}$ was achieved in Case 1, the Age_{ave} was much larger than that in Case 2, and vice versa. This is because $|PMVc|_{ave}$ and Age_{ave} are not subject to the same physical parameters. More specifically, $|PMVc|_{ave}$ is directly affected by the air-supply opening size, air-discharge direction, and air temperature, whereas Age_{ave} is more directly influenced by the flow. When a design process involves multiple objectives, each objective may correspond to a different solution set of design parameters. The intersection among different design objectives may lead to no solution, a single solution, or multiple solutions. The compatibility of different design objectives awaits further exploration.

Because the coefficients of the POD modes varied with the three design variables in this case, a three-dimensional interpolation was adopted. In such an interpolation, the coefficients of eight adjacent knots were used to construct a new one. More careful selection of representative snapshots is needed in a multiple-parameter design process.

Approximately 450 h was required to resolve the original 45 cases by means of full CFD simulations, and 3 h was spent in POD analysis and data construction for the additional 1,044 cases in the same computing facility. It can be seen that the proposed method is still very attractive for multiple-parameter design.

5.3.3.5 Summary

In this section, the POD method was used to determine the air-supply opening size, air-supply direction, and air-supply temperature in a five-row aircraft cabin. Multiple design targets were set to minimize the draft risk, ensure thermal comfort, and guarantee cabin air quality. The results reported here have led to the following findings:

- The POD-based method can be used effectively in multiple-parameter design to achieve the multiple objectives. The high computing efficiency of the method makes it quite attractive for the design of enclosed environments.
- The CFD-based POD design model is able to provide ranges of air-supply opening sizes, directions, and temperatures that satisfy multiple design targets. In addition, the air-supply parameters that correspond to the best performance for each specific design target can be obtained.

5.4 Conclusions

This chapter introduced the basic principles for extracting POD modes and some techniques for interpolating the coefficients of the POD modes in order to construct non-snapshot thermo-flow fields. The POD method was tested in an empty two-dimensional air cavity, an idealized three-dimensional office, and a three-dimensional aircraft cabin. The number of design variables can be varied, and the number of design targets ranges from one to multiple objectives. The following conclusions can be drawn:

- The extracted POD modes and the interpolated coefficients of the POD modes can be used to construct thermo-flow fields that are in good agreement with experimental data. The POD-based design method is effective for improving the design of an indoor environment, whether a single-objective design or a multiple-objective design.
- The POD-based method is able to provide the solution domains that satisfy a single design target or several design targets. It can also provide a specific solution that leads to the best performance for each individual design target.

The POD-based design process requires a relatively small amount of computing resources while achieving reasonably good accuracy. Most of the computing time for the POD method is devoted to preparation of the snapshot data by CFD, whereas the POD analysis, interpolation of the POD mode coefficients, and circumscription of the design solutions are more efficient.

Note

1 This chapter is partially reproduced from the paper entitled Prompt design of the air-supply opening size for a commercial airplane based on the proper orthogonal decomposition of flows, *Building and Environment*, 96: 131–141. The reprint is with permission from Elsevier.

References

Ahuja S, Cliff E, Narayanan S. Reduced-order modeling for control of indoor building airflows. *In: Proceedings of the 5th National Conference of IBPSA-USA (SIMBUILD 2012)*. 2012;1–8.

Allery C, Beghein C, Hamdouni A. Applying proper orthogonal decomposition to the computation of particle dispersion in a two-dimensional ventilated cavity. *Communications in Nonlinear Science and Numerical Simulation*. 2005;10(8):907–920.

ASHRAE. ASHRAE Standard 161-2007. *Air quality within commercial aircraft*. Atlanta, GA: ASHRAE Inc. 2007.

ASHRAE. ASHRAE Standard 55-2010. *Thermal Environmental Conditions for Human Occupancy*. Atlanta, GA: ASHRAE Inc. 2010.

ASHRAE. *ASHRAE Handbook – Fundamentals*. Atlanta, GA: ASHRAE Inc. 2013.

Bakewell HP, Lumley JL. Viscous sublayer and adjacent wall region in turbulent pipe flow. *Physics of Fluids*. 1967;10(9):1880–1889.

Blay D, Mergui S, Niculae C. Confined turbulent mixed convection in the presence of a horizontal buoyant wall jet. Fundamentals of mixed convection. *ASME HTD*. 1992;213:65–72.

Chen Q. Comparison of different k-ε model for indoor air flow computations. *Numerical Heat Transfer (Part B: Fundamentals)*. 1995;28(3):353–369.

Cui W, Zhu Y. 2014. *Evaluation of thermal comfort in aircraft cabins*. Private communication.

Ding P, Wu X, He Y, Tao W. A fast and efficient method for predicting fluid flow and heat transfer problems. *Journal of Heat Transfer*. 2008;130(3):1–17.

Elhadidi B, Khalifa HE. Application of proper orthogonal decomposition to indoor airflows. *ASHRAE Transactions*. 2005;111(1):625–634.

Fanger PO. *Thermal comfort: analysis and applications in environmental engineering*. Copenhagen: Danish Technical Press. 1970.

Fritsch FN, Carlson RE. Monotone piecewise cubic interpolation. *Siam Journal on Numerical Analysis*. 1980;17(2):238–246.

Glauser MN, George WK. Orthogonal decomposition of the axisymmetric jet mixing layer including azimuthal dependence. *In: Proceedings of the First European Turbulence Conference (Advances in Turbulence)*. 1987;357–366.

Herzog S. *The large scale structure in the near-wall region of turbulent pipe flow*. Ph. D. Thesis, Cornell University. 1986.

Holmes P, Lumley JL. Berkooz G. *Turbulence, coherent structures, dynamical systems and symmetry (Second Edition)*. Cambridge: Cambridge University Press. 2012.

Kahaner D, Moler C, Nash S. *Numerical methods and software*. New Jersey: Prentice Hall. 1989.

Li K, Su H, Chu J, Xu C. A fast-POD model for simulation and control of indoor thermal environment of buildings. *Building and Environment*. 2013b;60:150–157.

Li K, Xue W, Chao X, Su H. Optimization of ventilation system operation in office environment using POD model reduction and genetic algorithm. *Energy and Buildings*. 2013a;67:34–43.

Lumley JL. The structure of inhomogeneous turbulent flows. *In: Proceedings of Atmospheric Turbulence and Radio Wave Propagation*. 1967;166–178.

MATLAB. *MATLAB user's guide*. Natick, Massachusetts, USA: Mathworks Inc. 2010.

Sandberg M. What is ventilation efficiency? *Building and Environment*. 1981;16(2):123–135.

Sempey A, Inard C, Ghiaus C, Allery C. Fast simulation of temperature distribution in air conditioned rooms by using proper orthogonal decomposition. *Building and Environment*. 2009;44(2):280–289.

Sirovich L. Turbulence and the dynamics of coherent structures: I–III. *Quarterly of Applied Mathematics*. 1987;45(4):561–590.

Smith TR, Moehlis J, Holmes P. Low-dimensional modelling of turbulence using the proper orthogonal decomposition: A tutorial. *Nonlinear Dynamics*. 2005;41(1–3):275–307.

Tallet A, Allard F, Allery C. Numerical simulation of real-time air flow control by POD/ROM applied to an isothermal ventilated cavity. *In: Proceedings of the 10th International Conference on Healthy Buildings*. 2012;1960–1965.

Wei Y, Zhang T, Wang S. Prompt design of the air-supply opening size for a commercial airplane based on the proper orthogonal decomposition of flows. *Building and Environment*. 2015;96:131–141.

Yakhot V, Orzag S, Thangam S, Gatski T. Development of turbulence models for shear flows by a double expansion technique. *Physics of Fluids A: Fluid Dynamics*. 1992;4(7):1510–1520.

Yuan X, Chen Q, Glicksman LR. Measurement and computations of room airflow with displacement ventilation. *ASHRAE Transactions*. 1999;105(1):340–352.

Zhang T, Zhou H, Wang S. An adjustment to the standard temperature wall function for CFD modeling of indoor convective heat transfer. *Building and Environment*. 2013;68:159–169.

Zhang Z, Zhang W, Zhai Z, Chen Q. Evaluation of various turbulence models in predicting airflow and turbulence in enclosed environments by CFD: Part-2: Comparison with experimental data from literature. *HVAC&R Research*. 2007;13(6):871–886.

6 Enhancement of the inverse design methods[1,2]

The four inverse design methods introduced in the previous chapters have different advantages and disadvantages. For example, the CFD-based inverse design methods require large numbers of time-consuming CFD simulations; the CFD-based adjoint method converges faster than the CFD-based GA method, but it can identify only the local optima; and the POD method and CFD-based ANN method can be much faster than the other two methods, but their computational accuracy is not very high. To compensate for the existing shortcomings, this chapter introduces the integration of the GA method and adjoint/ANN method. Fast fluid dynamics (FFD) and a new numerical algorithm, SLPISO, have also been evaluated for their potential to reduce the computing time of CFD simulations. These efforts improve the applicability of inverse design methods to real engineering design problems.

6.1 Introduction

Among the four inverse design methods introduced in the previous chapters, which one is the most appropriate for the inverse design of an indoor environment? A method with fast computational speed and high accuracy is preferred. POD analysis and the CFD-based ANN method have much faster speeds than the other two inverse design methods. This is because POD analysis and the CFD-based ANN method work in an offline-online mode. The offline stage is computationally demanding in these two methods because of the large numbers of CFD simulations required for obtaining the spatial modes or training the ANN, respectively. In the online stage, however, both methods conduct algebraic calculations that require very little computing time. Unfortunately, the computational accuracy is not very good. The accuracy of the POD method is greatly affected by the representativeness of the spatial modes and the strategies used to interpolate the coefficients of these modes, while the accuracy of the ANN method is affected by the training samples and training parameters. In contrast, the CFD-based genetic algorithm method and CFD-based adjoint method always solve the Navier-Stokes equations directly with the use of a suitable turbulence model, and thus the accuracy of these methods is very high. The GA method is a gradient-free approach that emphasizes probability, and the adjoint method is a gradient-based approach that makes use of the gradient of the design objectives over the design variables. The adjoint method converges much faster than the GA method in a single design. However, the adjoint method can identify only the local optima, and multiple trials are required for identification of the global optima, which is a time consuming process. Therefore, we need to either improve the computational accuracy of POD analysis and the CFD-based ANN method, or accelerate the computational speed of the CFD-based GA method and CFD-based adjoint method. Here, we have taken the latter approach. This chapter describes our effort to accelerate CFD simulations by integration of the GA method and adjoint method, integration of the GA method and ANN method, and the use of faster numerical algorithms.

6.2 Integration of genetic algorithm and adjoint method

The goal of integrating the genetic algorithm method and adjoint method is to incorporate the advantages of both methods while compensating for their shortcomings. Theoretically, the GA method is able to identify the global optima, but it is more computationally demanding; the adjoint method can identify only the local optima, but it is more efficient. However, few studies have actually compared the two methods in terms of their accuracy and computing efficiency. It is also unknown to what extent the inverse design methods are more efficient and accurate than the conventional "trial-and-error" process.

As stated above, the CFD-based adjoint method is able to identify only the local optima. It requires multiple trials to obtain globally optimal solutions, which is a time-consuming process. To improve the computational efficiency, one may first take advantage of the CFD-based GA method to determine the initial air supply parameters, in which case there may be multiple solutions. Next, these pre-determined initial parameters can be fine-tuned by means of the CFD-based adjoint method. This external integration of the CFD-based GA and adjoint methods can be faster than the pure CFD-based GA and can provide the global optima, thus compensating for the shortcomings of the pure CFD-based adjoint method.

The CFD-based genetic algorithm method is able to identify the global optima, but its computational demand increases rapidly as the number of design variables increases. To improve the computational efficiency, one uses the adjoint method to obtain the gradients of the design objective with respect to the design variables, which will then be used to determine coupled-crossover individuals, crossover-point locations, and mutation directions. This internal integration of the CFD-based GA method and adjoint method ensures that the generation evolves towards global optima, and also increases the convergence speed.

In this section, we first compare the accuracy and efficiency of the CFD-based GA method, adjoint method, and trial-and-error method in the same inverse identification case. The external and internal integration of the GA method and adjoint method are then discussed.

6.2.1 Comparison of genetic algorithm and adjoint method

As a basis for comparison with the adjoint and genetic algorithm methods, this study used the conventional trial-and-error method. The trial-and-error method employs different sets of design variables ξ to obtain the objective function and selects the best set for the final design. We divided the presumed range for each variable into equal segments in order to generate different sets of ξ. For instance, if a design variable could vary between 0 and 1.0, the values used for the trial-and-error process were 0.25 and 0.75 under a resolution of 2. The CFD-based trial-and-error method for designing an indoor environment involves the following steps:

(1) Generate values for the design variables under a certain resolution;
(2) Obtain a set of $O(\xi)$ under this sampling resolution by means of CFD simulations;
(3) Check whether any of the $O(\xi)$ meet the desired objective. If yes, select the one with the best fit to the design objective as the final design. If not, generate new values of the design variables under a higher resolution;
(4) Repeat steps (2) and (3) until an acceptable design is found.

This study evaluated the ability of the three design methods to identify the air supply location (y_{in}, the y coordinate of the center point of the inlet) and other parameters ($\mathbf{U}_{in}$, T_{in}) ($\xi = (y_{in}, \mathbf{U}_{in}, T_{in})$) for a two-dimensional, non-isothermal case, by comparing the results with experimental data from Blay et al. (1992). In the original experiment, as shown in Fig. 6.1(a), the inlet location was at $x_{in} = 0$ and $y_{in}/H = 0.991$, the inlet height (h_{in}) was 18 mm, the inlet air velocity components were $U_{in,x} = 0.57$ m/s and $U_{in,y} = 0.0$ m/s, the inlet air temperature (T_{in}) was 15°C, and the outlet height was 24 mm. The temperature of the walls (T_{wall}) was 15°C, and that of the floor (T_{floor}) was 35.5°C. The case contained flow characteristics that are representative for an indoor environment, such as jets and thermal plumes. The experiment measured the air velocity and temperature at the locations indicated by the red dashed lines in Fig. 6.1(a). We set the measured air velocity and temperature as the design objectives. Then the objective function for this case was:

$$O(\xi) = \int_{\Omega} \left[(U_x^* - U_{\exp,x}^*)^2 + (U_y^* - U_{\exp,y}^*)^2 + (T^* - T_{\exp}^*)^2 \right] d\Omega \tag{6.1}$$

where * represents a value that has been normalized by the experimental data. This design objective was clearly attainable because it used measured data from the experiment with the corresponding boundary conditions. Fig. 6.1(b) shows the resolution of the mesh when the inlet was located at $x_{in} = 0$ and $y_{in}/H = 0.991$. The mesh resolution at the inlet, outlet, and near wall region was finer because of the high gradients of velocity and temperature in these locations. This principle was applied during the design process to regenerate the mesh whenever the air supply location changed.

The design process required the initialization of the design variables ξ, and this study assumed a range of ξ. We also assumed that the left wall in Fig. 6.1(a) was a possible location for the inlet. Furthermore, we assumed that the magnitude of the air-supply velocity was between 0 and 1 m/s and that its direction was $\alpha \in [-45°, +45°]$ (where the air velocity direction at 0° was normal to the inlet face). The range air supply temperature ranged from 10 to 20°C.

For the adjoint method, the initialized air supply conditions were $y_{in}/H = 0.5$, $|\mathbf{U}_{in}| = 0.5$ m/s, $\alpha = 0°$, and $T_{in} = 12.5$°C. For the GA method, the initialized ξ in the first generation of eight individuals were generated stochastically within the specified ranges. As a basis for comparison, this investigation used the trial-and-error method with prescribed design variables as shown in Table 6.1.

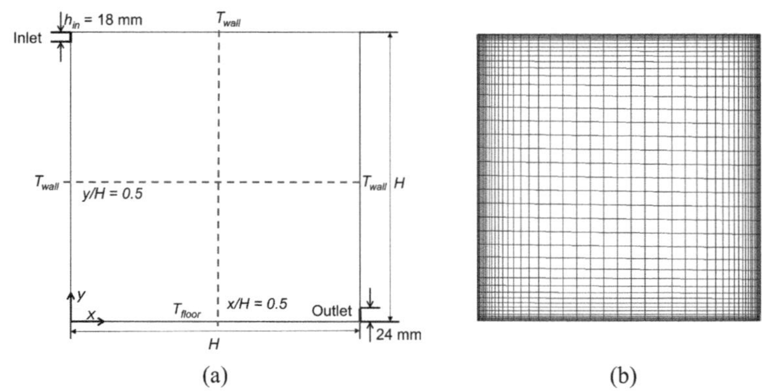

Figure 6.1 (a) Sketch and (b) mesh for the 2D non-isothermal cavity

Table 6.1 Prescribed ξ for the trial-and-error method

Resolution	*Inlet location* (y_{in})	$\lvert U_{in} \rvert$ *(m/s)*	*α (°)*	T_{in} *(°C)*
2×2×2×2	0.26; 0.78	0.25; 0.75	−22.5; 22.5	12.5; 17.5
3×3×3×3	0.173; 0.52; 0.867	0.16; 0.5; 0.83	−30; 0; 30	11.6; 15; 18.3
4×4×4×4	0.13; 0.39; 0.65; 0.91	0.125; 0.375; 0.625; 0.875	−33.75; −11.25; 11.25; 33.75	11.25; 3.75; 16.25; 18.75

Fig. 6.2 shows the objective function versus the number of cases calculated for each method. The objective function for the adjoint method decreases very quickly, and it was possible to obtain the local optima within five design cycles (the calculation stopped when $|O_{new} - O_{old}| < 0.01$), which means that a total of ten cases were calculated. This is because the adjoint method calculated both the Navier-Stokes equations and adjoint equations in one design cycle, and the time required to solve the adjoint equations was almost the same as that required to solve the Navier-Stokes equations.

As shown in Fig. 6.2, the GA method led to the smallest objective function. This result is reasonable because the GA method has a greater chance of identifying the global optima. However, both the GA and trial-and-error approaches required the calculation of about 200 cases in order to obtain converged solutions, and the computing effort was about 20 times higher than that of the adjoint method. For the GA method, we continued the calculation after convergence was reached. A smaller objective function was obtained after more than 600 cases. This result indicates that the GA method cannot guarantee the identification of global optima. The optimal solution according to the adjoint method was $|\mathbf{U}_{in}| = 0.561$ m/s, $\alpha = -2°$, $T_{in} = 286.5$ K, and $y_{in} = 1.031$ m; according to the GA method, it was $|\mathbf{U}_{in}| = 0.91$ m/s, $\alpha = 45°$, $T_{in} = 287.4$ K, and $y_{in} = 0.977$ m.

For the trial-and-error method, the computing effort increased dramatically with the use of the refined sampling resolution for the design variables. However, the higher resolution did not necessarily improve the design results, as shown in the figure. Furthermore, the method did not lead to an objective function that was as small as that obtained by the adjoint or GA method.

6.2.2 External integration of genetic algorithm and adjoint method

The results in the previous section confirm that the adjoint method is fast, although it can only identify the local optima. The convergence speed depends on the initial values of the design variables. The convergence speed of the GA method is slow, and the method requires a large number of CFD calculations, but it can provide better optimal solutions. To increase the design speed, this study proposes the use of the GA method for initial design and the adjoint method for fine tuning. Integration of the two methods could provide better initial design variables for the adjoint method, while decreasing the computing effort by the GA method.

We tested the integrated method for the inverse design of the thermal environment in a single-aisle, fully occupied aircraft cabin, where the goal was the optimal air supply conditions for the diffusers. To decrease the computing load, we performed calculations for only half of one row of the cabin, as illustrated in Fig. 6.3. The design domain Ω was set as the area 0.1 m away from the passengers. To avoid longitudinal flow, we fixed the U_z as 0 at the inlet. The airflow rate was constrained at 9.4 L/s per passenger according to ANSI/ASHRAE Standard 161 (2007), and we assumed that the inlet air temperature ranged from

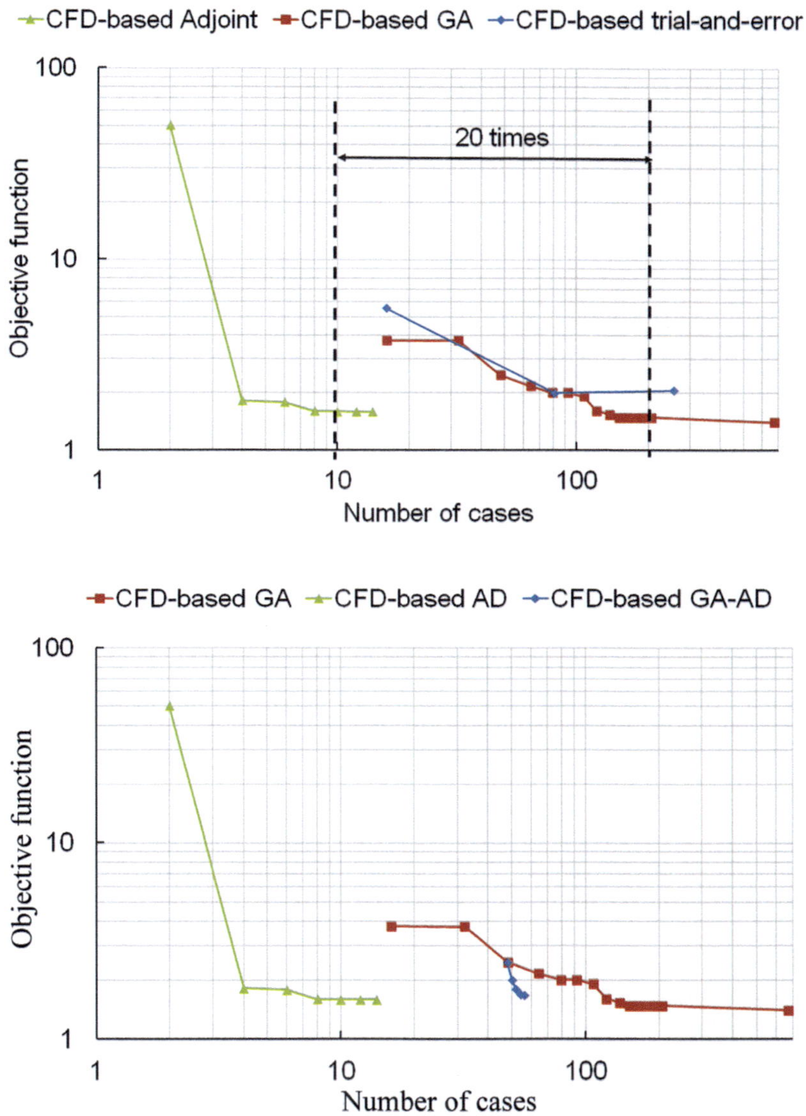

Figure 6.2 Objective function vs. number of cases for different design methods

16°C to 26°C. The inlet location y_{in} was in the range of 0–y_{mix} (0.409 m) for mixed ventilation. The lower bound was 10 cm higher than the passenger's head to prevent drafts. The angle between the air supply direction and the direction normal to the inlet surface was maintained at less than 45°. Table 6.2 lists the thermal boundary conditions that did not vary during the optimization process, according to the measured data from Zhang and Chen (2007).

To evaluate the thermal comfort level in the aircraft cabin, this study used a modified predicted mean vote for air cabins (PMVc), developed by Cui and Zhu (2015), that accounts for low pressure during flight. The design objective was to achieve a thermally comfortable

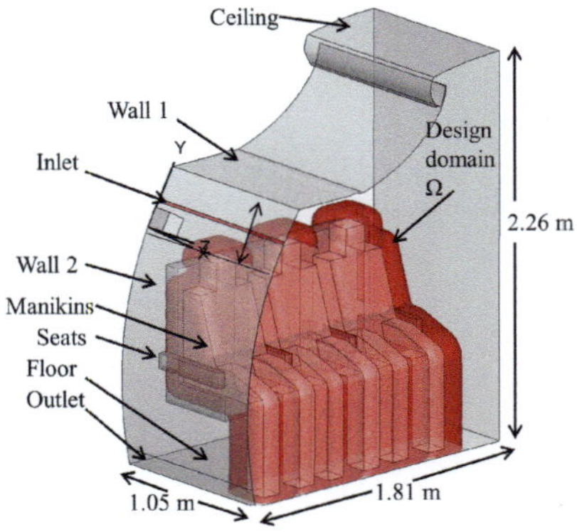

Figure 6.3 Schematic of the flow domain and design domain for the aircraft cabin

Table 6.2 Thermal boundary conditions that were fixed during the optimization process

Surface	*T*	*Surface*	*T*	*Surface*	*T*	*Surface*	*T*
Wall 1	21°C	Outlet	Zero gradient	Ceiling	22°C	Floor	23°C
Wall 2	22°C	Seats	Adiabatic	Passengers	30.3°C		

cabin environment, where PMVc should approach zero. Thus, the corresponding objective function was:

$$O(\xi) = \int_{\Omega} (PMV_c)^2 d\Omega \, / \int_{\Omega} d\Omega \tag{6.2}$$

Fig. 6.4 shows the absolute averaged PMVc versus the number of cases for the different methods. The design variables for the best individual in the second generation of the GA method were used as initial variables for the adjoint method. The adjoint method converged quickly, but improved the thermal comfort level by only 5%. A possible reason for this minimal improvement was that the best |PMVc| in the second generation of the GA method was already very small at 0.0968. However, the GA method was able to provide good initial design variables for the adjoint method. The GA method converged within 46 RANS (Reynolds-averaged Navier–Stokes) simulations, and the improvement in the thermal comfort level was only slightly better than that with the adjoint method. Therefore, the integrated method cannot guarantee the identification of global optima unless appropriate timing for introduction of the adjoint method can be identified, which requires further investigation.

Figs. 6.5(a) and (b) show the PMVc distributions in the cabin for the worst and best individuals, respectively, in the first generation of the GA method. Although the thermal comfort level for the best individual in the first generation was very good, the integrated GA and adjoint methods can further improve the design if the number of generations is increased in the GA method as shown in Fig. 6.5(c), or if the GA method is integrated with the adjoint method as shown in Fig. 6.5(d). However, the improvements were not significant.

6.2.3 Internal integration of genetic algorithm and adjoint method

Indoor environmental design and optimization is a multi-variable, multi-solution problem. The genetic algorithm method performs well in identifying global optima. Therefore, the

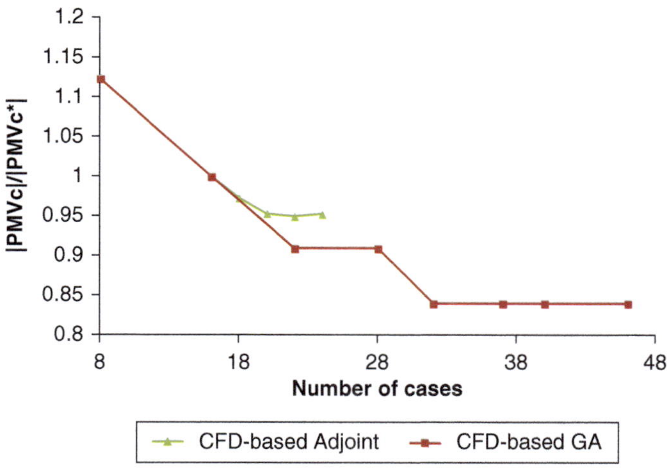

Figure 6.4 Normalized absolute averaged PMVc vs. the number of cases for different methods ($|PMVc^*| = 0.0968$ was the best individual in the second generation of the GA method)

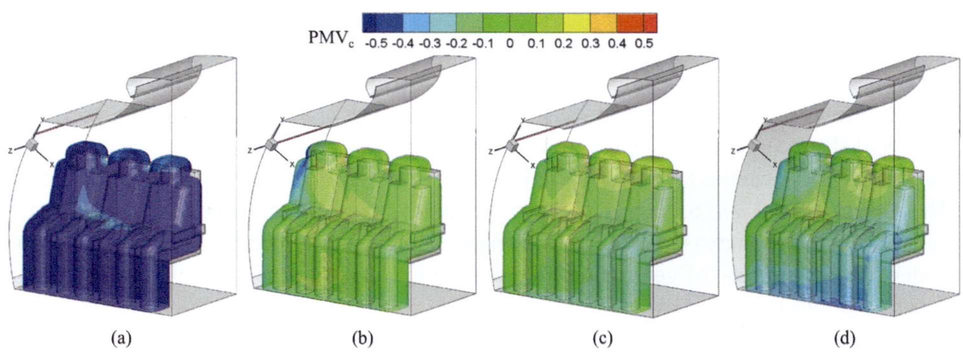

Figure 6.5 PMVc distributions for (a) the worst individual in the 1st generation of the GA method, (b) the best individual in the 1st generation of the GA method, (c) the optimal design by the GA method, and (d) the optimal design by the adjoint method

basis of the integrated method is the GA method, and the adjoint method is incorporated to reduce the calculation effort. The internal integration of the GA method and adjoint method (ADGA) makes use of the latter method to obtain the gradients of the objective function with respect to design variables as shown in Fig. 6.6. Each gradient is then used to determine coupled-crossover individuals, crossover-point locations, and mutation directions, which ensures that the generation evolves towards the global optima and increases the speed of convergence.

This study used a 2D mixing convection case (Blay et al., 1992), as shown in Fig. 6.1(a), to evaluate the performance of the integrated method. The measured air velocity and temperature along the horizontal red dashed line in Fig. 6.1(a) were used as the design objective. We again used Eq. (6.1) as the objective function. The goal of this inverse design process

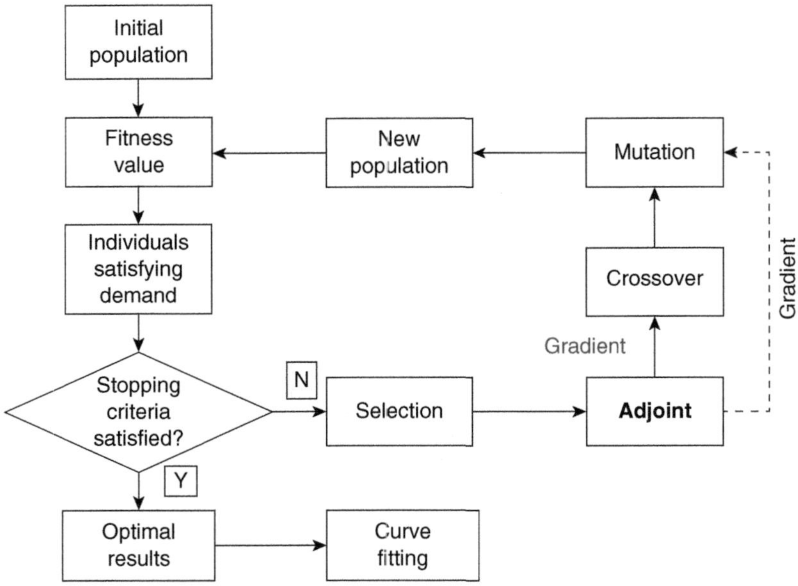

Figure 6.6 Flowchart of integrated GA method and adjoint method

was to identify the following air supply parameters: inlet air velocity magnitude (V, range: 0–1 m/s, accuracy: 0.01 m/s), inlet air temperature (T, range: 278.15–298.15 K, accuracy: 1 K), inlet flow direction (a, range: −20°–20°, accuracy: 5°), and inlet effective area ratio (A, range: 0–1, accuracy: 0.2). We conducted the inverse identification by means of the GA method and ADGA method with different numbers of design variables.

The GA lines in Figs. 6.7 and 6.8 show that as the number of variables increased, the computational effort increased rapidly. The improvement introduced by the ADGA method is evident in the number of generations required (Fig. 6.7). In addition, this method had a lower growth rate than the GA. Because of double-counting of cases in the adjoint method, the ADGA method required a larger number of calculations than the GA method alone, but the advantage of the integrated method is evident as the number of variables increases. Thus, the ADGA method is appropriate for problems with large numbers of variables and large computational demands. If there are fewer than three variables, the pure GA or adjoint method may be more suitable.

6.3 Integration of genetic algorithm method and artificial neural network

The use of the genetic algorithm (GA) method in inverse design entails a large computational cost. Surrogate models are among the most promising approaches to reducing this cost. A surrogate model mimics the behavior of the original model, producing the same responses at reduced computational cost (Nguyen, et al., 2014). An artificial neural network (ANN) is a typical surrogate model and can be used in the inverse design process. Coupling the GA method with an ANN allows one to make full use of the ANN's ability to learn complex nonlinear mapping and the GA's ability to find the global optimum in a bounded parametric search space (Chow, et al., 2002; Dounis and Caraiscos, 2009). Fig. 6.9 illustrates the trends

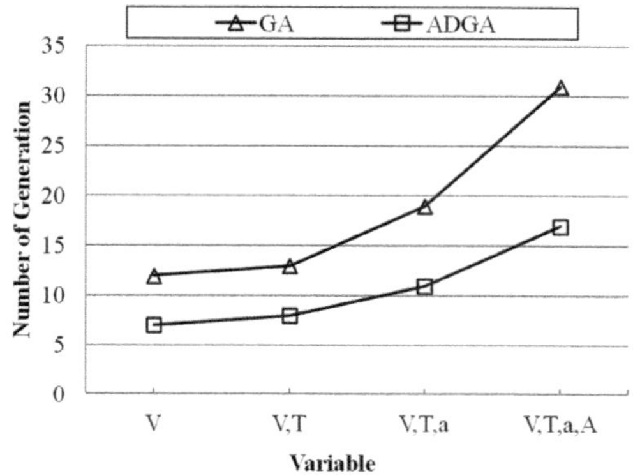

Figure 6.7 Comparison of the number of generations for the GA and ADGA methods with different numbers of variables

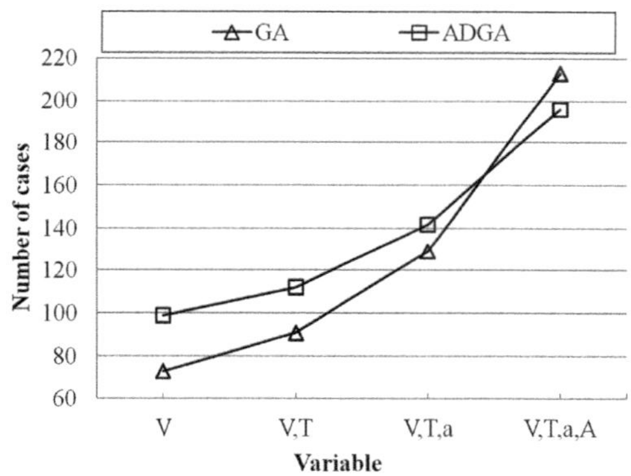

Figure 6.8 Comparison of the number of cases calculated for the GA and ADGA methods for different numbers of variables

in the use of the GA method and ANNs in building environment research in the last two decades. The literature search was performed on Scopus, the world's largest abstract and citation database of peer-reviewed literature. As shown in the figure, the number of publications on GA-ANN coupling has increased in recent years, along with an increase in the number that address the GA method and ANNs separately. This chapter introduces the combination of the GA method with an artificial neural network in the inverse design of an indoor environment.

6.3.1 Classification of combination modes

Two methods have been proposed for integrating an ANN with a genetic algorithm for the inverse design of an indoor environment: an ANN-based method and a GA-based method. In the ANN-based method, an artificial neural network is used as the engine of the inverse design process, and a genetic algorithm is used to optimize the structure of the ANN with the purpose of improving the prediction accuracy. Ferreira et al. (2012) used a multi-objective genetic algorithm to train an ANN so as to minimize energy consumption and maintain thermal comfort. The estimated energy saving was greater than 50%. Zhang et al. (2013) used a

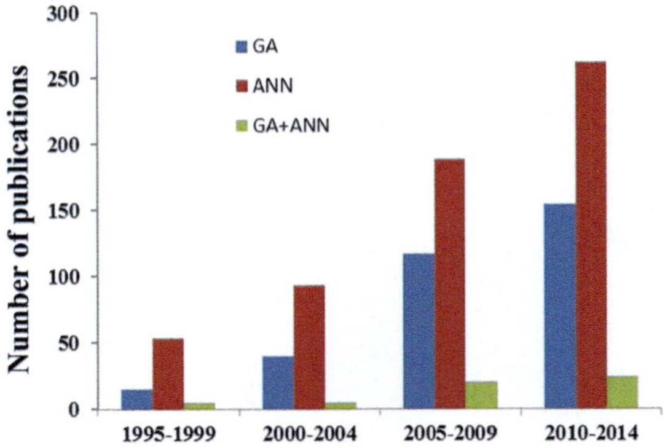

Figure 6.9 Trends in the use of the GA method and ANNs in building environment research

genetic algorithm to train an ANN to reconstruct the temperature field from the temperature data of discrete observation positions. Boithias et al. (2012) used a GA to identify the optimal parameters of ANN. Then the ANN was able to predict energy consumption and indoor discomfort. The validation process for the optimized ANN demonstrated that it had good prediction accuracy. However, our results in Chapter 4 have shown that there are many limitations in the application of the ANN-based method to the inverse design of an indoor environment.

In the GA-based method, a genetic algorithm is used to search for solutions to an inverse design problem, and an ANN is used to obtain values for the design objectives with the purpose of reducing the computational requirements (Gossard et al., 2013). The flow chart for this method is shown in Fig. 6.10. The training data for the ANN are obtained from CFD simulations. The inputs and outputs of the ANN are the design variables and design objectives, respectively. This integration mode has been widely used in the inverse design of indoor environments to reduce the computational cost of the CFD-based GA method. The present chapter discusses the feasibility of the GA-based method.

Nassif (2014) used a genetic algorithm and an artificial neural network to reduce energy consumption for HVAC systems. The ANN was used to predict the energy cost, and the relative prediction error was less than 10%. The test results showed that the models successfully captured the system performance, and the optimization process was able to reduce cooling

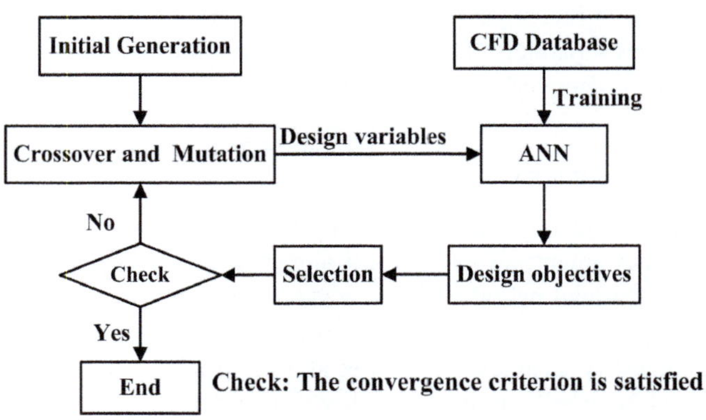

Figure 6.10 Flow chart of inverse design method based on GA-ANN integration

energy consumption by about 11% in comparison with traditional operating strategies. Yu et al. (2015) used an improved multi-objective GA and ANN to design energy consumption and indoor thermal comfort on the basis of simulation data. The relative prediction error of the ANN was less than 2%. The authors used the multi-objective approach to determine dozens of potential designs for a typical building in China, with a wide range of trade-offs between thermal comfort and energy consumption. Zhou and Haghighat (2009a, 2009b) used an ANN and a GA to enhance indoor thermal comfort and air quality without increasing the ventilation cost. The relative prediction error of the ANN was less than 6%. According to the design results, the performance of the ventilation system was efficiently enhanced. The computational cost was reduced from 10 years to 15 weeks when the ANN was used in place of CFD. Magnier and Haghighat (2010) integrated an ANN and a GA to design a building environment. The relative prediction error of the ANN was less than 4%, and the researchers obtained a set of solutions with a trade-off of different objectives. Energy consumption was significantly reduced, while thermal comfort improved. Furthermore, the use of the ANN instead of CFD reduced the computational cost from 10 years to 3 weeks. Zhang and You (2014a) coupled an ANN with a genetic algorithm to design an aircraft cabin environment and found several groups of optimal air supply parameters with different trade-offs between thermal comfort and energy consumption. The computational cost was reduced by 97% when the ANN was used in place of CFD. Song et al. (2012) coupled an ANN with a GA to design the thermal environment and energy consumption in a building. They used CFD to generate a database for training and cross validation of a primary ANN. Good agreement was achieved between the CFD- and ANN-based predictions. The total simulation time was significantly less than that required by a fully CFD-based optimization method.

A disadvantage of the use of an ANN instead of a CFD model is that it increases the uncertainty and risk of cumulative error, with the existence of prediction error. To solve this problem, Zhang and You (2014b) used both an ANN and CFD to calculate the design objectives for new individuals, and the design results were obtained by CFD. To identify a suitable scale for the samples used to train an artificially neural network, Zhang and You (2015) have proposed an inverse design method with a self-updating ANN. In this method, the CFD results obtained in the design process are added to the CFD database to enable the self-adaption of training samples, and the ANN is trained with the current CFD database in each generation. The scale of the database is determined automatically, and the need for human adjustment is eliminated. The flow chart of this inverse design method can be seen in Fig. 6.11. The initial CFD database is used to train the ANN. The ANN is then used to judge whether the current individual satisfies the preset requirement. If the predicted value satisfies the requirement, the CFD simulation is conducted to obtain the design objectives.

6.3.2 Multi-objective problems

Most indoor environment designs problems involve multiple objectives. When a GA is integrated with an ANN the engine for the inverse design process is the GA. For a multi-objective problem, two approaches are usually applied within the GA method (Caldas and Norford, 2003):

(1) Simple aggregation and weighting of individual objectives

The objective function for a multi-objective problem can be constructed by using weighting factors that reflect the overall performance of the solution. The GA is then used to obtain the design variables that lead to the minimum objective function. Thus, a multi-objective problem is solved in the same way as a single-objective problem.

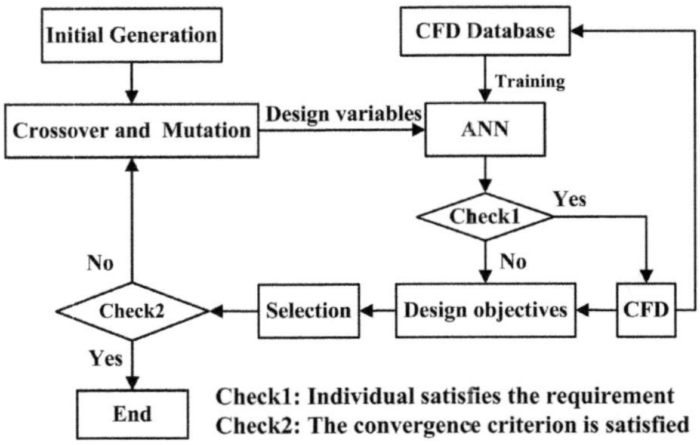

Figure 6.11 Flow chart of inverse design method based on GA and self-updating ANN

(2) Pareto-based approaches

A set of solutions can be obtained by Pareto-based approaches, with the trade-off of different objectives. Such approaches apply the concept of dominated and non-dominated solutions. Solution X is said to dominate solution Y when X is better than Y for at least one objective function and at least as good as Y for all the other objective functions. A solution is "Pareto optimal" if it is not dominated by any other solution. Fig. 6.12 illustrates a common strategy for selecting the Pareto solutions (the Pareto front) from the candidate solutions when two objective functions (f_1, f_2) exist.

In the selection process of a GA, the dominant relationship is used to determine the dominance layer of an individual, and the crowding degree is used to determine the order of the individuals in each dominance layer. The crowding degree can be obtained as follows:

$$CD_{hh} = \sum_{gg=1}^{GG} \left(\left| f_{gg}(hh+1) - f_{gg}(hh-1) \right| \right) \tag{6.3}$$

where CD_k is the crowding degree of individual hh, $f_{gg}(k+1)$ is the value of the gg^{th} objective of individual $hh+1$, $f_{gg}(k-1)$ is the value of the gg^{th} objective of individual $hh-1$, and GG is the number of objectives.

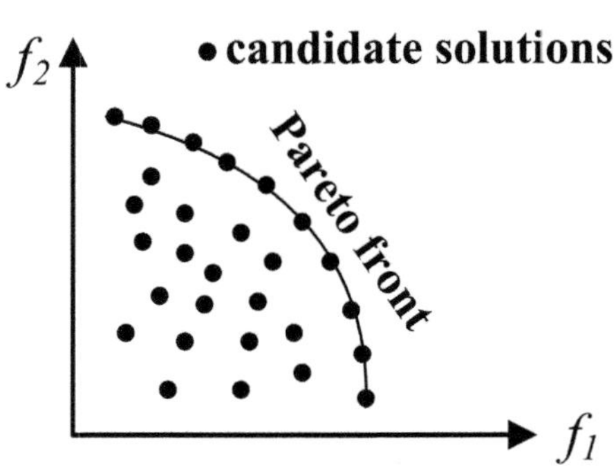

Figure 6.12 Selection of the Pareto solutions from the candidate solutions

In these two approaches to multi-objective problems, an ANN can be used to obtain the design objectives for new individuals, and a GA can be used with the ANN results to select the generations.

In a multi-objective design problem, there are usually many solutions, and it can be difficult to obtain a globally optimal solution. The hypervolume (Zitzler et al., 2010) can be used to solve this problem, and it is expressed as follows:

$$Hy\left(B, y_{ref}\right) = Le\left(\underset{y_{pareto} \in B}{U}\left\{z \middle| y_{pareto} \succ z \succ y_{ref}\right\}\right), \qquad B \in R^{m} \tag{6.4}$$

where λ *Le* is the Lebesgue measure, R^m is the m-dimension real number space, B is the spatial vector, y_{pareto} is the Pareto front, and y_{ref} is the reference point. A schematic diagram of the hypervolume for two objectives (f_1, f_2) can be seen in Fig. 6.13.

In the design stage, when a solution is obtained by the GA method, potential solutions near this solution are also studied. If these potential solutions satisfy the design requirement, they can be considered as the Pareto front, and the hypervolume can be obtained from the design objectives of these potential solutions. Thus, the hypervolume can be used to evaluate the stability of solutions. In the selection process of the GA, the hypervolume is used to evaluate the fitness of individuals instead of the values of design objectives. The calculation of the hypervolume involves a large quantity of individuals, and an ANN can be used to obtain the hypervolume at reduced computational cost.

6.3.3 Factors affecting convergence and computational cost

When a GA and an ANN are integrated in the inverse design process, convergence is affected by both the GA and the ANN simultaneously. In a single-objective design problem, the convergence criterion is usually the obtaining of a single globally optimal solution. Thus, the global search ability of the inverse design method is very important. A GA is a global optimization method, and it has a high probability of exploring the entire design space and reaching the global optimum. However, the use of an ANN increases the uncertainty and risk of accumulative error with the existence of prediction error (Nguyen, et al., 2014). The searching process of GA is

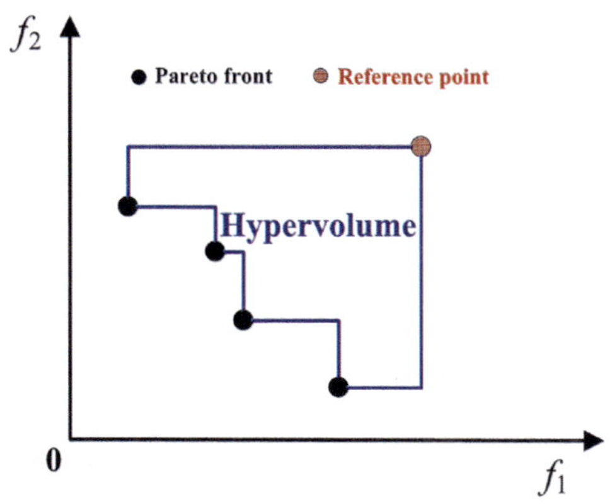

Figure 6.13 Schematic diagram of hypervolume

difficult to converge with low prediction accuracy of ANN. In a single-objective design problem, the convergence criterion is usually the obtaining of several solutions that satisfy the design requirements. In this case, the local search ability of the inverse design method is very important, and the convergence speed is affected by the design requirements. If the design requirements are easily satisfied, convergence occurs quickly and the effect of the ANN can be neglected.

The computational cost of an ANN is negligible in comparison with that of a CFD simulation. So the computational cost for the inverse design is determined by the total CFD simulation time. The number of CFD cases can be used to estimate the computational cost because the simulation time for each case is similar. The CFD cases are generated in two parts: the generation of training samples for ANN and the searching process of GA. When a GA and self-updating ANN are combined, only a small number of training samples is needed. Thus, the computational cost is determined mainly by the search ability of the GA.

6.3.4 Methods for improving GA and ANN performance

In a multi-objective design problem, the globally optimal solution is usually hard to identify. Instead, the design target is to obtain several locally optimal solutions. The initial generation has significant effect on the local convergence rate of GA. The training samples for an ANN are generated before the start of the inverse design process. Selection of the initial generation from the training samples would provide individuals of higher quality than those that result from random selection, and the local search speed would be accelerated. Furthermore, the CFD simulation results could be used sufficiently.

A means of improving the efficiency of GA- and ANN-based inverse design is fuzzy logic control. Fuzzy control of the computing probability and evolution space enhances the prediction accuracy of the ANN and the search ability of the GA, respectively.

The computing probability is the probability of conducting the CFD process for the current individual. If the computing probability is larger than a threshold value (0~1), the CFD process will be conducted. The control of computing probability is shown in Fig. 6.14. The inputs to the fuzzy logic controller (FLC) are the objective distance and prediction accuracy. The objective distance is the distance between the predicted value and the design requirement, and the prediction accuracy is the success rate of the ANN. The output of the FLC is computing probability, which ranges from 0 to 1.

The evolution space is the variable space for new individuals. Better individuals have a smaller evolution space. The control of evolution space is shown in Fig. 6.15.

6.3.5 Application examples

6.3.5.1 Nielsen case

Integration of a GA and an ANN was first validated by the Nielsen case (Nielsen, 1978), with the well-trained ANN that was obtained in Chapter 4. The value of $O(U_{inlet})$ in Chapter 4 was used as the fitness value in the GA. The convergence criterion was $O(U_{inlet}) = 0$. This was a

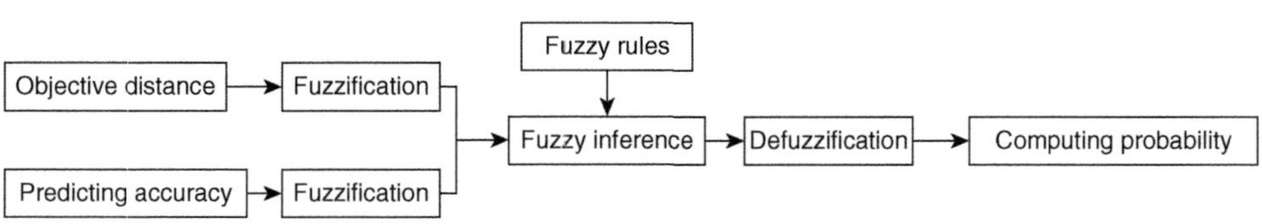

Figure 6.14 Control of computing probability

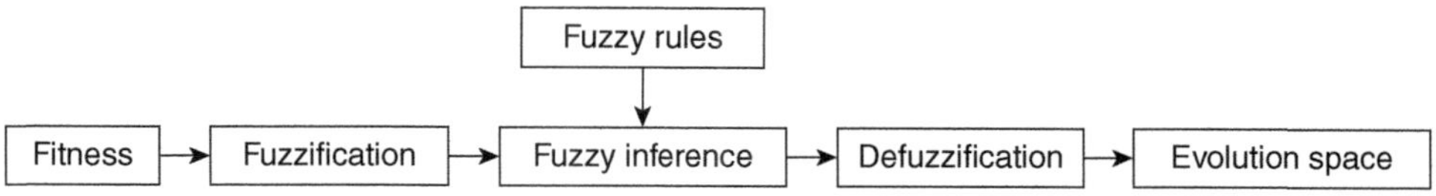

Figure 6.15 Control of evolution space

single-solution design problem. Convergence occurred when, and only when, $U_x = 0.455$ m/s and $U_y = 0$ m/s. The crossover and mutation rates for the GA were 0.8 and 0.1, respectively. The maximum generation size was 200, and each generation consisted of 10 individuals. The convergence history for $O(U_{inlet})$ is shown in Fig. 6.16. When the number of generations reached 31, $O(U_{inlet})$ converged to 0. As shown in Fig. 6.17, U_x converged to 0.455 m/s, and U_y converged to 0 m/s.

The increase in computational cost with the number of CFD cases is shown in Fig. 6.18. When the inverse design process ended, 242 new individuals had been generated, but only 104 of them were involved in the CFD process. Thus, the computational cost was reduced by 57.0% (the number of individuals decreased from 242 to 104) when the ANN was incorporated into the design process.

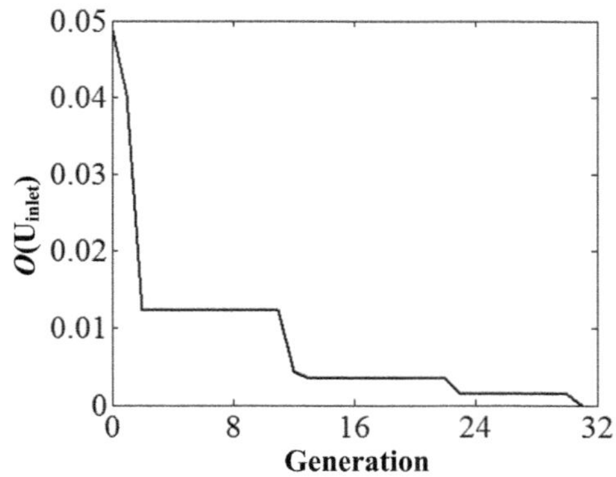

Figure 6.16 Convergence of the inverse design for the Nielsen case

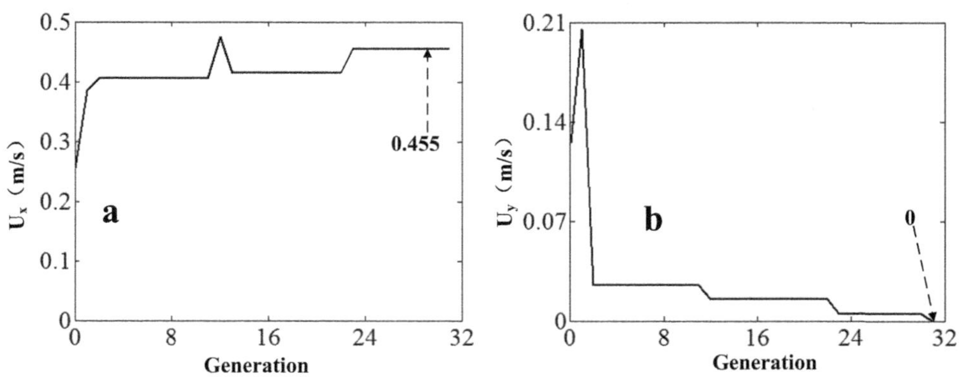

Figure 6.17 Convergence histories of (a) U_x and (b) U_y

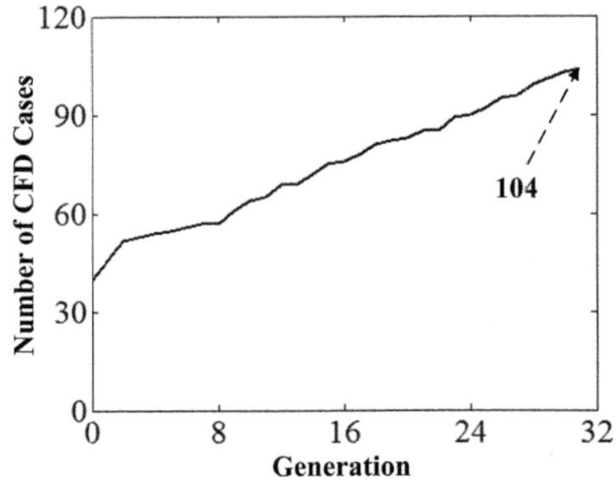

Figure 6.18 Increase in computational cost

6.3.5.2 Blay case

Next, the GA-ANN integration was validated by the Blay case (Blay et al., 1992), again with the use of the well-trained ANN from Chapter 4. The value of F in Chapter 4 was used as the fitness value in the GA. The convergence criterion was $F_{Blay} = 0$. This single-solution design problem converged when, and only when, $U_{in} = 0.57$ m/s and $T_{in} = 15°C$. The crossover and mutation rates for the GA were 0.8 and 0.1, respectively. The maximum generation size was 200, and each generation consisted of 10 individuals. The convergence history for F_{Blay} is shown in Fig. 6.19. When the number of generations reached 21, F_{Blay} converged to 0.

The convergence histories for U_{in} and T_{in} are shown in Fig. 6.20. When the number of generations reached 8, the value of U_{in} converged to 0.57 m/s. At 21 generations, T_{in} converged to 15.0°.

The increase in computational cost with the number of CFD cases is shown in Fig. 6.21. When the inverse design process ended, 162 new individuals had been generated, but only 92 of them were involved in the CFD process. Thus, the computational cost was reduced by 43.2% (the number of individuals decreased from 162 to 92) when the ANN was incorporated into the design process.

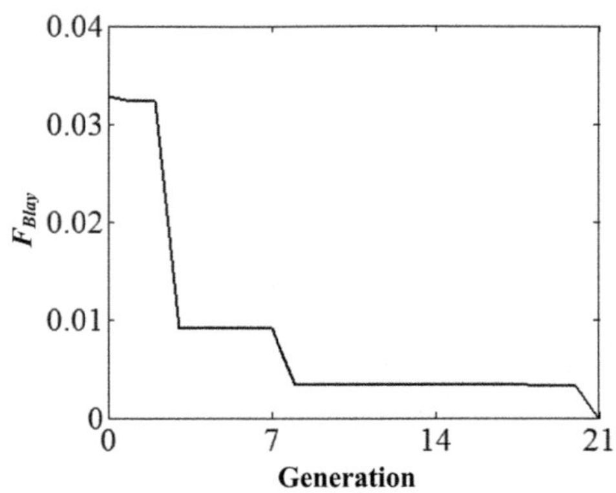

Figure 6.19 Convergence of the inverse design for the Blay case

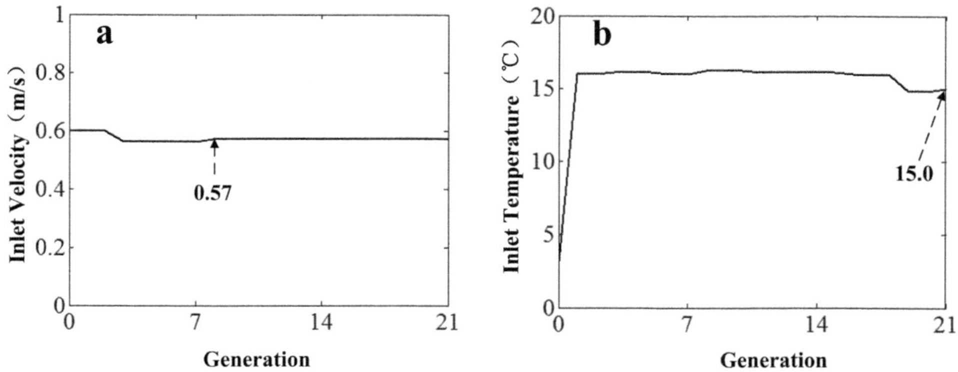

Figure 6.20 Convergence histories of (a) U_{in} and (b) T_{in}

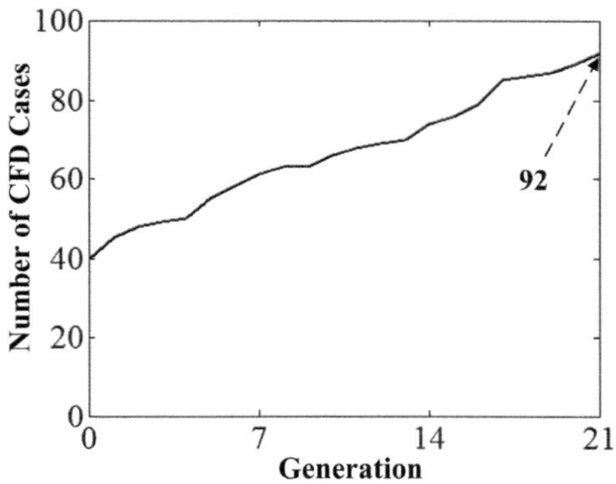

Figure 6.21 Increase in computational cost

6.3.5.3 Yuan case

The GA-ANN integration was then validated by the Yuan case (Yuan et al., 1999), with PMV as the design objective. The validation process again used the well-trained ANN from Chapter 4, and the initial generation for the GA was selected from the training samples of the ANN. An FLC was used to obtain the computing probability and evolution space. The proportion of training samples that satisfied the PMV requirement was 32.5%. In the inverse design process, the convergence criterion was the obtaining of 13 solutions (the same as in Chapter 4), which occurred after 25 CFD cases. Thus, the efficiency of the design process was 52% (13/25). This efficiency was significantly better than in the sampling stage.

Because the optimal method was involved in this design process, better solutions were obtained than in Chapter 4. Four of the solutions had |PMV| values that were better than the optimal value (0.094) obtained in Chapter 4. The minimum value of |PMV| in the present case was found to be 0.088. The PMV distributions are shown in Fig. 6.22.

6.3.5.4 Cabin case

The integrated GA and ANN were also used to design a model for an aircraft cabin. The |PMVc|, U_{head}, U_{feet}, and age of air were considered in this multi-objective design problem.

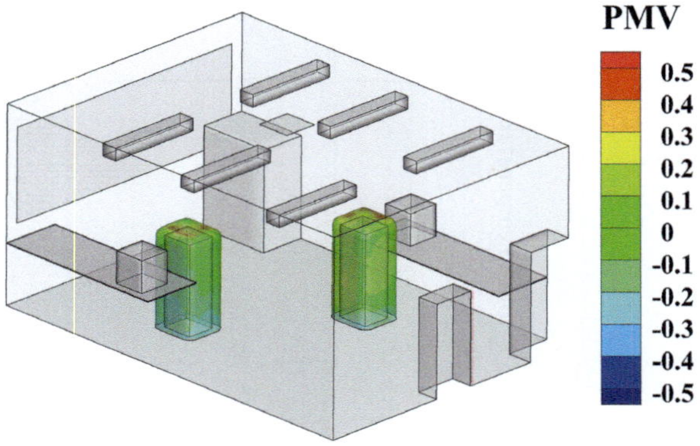

Figure 6.22 Design objective of the optimal solution obtained by inverse design

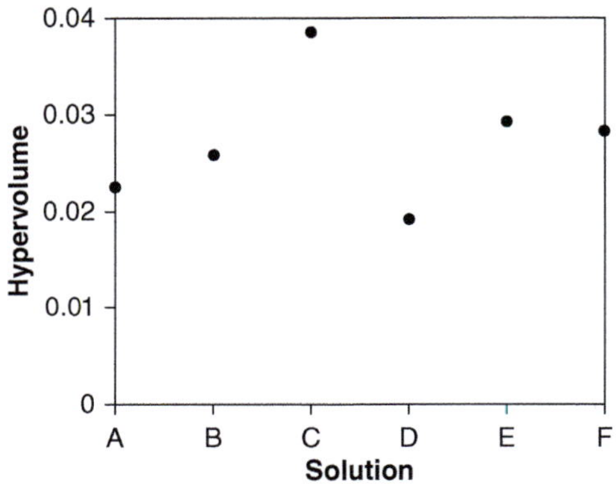

Figure 6.23 Hypervolumes of the solutions

The well-trained ANN from Chapter 4 was adopted, and the initial generation for the GA was selected from the training samples of the ANN. The FLC was used to obtain the computing probability and evolution space. Among the training samples, the proportions that satisfied the requirements for |PMVc|, U_{head}, U_{feet}, and age of air were 37.5%, 52.5%, 82.5%, and 17.5%, respectively. Only 2.5% of the samples satisfied all the design requirements. In the inverse design process, the convergence criterion was the obtaining of five solutions, and convergence was reached after 18 CFD cases had been conducted. Thus, the efficiency of the design process was 35.7%, which was significantly better than that of the sampling stage. There were a total of six solutions, labeled from A to F. The first solution was obtained by sampling, and the others were obtained in the inverse design process. There is no consensus as to whether large or small U_{head} is better. Only |PMVc|, U_{feat}, and the age of air were used to evaluate the solutions when U_{head} satisfied the preset requirement. No one solution was found to dominate all the other solutions. The hypervolumes, shown in Fig. 6.23, can be used to evaluate these solutions. It can be seen in the figure that solutions with better hypervolumes than solution A were obtained, and it is possible that additional better solutions would have been obtained if the design process had continued. However, greater computational cost would have been involved, and it would not have been possible to identify the optimal solution.

Among the six solutions, solution C had the best hypervolume (0.0385). The design variables for solution C were T_{in} = 23.6°C, A_y = −24°, S_{in} = 0.018 m, and L_{in} = 0.15 m. The distributions of the different design objectives for solution C are shown in Fig. 6.24. Although none of the design objectives for solution C were optimal, this solution was more stable than the others.

6.4 Enhancement of computing speed by fast fluid dynamics

Inverse design by means of the CFD-based adjoint method requires tens of design cycles (Jameson, 1995). In each cycle, this method solves the Navier-Stokes equations and adjoint equations in sequence by the SIMPLE (Semi-Implicit Method for Pressure-Linked Equations) algorithm (Patankar and Spalding, 1972). Therefore, a single design requires tens of Reynolds-averaged Navier-Stokes (RANS) simulations. In the design of an office environment, a CFD simulation normally requires mesh with 0.5 million cells, and the computing time for a single RANS simulation is about two hours with the use of a personal computer. Therefore, the inverse design process still takes at least tens of hours. For a larger indoor space such as an aircraft cabin, this method requires a powerful computer that can handle mesh with large grid numbers, and the computing time can be very long (Lin et al., 2005).

A method is desired that is more efficient and that can solve the Navier-Stokes/adjoint equations as accurately and informatively as does CFD simulation with the SIMPLE algorithm. A good option for solving the Navier-Stokes equations is fast fluid dynamics (FFD), which applies a time-advancement scheme and a semi-Lagrangian (SL) scheme (Staniforth and Côté, 1991). After a major development by Stam (1999) in the area of computer games, FFD has been further developed for simulating indoor airflow (Zuo and Chen, 2009; Jin et al., 2012; Liu et al., 2016; Xue et al., 2016). These efforts have made FFD a viable alternative to CFD for simulating indoor airflow, achieving similar accuracy with low computing time and/or computational capacity. However, the current FFD solvers consider the airflow to be laminar, which is not true for indoor environments. Furthermore, a turbulence model cannot be directly integrated with FFD because the SL scheme in FFD introduces significant numerical viscosity. Therefore, the integration of FFD with a turbulence model requires further investigation.

Because the adjoint equations have a similar form to that of the Navier-Stokes equations, they can be solved in the same way. Othmer et al. (2007) applied a SIMPLE-type algorithm to solve the adjoint equations for shape optimization. Our previous study (Liu and Chen, 2015) also used a SIMPLE-type algorithm to solve the adjoint equations for the inverse design of an indoor environment. Although the algorithm has proven to be robust, its computational demand is the same as that for solving the Navier-Stokes equations by the SIMPLE

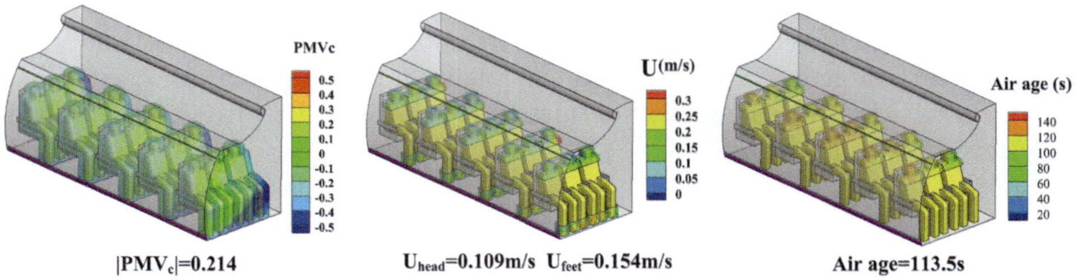

Figure 6.24 Distributions of different design objectives for solution C

algorithm. However, if FFD were further improved and could solve the Navier-Stokes equations quickly and accurately, it is quite possible that the FFD models could be used to solve the adjoint equations.

Therefore, this study developed an FFD technique with a turbulence model to achieve fast and accurate simulation of the air distributions in an indoor environment. The FFD models were further developed to solve the adjoint equations. This study implemented the FFD solvers for solving Navier-Stokes/adjoint equations in an open-source CFD program, OpenFOAM (Open Field Operation and Manipulation) (2007), and this section describes the validation process for indoor airflow predictions, inverse identification problems with experimental air-distribution data from the literature, and the inverse design of indoor environments.

6.4.1 Fast fluid dynamics

6.4.1.1 Use of FFD to solve the Navier-Stokes equations

FFD solves the Navier-Stokes equations in a similar manner to CFD. The Navier-Stokes equations for incompressible Newtonian flow are:

$$N_1 = -\frac{\partial U_i}{\partial x_i} = 0 \tag{6.5}$$

$$N_{i+1} = \frac{\partial U_i}{\partial t} + U_j \frac{\partial U_i}{\partial x_j} + \frac{1}{\rho}\frac{\partial p}{\partial x_i} - \nu \frac{\partial^2 U_i}{\partial x_j \partial x_j} - \frac{1}{\rho} F_i = 0 \tag{6.6}$$

$$N_5 = \frac{\partial T}{\partial t} + U_j \frac{\partial T}{\partial x_j} - \kappa \frac{\partial^2 T}{\partial x_j \partial x_j} - S_T = 0 \tag{6.7}$$

where $\mathbf{N}(\chi) = (N_1, N_2, N_3, N_4, N_5)^T$ are the Navier-Stokes equations in vector form; $\chi = (p, \mathbf{U}, T)$ the state variables; $i, j = 1, 2,$ and 3; U_i the ith component of the velocity vector; p the pressure, ρ the density; F_i the ith component of the body forces; ν the effective viscosity; κ the effective thermal conductivity; and S_T the energy source.

Previous FFD models used the non-incremental pressure-correction scheme with the SL scheme (NIPC-SL) to solve the Navier-Stokes equations (Zuo and Chen, 2009). The SL scheme stabilizes the calculations for a large time step size, but it introduces extra numerical diffusion that is dependent on the time step size (Zuo et al., 2010). The numerical diffusion could be treated as the substitution of turbulent diffusion, but its significance is not controllable. It is thus necessary to employ a turbulence model for simulating real indoor airflow. However, it is quite possible that the effective diffusion would be overestimated. Therefore, this study also investigated three pressure-correction schemes without the SL scheme: the non-incremental pressure-correction (NIPC) scheme, standard incremental pressure-correction (SIPC) scheme, and rotational incremental pressure-correction (RIPC) scheme. The following subsection introduces the NIPC, SIPC, RIPC, and NIPC-SL in sequence.

6.4.1.1.1 NON-INCREMENTAL PRESSURE-CORRECTION SCHEME

The NIPC scheme is the simplest pressure-correction scheme and was originally proposed by Chorin (1968) and Temam (1969). It applies a two-step time-advancement scheme that splits the momentum equation (Eq. (6.6)) into two discretized equations:

$$\frac{U_i^* - U_i^n}{\Delta t} = -U_j^n \frac{\partial U_i^*}{\partial x_j} + \nu \frac{\partial^2 U_i^*}{\partial x_j \partial x_j} + \frac{1}{\rho} F_i \tag{6.8}$$

$$\frac{U_i^{n+1} - U_i^*}{\Delta t} = -\frac{1}{\rho} \frac{\partial p}{\partial x_i} \tag{6.9}$$

where U^n and U^{n+1} represent the air velocity at the previous and current time steps, respectively, and U^* is the intermediate air velocity. To resolve the coupled pressure and velocity, the scheme uses the pressure projection method (Chorin, 1967), which substitutes Eq. (6.9) into Eq. (6.5) to produce:

$$\frac{\partial^2 p}{\partial x_i \partial x_i} = \frac{\rho}{\Delta t} \frac{\partial U_i^*}{\partial x_i} \tag{6.10}$$

The NIPC scheme first solves Eq. (6.8) with the implicit Euler scheme for the temporal term; the semi-implicit scheme for the convection term; and the implicit scheme for the diffusion term to obtain intermediate velocity U^*. Next, the NIPC scheme solves the Poisson equation, Eq. (6.10), to obtain the pressure p. Finally, the scheme calculates the air velocity at the next step U^{n+1} by solving Eq. (6.9) as follows:

$$U_i^{n+1} = U_i^* - \frac{\Delta t}{\rho} \frac{\partial p}{\partial x_i} \tag{6.11}$$

6.4.1.1.2 STANDARD INCREMENTAL PRESSURE-CORRECTION SCHEME

As observed by Goda (1979), the pressure term is missing in Eq. (6.8), and adding the pressure gradient of the previous step would increase the accuracy. Then the two discretized equations for the momentum equations are:

$$\frac{U_i^* - U_i^n}{\Delta t} = -\frac{1}{\rho} \frac{\partial p^n}{\partial x_i} - U_j^n \frac{\partial U_i^*}{\partial x_j} + \nu \frac{\partial^2 U_i^*}{\partial x_j \partial x_j} + \frac{1}{\rho} F_i \tag{6.12}$$

$$\frac{U_i^{n+1} - U_i^*}{\Delta t} = -\frac{1}{\rho} \frac{\partial (p^{n+1} - p^n)}{\partial x_i} \tag{6.13}$$

where p^n and p^{n+1} represent the air pressure at the previous and current time steps, respectively. Using the pressure projection again, we have:

$$\frac{\partial^2 (p^{n+1} - p^n)}{\partial x_i \partial x_i} = \frac{\rho}{\Delta t} \frac{\partial U_i^*}{\partial x_i} \tag{6.14}$$

Like the NIPC scheme, the SIPC scheme first solves Eq. (6.12) with the implicit Euler scheme for the temporal term; the semi-implicit scheme for the convection term; and the implicit scheme for the diffusion term to obtain intermediate velocity U^*. The SIPC scheme then solves the Poisson equation, Eq. (6.14), to obtain the pressure p^{n+1}. Finally, the SIPC scheme calculates the air velocity at the next time step U^{n+1} by solving Eq. (6.13) as follows:

$$U_i^{n+1} = U_i^* - \frac{\Delta t}{\rho} \frac{\partial (p^{n+1} - p^n)}{\partial x_i} \tag{6.15}$$

6.4.1.1.3 ROTATIONAL INCREMENTAL PRESSURE-CORRECTION SCHEME

The SIPC scheme uses a non-physical Neumann boundary condition enforced on the pressure that is applied to the numerical boundary layer. This limits the accuracy of the scheme (Guermond et al., 2005; Timmermans et al., 1996) and slightly modifies Eq. (6.15), as follows:

$$\phi^{n+1} = p^{n+1} - p^n + \nu \frac{\partial U_j^*}{\partial x_j} \tag{6.16}$$

$$U_i^{n+1} = U_i^* - \frac{\Delta t}{\rho} \frac{\partial \phi^{n+1}}{\partial x_i} \tag{6.17}$$

The solution procedure is exactly the same as for the SIPC scheme.

6.4.1.1.4 NON-INCREMENTAL PRESSURE-CORRECTION SCHEME WITH SEMI-LAGRANGIAN SCHEME

The NIPC-SL scheme applies a three-step time-advancement scheme (Ferziger and Perić, 2002) that splits the momentum equation (Eq. (6.6)) into three discretized equations:

$$\frac{U_i^* - U_i^n}{\Delta t} = -U_j^n \frac{\partial U_i^n}{\partial x_j} \tag{6.18}$$

$$\frac{U_i^{**} - U_i^*}{\Delta t} = \nu \frac{\partial^2 U_i^{**}}{\partial x_j \partial x_j} + \frac{1}{\rho} F_i \tag{6.19}$$

$$\frac{U_i^{n+1} - U_i^{**}}{\Delta t} = -\frac{1}{\rho} \frac{\partial p}{\partial x_i} \tag{6.20}$$

where U^* and U^{**} are intermediate air velocities. Using the pressure projection, we have:

$$\frac{\partial^2 p}{\partial x_i \partial x_i} = \frac{\rho}{\Delta t} \frac{\partial U_i^{**}}{\partial x_i} \tag{6.21}$$

The NIPC-SL scheme first solves Eq. (6.18) with the SL scheme to obtain intermediate air velocity U^*. The NIPC-SL scheme then solves the two Poisson equations, Eqs. (6.19) and (6.21), in sequence to obtain intermediate air velocity U^{**} and air pressure p, respectively. Finally, the algorithm calculates the air velocity at the next time step U^{n+1} by solving Eq. (6.20). The following paragraphs discuss the use of the SL scheme to solve the advection term, Eq. (6.18).

The SL scheme applies the Lagrangian advection to a grid so that it will remain stable at large time steps. The scheme traces the flow particle trajectory at each cell back in time to its former position, which is called the departure point X_d, and copies the quantities at that position to the present cell, which is called the arrival point X_a. The SL scheme actually solves Eq. (6.18) by:

$$U^*(X_a) = U^n(X_d) = U^n(X_a - \Delta t U^n(X_a)) \tag{6.22}$$

where $U^*(X_a)$ is the value of U^* at point X_a, and $U^n(X_d)$ is the value of U^n at point X_d. Since the departure point represents the position of the arrival point in the previous time step, $X_d = X_a - \Delta t U^n(X_a)$. To obtain $U^n(X_d)$, the SL scheme first identifies the index of the departure cell. As illustrated in Fig. 6.25(a), the departure point X_d is not necessarily located at the cell center. Therefore, $U^n(X_d)$ is estimated by interpolation from the values in the surrounding cells, and the result is set as $U^*(X_a)$ in Eq. (6.16).

For a structured mesh, the cell index can be defined on the basis of the coordinates in the three directions, and thus one can easily find the departure cell. For an unstructured mesh, there is no direct way to define the cell index for the departure point. Giraldo (1998) used the QuadTree algorithm to find the closest cell to the departure point for a quadrilateral mesh. This method is fast, but it would fail with a highly distorted mesh and non-convex geometry because the cell closest to the departure point may not be the departure cell. However, Xiu and Karniadakis (2001) have presented a fast search algorithm for the semi-Lagrangian scheme with a two-dimensional mesh. The algorithm can identify the departure cell by checking whether or not the departure point was inside the cells between the arrival and departure points. Fig. 6.25(b) is an example of this process. The search starts with the arrival cell I_a (cell index). To determine whether or not the departure point X_d is inside cell I_a, we can identify the crossover point X_{cr}^a at the cell face f^a by linking X_d and X_a. If point X_d is inside cell I_a and $X_{da} = X_d - X_a$, we have

$$X_{da} \cdot (X_{cr}^a - X_d) \geq 0 \tag{6.23}$$

Otherwise, the search moves to the neighboring cell (I_1) of cell face f^a. The same method is applied to determine whether or not the departure point is in cell I_1. If not, the search moves to the next neighboring cell, until

$$X_{da} \cdot (X_{cr}^n - X_d) < 0 \tag{6.24}$$

This result indicates that the departure cell is cell I_n. If cell I_n is outside the flow domain, then the departure cell is considered to be in cell I_{n-1}. The present investigation has extended the concept to three-dimensional meshes. Thus, the index number of the departure cell can be determined.

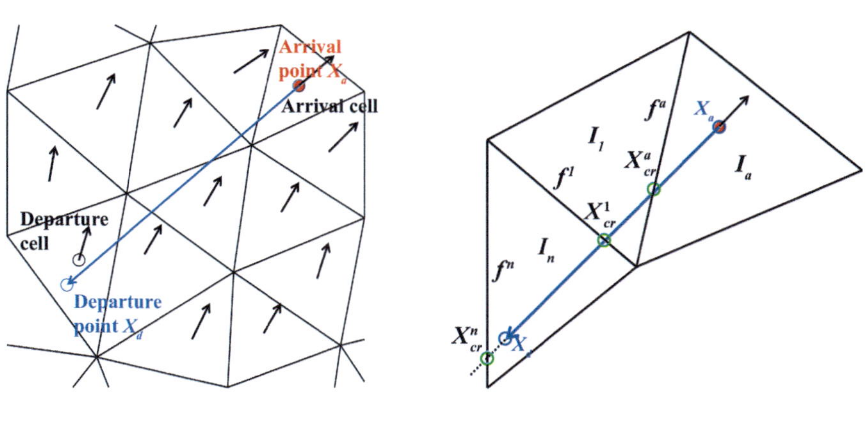

Figure 6.25 Illustration of (a) the semi-Lagrangian scheme and (b) the search algorithm

Once the index number of the departure cell has been identified, the next step is to obtain $U^n(X_d)$. Since X_d is not necessarily at the cell center, $U^n(X_d)$ is estimated by interpolation from the values in the surrounding cells.

For an indoor environment, the transport of energy and species must often be predicted. Therefore, it is also necessary to solve the scalar transport equation:

$$\frac{\partial \Phi}{\partial t}+U_j\frac{\partial \Phi}{\partial x_j}=\Upsilon_\Phi\frac{\partial^2\Phi}{\partial x_j\partial x_j}+S_\Phi \tag{6.25}$$

where Φ is the transport scalar to be solved, Υ_Φ the diffusive coefficient, and S_Φ the source term. Because Eq. (6.25) is a linear partial differential equation, it can be solved iteratively. Therefore, the discretized form of Eq. (6.25) by the implicit scheme is:

$$\frac{\Phi^{n+1}-\Phi^n}{\Delta t}+U_j^n\frac{\partial \Phi^{n+1}}{\partial x_j}=\Upsilon_\Phi\frac{\partial^2\Phi^{n+1}}{\partial x_j\partial x_j}+S_\Phi \tag{6.26}$$

The implicit scheme was used to ensure stability when the time step size was large.

6.4.1.2 Use of FFD to solve the adjoint equations

According to Chapter 3, the final forms of the adjoint equations are:

$$\frac{\partial U_{a,i}}{\partial x_i}=0 \tag{6.27}$$

$$\frac{\partial U_{a,i}}{\partial t}-U_j\frac{\partial U_{a,i}}{\partial x_j}=-\frac{\partial p_a}{\partial x_i}+\nu\frac{\partial^2 U_{a,i}}{\partial x_j\partial x_j}+F_{a,i} \tag{6.28}$$

$$\frac{\partial T_a}{\partial t}-U_j\frac{\partial T_a}{\partial x_j}=\kappa\frac{\partial^2 T_a}{\partial x_j\partial x_j}-\frac{\partial f}{\partial T} \tag{6.29}$$

where

$$F_{a,i}=\frac{\partial U_{a,j}}{\partial x_i}U_j-T_a\frac{\partial T}{\partial x_i}-\frac{\partial f}{\partial U_i} \tag{6.30}$$

The adjoint equations have the same form as the Navier-Stokes equations, so we can apply the previously mentioned FFD methods to solve the adjoint equations. For example, applying the SIPC scheme, this study again used a two-step time-advancement scheme that splits the adjoint momentum equation (Eq. (6.28)) into two discretized equations:

$$\frac{U_{a,i}^*-U_{a,i}^n}{\Delta t}=-\frac{\partial p_a}{\partial x_i}+U_j\frac{\partial U_{a,i}^*}{\partial x_j}+\nu\frac{\partial^2 U_{a,i}^*}{\partial x_j\partial x_j}+F_{a,i} \tag{6.31}$$

$$\frac{U_{a,i}^{n+1}-U_{a,i}^*}{\Delta t}=-\frac{\partial (p_a^{n+1}-p_a^n)}{\partial x_i} \tag{6.32}$$

where p_a^{n+1} and p_a^n represent the adjoint pressure at the previous and current time steps, respectively. U_a^* is the intermediate adjoint velocity. Using the pressure projection, we have:

$$\frac{\partial^2 (p_a^{n+1} - p_a^n)}{\partial x_i \partial x_i} = \frac{1}{\Delta t}\frac{\partial U_{a,i}^*}{\partial x_i} \tag{6.33}$$

The SIPC scheme first solves Eq. (6.31) with the implicit Euler scheme for the temporal term; the semi-implicit scheme for the convection term; and the implicit scheme for the diffusion term to obtain intermediate adjoint velocity U_a^*. Next, the SIPC scheme solves the Poisson equation, Eq. (6.33), to obtain the adjoint pressure p_a^{n+1}. Finally, the SIPC scheme calculates the adjoint velocity at the next time step U_a^{n+1} by solving Eq. (6.32) as follows:

$$U_{a,i}^{n+1} = U_{a,i}^* - \Delta t \frac{\partial (p_a^{n+1} - p_a^n)}{\partial x_i} \tag{6.34}$$

6.4.2 Validation of fast fluid dynamics

After developing the FFD solvers for the Navier-Stokes/adjoint equations, this study implemented them in OpenFOAM and validated their accuracy and efficiency step by step. In the first step, we validated the FFD solvers for predicting steady-state and transient flow in indoor environments. Next, we assessed the accuracy of the FFD solvers in solving the adjoint equations. Finally, we compared the CFD-based and FFD-based adjoint methods for the inverse design of indoor environments.

6.4.2.1 FFD to predict the indoor airflow

This study used two cases with experimental data from the literature, mixed convection flow in a simplified room and an office with displacement ventilation, to test the FFD solvers in predicting steady-state and transient indoor airflow. CFD simulations with the SIMPLE algorithm were used as a reference.

6.4.2.1.1 STEADY-STATE MIXED CONVECTION FLOW IN A SIMPLIFIED ROOM

We study tested the FFD solvers with data for mixed convection flow in a simplified room from Wang and Chen (2009). Fig. 6.26 shows a 2.44 m × 2.44 m × 2.44 m room with a plane jet from the upper left corner. The inlet height was 0.03 m. The inlet air velocity was 0.455 m/s with a corresponding Reynolds number of approximately 2,000. The supply air temperature was 22.2°C. The inlet jet developed along the ceiling and reached the far right side of the room, and then the flow moved downward because of the presence of the right-side wall, forming a circulation pattern in the room. The outlet height was 0.08 m. A box in the middle of the room, with dimensions of 1.22 m × 1.22 m × 1.22 m and a 700 W heat source inside, was used to represent a mass such as furniture or occupants. The temperatures of the box surface, ceiling, surrounding walls, and floor were 36.7, 25.8, 27.4, and 26.9°C, respectively. The heated box formed a thermal plume, and thus the type of airflow in the room was mixed convection.

This investigation applied the zero equation (0-eqn) model (Chen and Xu, 1998) to simulate the turbulence, since Zhang et al. (2007) had concluded that this model performs well in predicting the mixed convection flow in an indoor environment. The turbulent viscosity

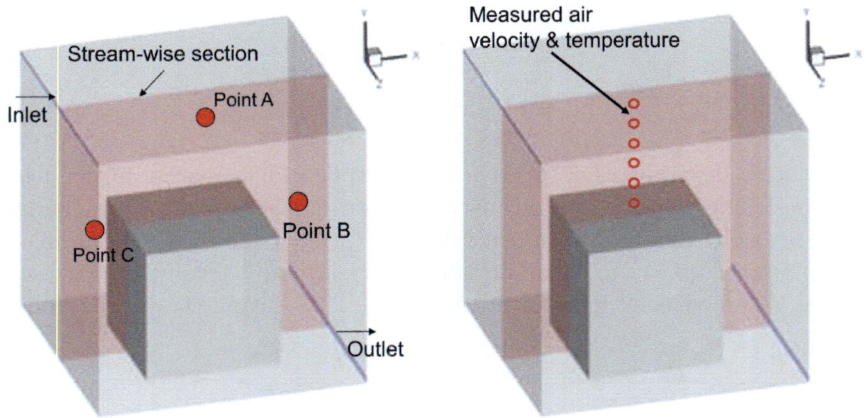

Figure 6.26 Sketch of the room with a box, the monitored positions, and the design objective

was determined from the local mean air velocity $|\mathbf{U}|$ and a length that represented the shortest distance l to the wall:

$$\nu_t = 0.03874\,|\,\mathbf{U}\,|\,l \tag{6.35}$$

We compared the simulated and measured velocity distributions in Fig. 6.27 to assess the performance of the FFD solvers. The CFD simulation used the SIMPLE algorithm, and the 0-eqn model was used as a reference. In general, the velocity distributions, except the one predicted by NIPC-SL, were very similar and agreed well with the measured distribution. The poor performance of NIPC-SL was due to the significant numerical diffusion contributed by the SL scheme.

To allow a quantitative comparison, Fig. 6.28 shows the calculated air velocity profiles along with the measured data at four representative positions (vertical lines) in the room. These positions are in the jet upstream (position 1), in the jet downstream (position 5), at the center of the room (position 3), and at a location close to the right-side wall (position 6). The air velocity was normalized by $|\mathbf{U}|_{max}$ = 1.5 m/s. The CFD simulation used the SIMPLE algorithm, and the 0-eq model provided a fairly good prediction of the air velocity. The NIPC-SL with the 0-eqn model performed poorly, as it under-predicted the jet flow and the re-circulated flow. The NIPC, SIPC, and RIPC schemes with the 0-eqn model had very similar performance, and the predicted air velocity profiles agreed well with the experimental data. However, there was still a significant difference in the lower part of position 5. This difference occurred because the flow in that region was very complicated; the box and the jet interacted and generated a secondary recirculation. Wang and Chen (2009) had tried more sophisticated models such as large eddy simulation, which were not able to predict the measured velocity profile at this location either.

This case was non-isothermal, and Fig. 6.29 compares the predicted temperature profiles with the experimental data in the same positions as in Fig. 6.28. The temperature was normalized by using T_{min} = 22.2°C and T_{max} = 36.7°C, which were the measured minimum and maximum air temperatures, respectively, in this case. All the numerical simulations of air temperature agreed well with the experimental data.

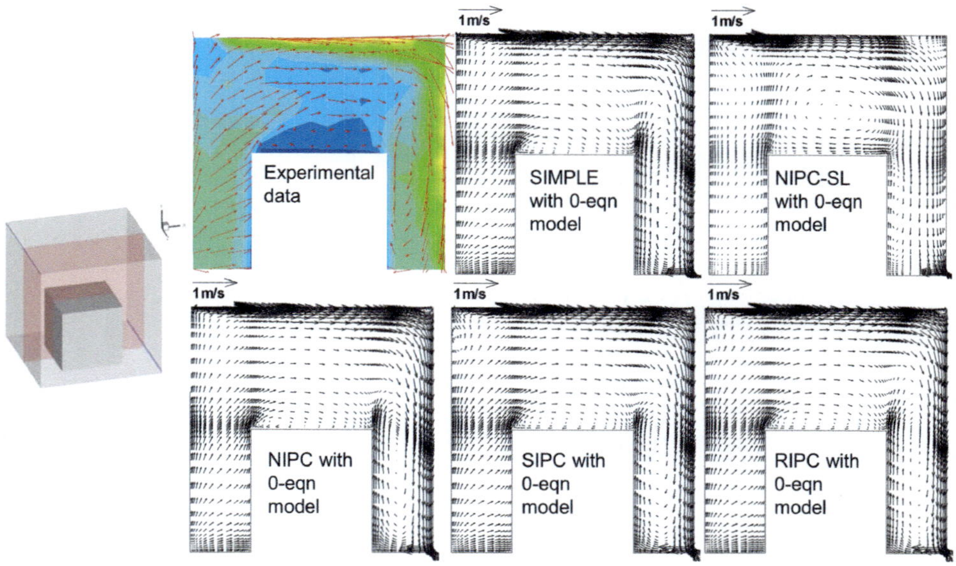

Figure 6.27 Airflow patterns measured by Wang and Chen (2009) and predicted by CFD and FFD with the 0-eqn model

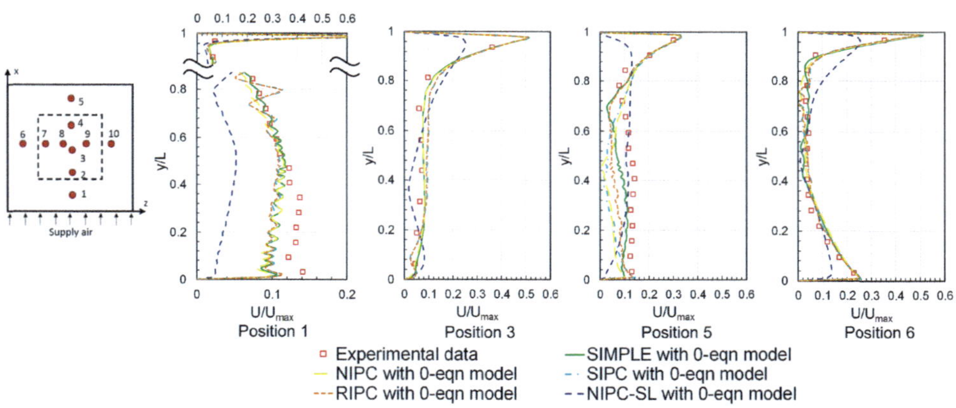

Figure 6.28 Comparison of air velocity profiles predicted by CFD and FFD with the 0-eqn model, and the experimental data from Wang and Chen (2009) at positions 1, 3, 5, and 6

In the above calculations, the FFD simulations with the 0-eqn model solved the transient Navier-Stokes equations with a time step size of 0.05 s for a physical flow time of 200 s, to obtain steady-state flow. The corresponding mean Courant–Friedrichs–Lewy (CFL) number was as small as 0.19, which ensured computational accuracy. To test the effect of time step size on the predictions, this study increased the step size as shown in Table 6.3. When Δt = 0.2 s, the corresponding mean CFL number was 0.75. To ensure that the mean CFL number was less than 1, the time step size was not further increased.

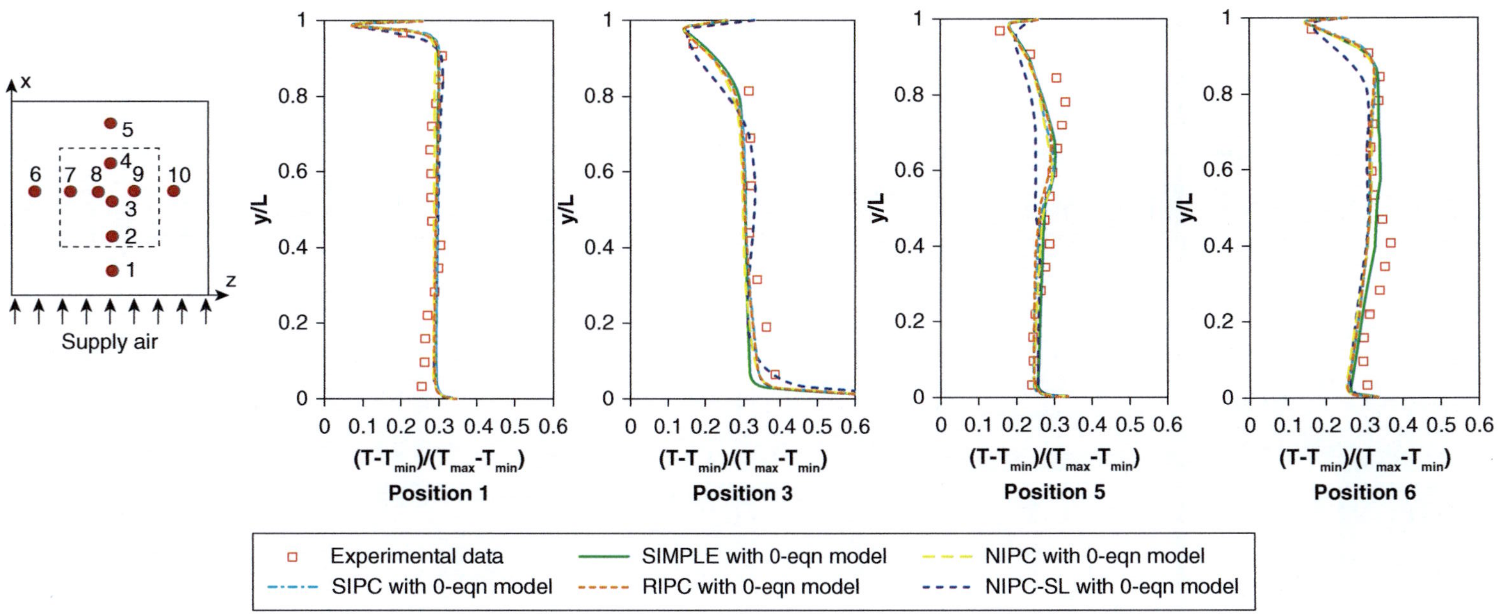

Figure 6.29 Comparison of air temperature profiles predicted by CFD and FFD with the 0-eqn model, and the experimental data from Wang and Chen (2009) at positions 1, 3, 5, and 6

Table 6.3 Time step size vs. CFL number.

Δt	*0.05 s*	*0.10 s*	*0.20 s*
CFL_{mean}	0.19	0.38	0.75
CFL_{max}	6.8	13.6	27.3

This study conducted the FFD simulations using the SIPC scheme with different time step sizes, again for a physical flow time of 200 s. Fig. 6.30 compares the predicted air velocity profiles with the measured data at the same four representative positions in the room. At position 5, the predicted air velocity profile with $\Delta t = 0.2$ s differed somewhat from the other predictions. It must have been a coincidence that the prediction with the largest time step size had better agreement with the experimental data. Fig. 6.31 compares the predicted profiles for air temperature with the measured data, and the predictions with different time step sizes exhibit almost no difference.

6.4.2.1.2 TRANSIENT MIXED CONVECTION FLOW IN A SIMPLIFIED ROOM

In addition to steady-state flow, it was necessary to validate the performance of the FFD solvers in predicting transient flow in an indoor environment. Since the above-mentioned experiment by Wang and Chen (2009) did not measure the transient flow fields, this study used the airflow predicted by CFD simulation with the SIMPLE algorithm as a reference. The NIPC-SL scheme was abandoned as it could not accurately predict even the steady-state indoor airflow. To eliminate the possibility of error introduced by the turbulence model, this study adopted the RNG k-ε model (Yakhot et al., 1992), which is much more sophisticated than the 0-eqn model, to simulate the turbulence. We calculated the flow for 100 s with a time step size of 0.05 s. The initialized temperature field and flow field were uniform with $T = 301.5$ K and $\mathbf{U} = \mathbf{0}$, respectively.

We monitored the transient air velocity and temperature at three typical locations, points A, B, and C, as shown in Fig. 6.26. These locations were in the jet downstream, at the outlet, and in the flow recirculation area. As shown in Fig. 6.32, the transient air velocity and temperature predicted by the FFD simulations had the same features as those predicted by the SIMPLE algorithm. There was a noticeable difference in velocity magnitude, although the difference was as small as 0.02 m/s. For the monitored air temperature, the predictions by the projection methods differed only minimally from the prediction by the SIMPLE algorithm. Overall, the SIPC and RIPC performed in a very similar manner, and their predictions were in better agreement with those predicted by SIMPLE algorithm.

Figs. 6.33 and 6.34 further compare the flow fields and temperature fields in the streamwise section (refer to Fig. 6.26) predicted by the SIMPLE algorithm and FFD simulations at $t = 5$ s, 10 s, 15 s, 20 s, 25 s, and 30 s, when the flow and temperature fields changed significantly. It is clear that all the FFD simulations were as capable as the SIMPLE algorithm in capturing the flow and thermal features. At each time point, the flow and temperature fields predicted by the FFD simulations agreed well with those predicted by the SIMPLE algorithm. Both the FFD and CFD simulations were conducted with the same structured mesh on a personal computer using a single CPU core at 3.50 GHz. The computing time for the FFD simulations was 0.4 hour, and that for the CFD simulations was 7.7 hours. FFD was almost 20 times faster than CFD in predicting the transient flow, while maintaining similar prediction accuracy. We also found that the computing time for FFD simulation with the 0-eqn model was 0.32 hour, and thus the integration with the RNG k-ε model would not significantly increase the computational effort.

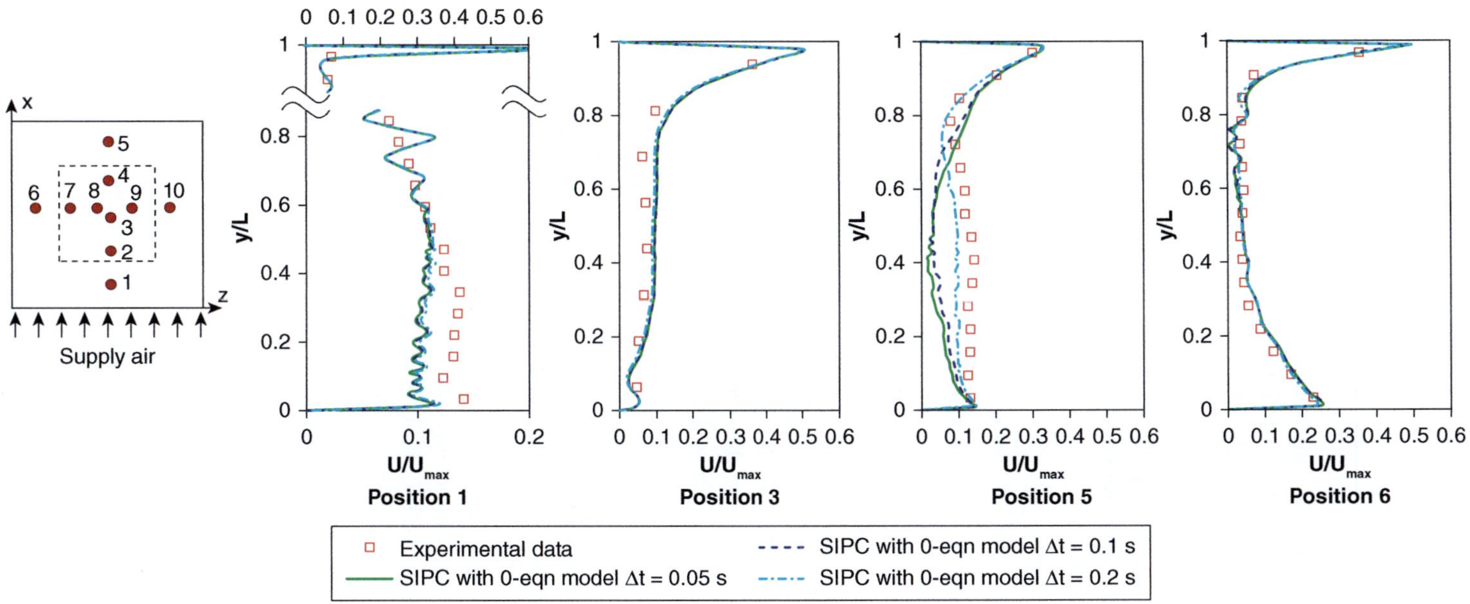

Figure 6.30 Comparison of air velocity profiles predicted by the SIPC scheme with the 0-eqn model, and the experimental data from Wang and Chen (2009) at positions 1, 3, 5, and 6

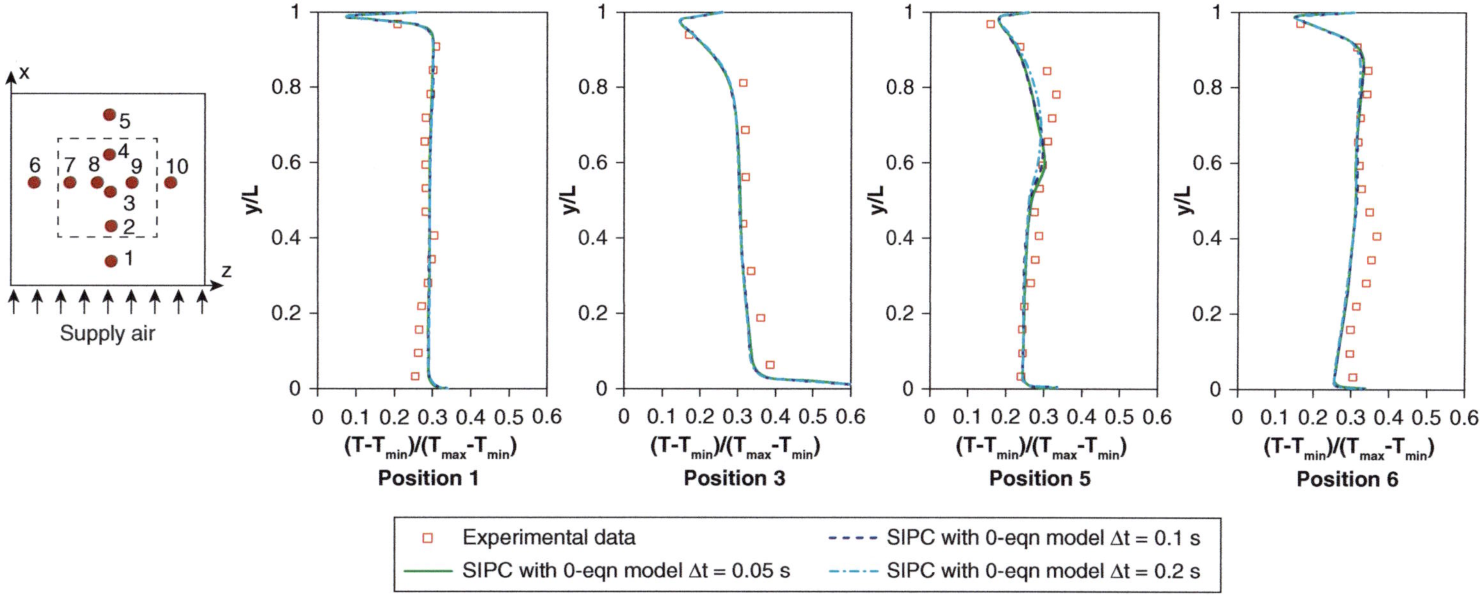

Figure 6.31 Comparison of air temperature profiles predicted by the SIPC scheme with the 0-eqn model, and the experimental data from Wang and Chen (2009) at positions 1, 3, 5, and 6

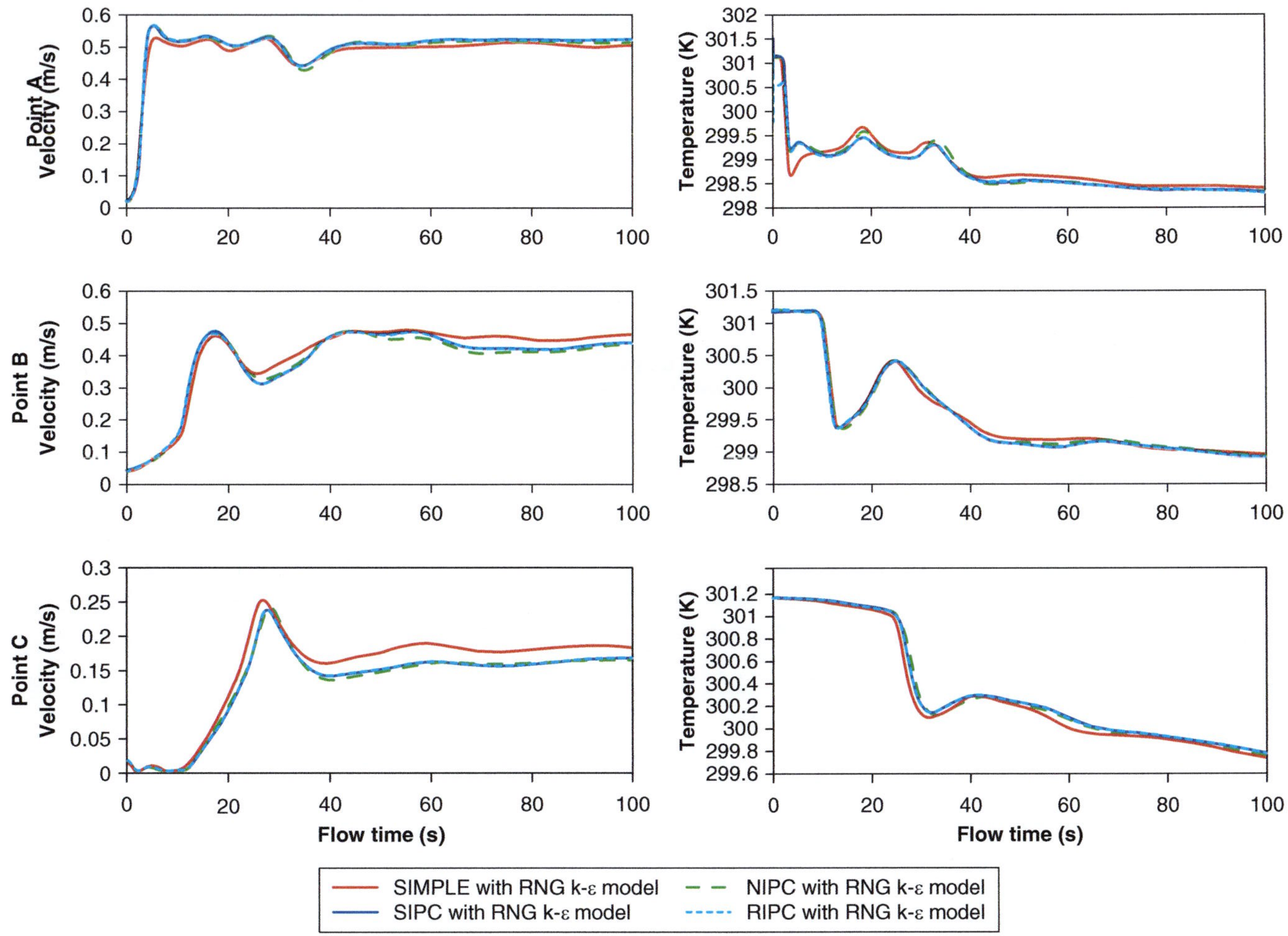

Figure 6.32 Monitored transient air velocity and temperature at points A, B, and C (refer to Fig. 6.26 for the locations of the monitoring points)

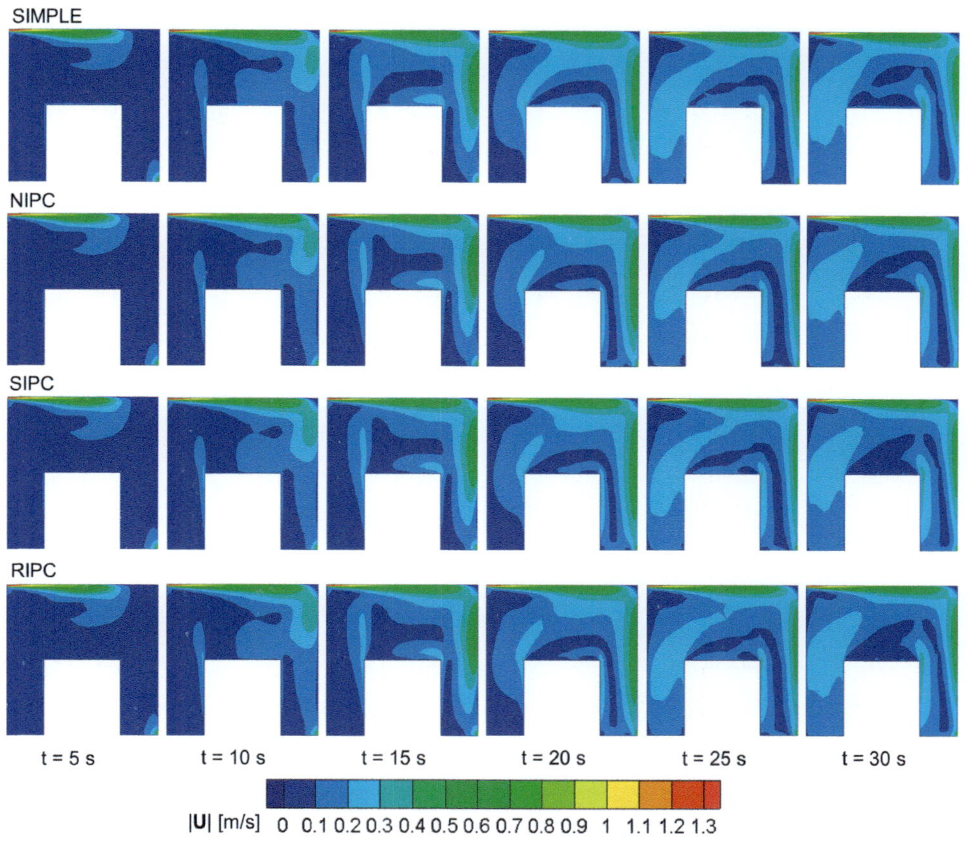

Figure 6.33 Predicted flow fields in the stream-wise section (refer to Fig. 6.26) at t = 5 s, 10 s, 15 s, 20 s, 25 s, and 30 s

For the transient airflow, this study again conducted FFD simulations using the SIPC scheme and RNG k-ε model with the various time step sizes shown in Table 6.3. Figs. 6.35 and 6.36 show the flow fields and temperature fields on the stream-wise section predicted by the SIPC scheme at $t = 5$ s, 10 s, 15 s, 20 s, 25 s, and 30 s with different time step sizes. There was a noticeable difference among the predictions by various time step sizes at the beginning of the simulation (when $t < 15$ s). Later, as the airflow developed, the predictions were almost the same for the different time step sizes. This result agreed with the observed similarity among the predicted steady-state air distributions with different time step sizes. To further decrease the computational effort, a smaller time step size could be used at the beginning of the simulation and a larger step size later.

6.4.2.1.3 STEADY-STATE FLOW IN AN OFFICE WITH DISPLACEMENT VENTILATION

This investigation also tested the FFD simulation with a more realistic indoor airflow case. Fig. 6.37 shows an occupied office with displacement ventilation. The office was simulated by an experimental chamber with dimensions of 5.16 m × 3.65 m × 2.43 m. A displacement

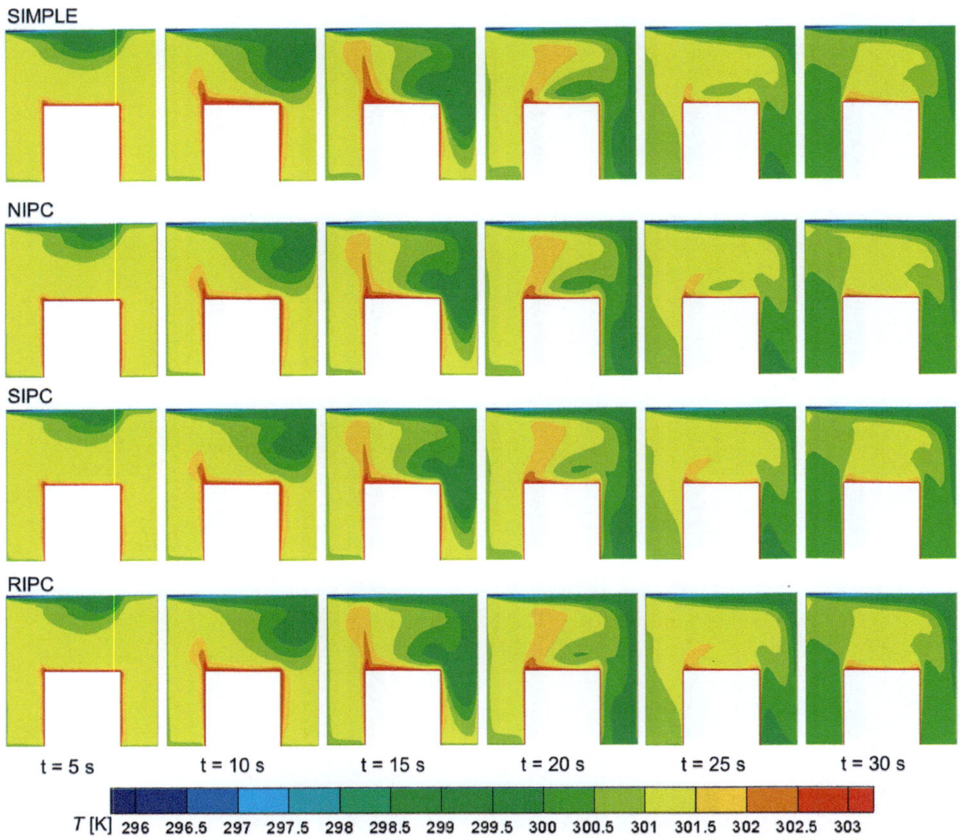

Figure 6.34 Predicted temperature fields in the stream-wise section (refer to Fig. 6.26) at t = 5 s, 10 s, 15 s, 20 s, 25 s, and 30 s

diffuser on the side wall supplied air with a ventilation rate of four air changes per hour. The supply-air temperature was 17°C. There were two occupants, two computers, two tables, two boxes, and six lamps in the room, as shown in Fig. 6.37(a). For detailed sizes, locations, and heat release of these items and other thermo-fluid boundary conditions, one can refer to Yuan et al. (1999). The experiment measured the air temperature and air velocity along nine vertical poles distributed in the stream-wise center plane and the cross-sectional center plane, as indicated by the red lines in Fig. 6.37(b). This figure also shows the mesh that was chosen on the basis of our grid independence test.

Since this was a much more realistic case, we study used the RNG k-ε model to simulate the turbulence. Because of the poor performance of the NIPC-SL algorithm in the previous two cases, we did not consider it for this case. The CFD simulation with the SIMPLE algorithm was again used as a reference.

This study compared the calculated air velocity profiles with the measured data at the nine positions in the room, as shown in Fig. 6.38. The air velocity was normalized by the inlet air velocity $|\mathbf{U}|_{in} = 0.09$ m/s. There was good quantitative agreement between the predicted and measured air velocity. All the FFD simulations had almost the same performance, and it was hard to determine whether the SIMPLE algorithm or the FFD models performed better.

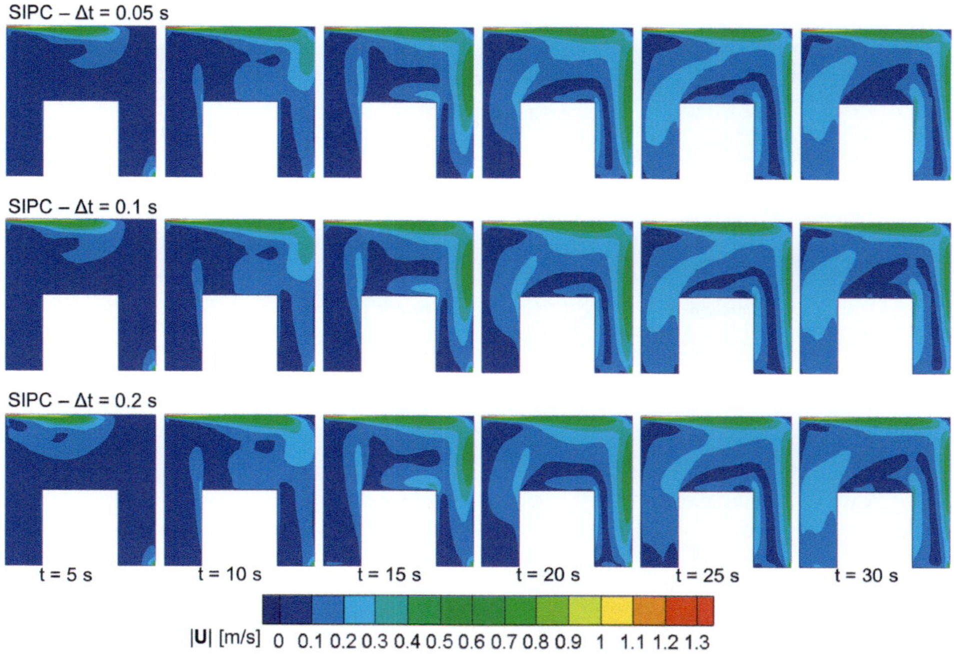

Figure 6.35 Flow fields predicted by the SIPC scheme with different time step sizes in the stream-wise section (refer to Fig. 6.26) at t = 5 s, 10 s, 15 s, 20 s, 25 s, and 30 s

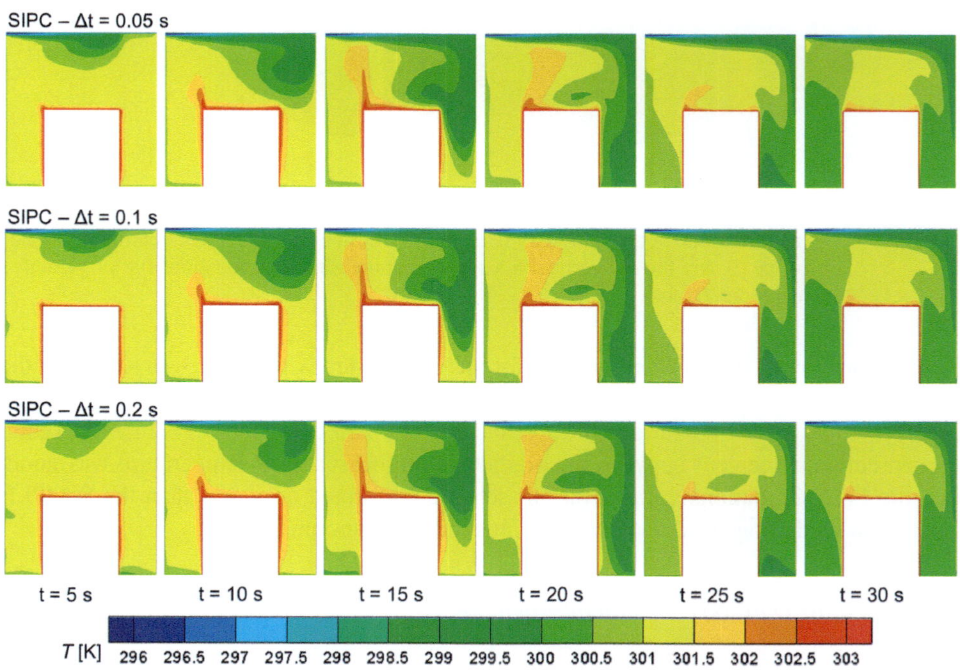

Figure 6.36 Temperature fields predicted by the SIPC scheme with different time step sizes in the stream-wise section (refer to Fig. 6.26) at t = 5 s, 10 s, 15 s, 20 s, 25 s, and 30 s

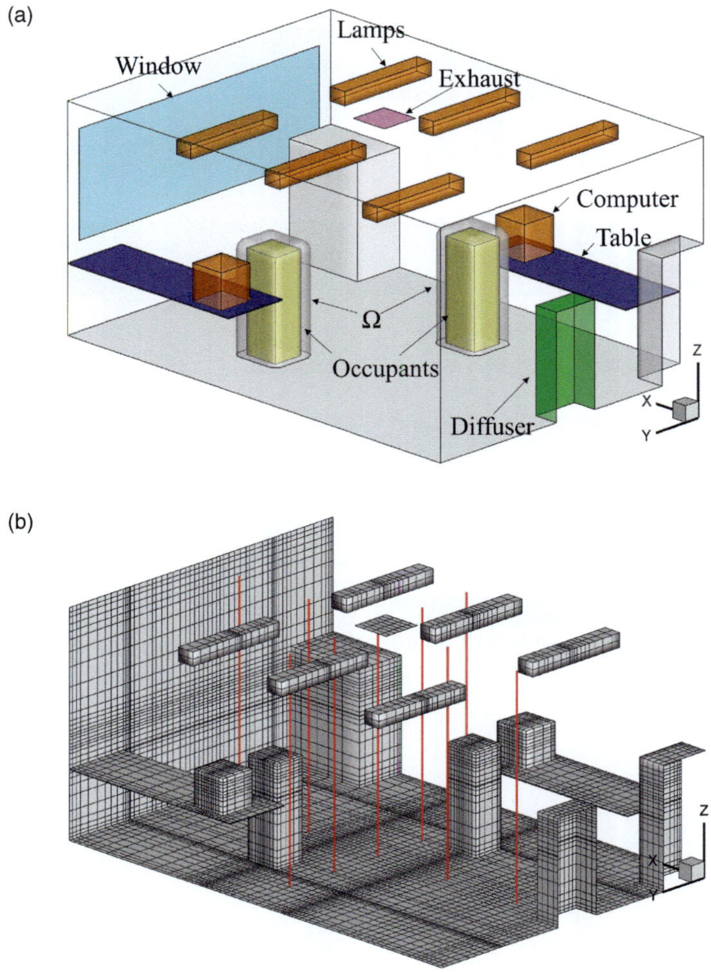

Figure 6.37 (a) Schematic of an occupied office and (b) mesh of displacement ventilation in the office (where the red vertical lines represent measurement locations for air velocity and temperature)

Fig. 6.39 compares the calculated profiles for air temperature with the measured data at the nine positions in the room. The normalized air temperature θ was defined as $(T - T_s)/(T_r - T_s)$, where T_s (supply air temperature) = 17.0°C and T_r (return air temperature) = 26.7°C. Again, the quantitative agreement between the predicted and measured air temperature was good. Thus, the FFD simulations without the SL scheme performed no worse than the SIMPLE algorithm in predicting steady-state flow in this indoor environment.

6.4.2.2 Use of FFD to solve the adjoint equations

To validate the accuracy of FFD in solving the adjoint equations, this study compared the adjoint velocity fields predicted by the SIMPLE algorithm and SIPC scheme in the first design cycle of an inverse identification problem. The goal of this problem was to inversely

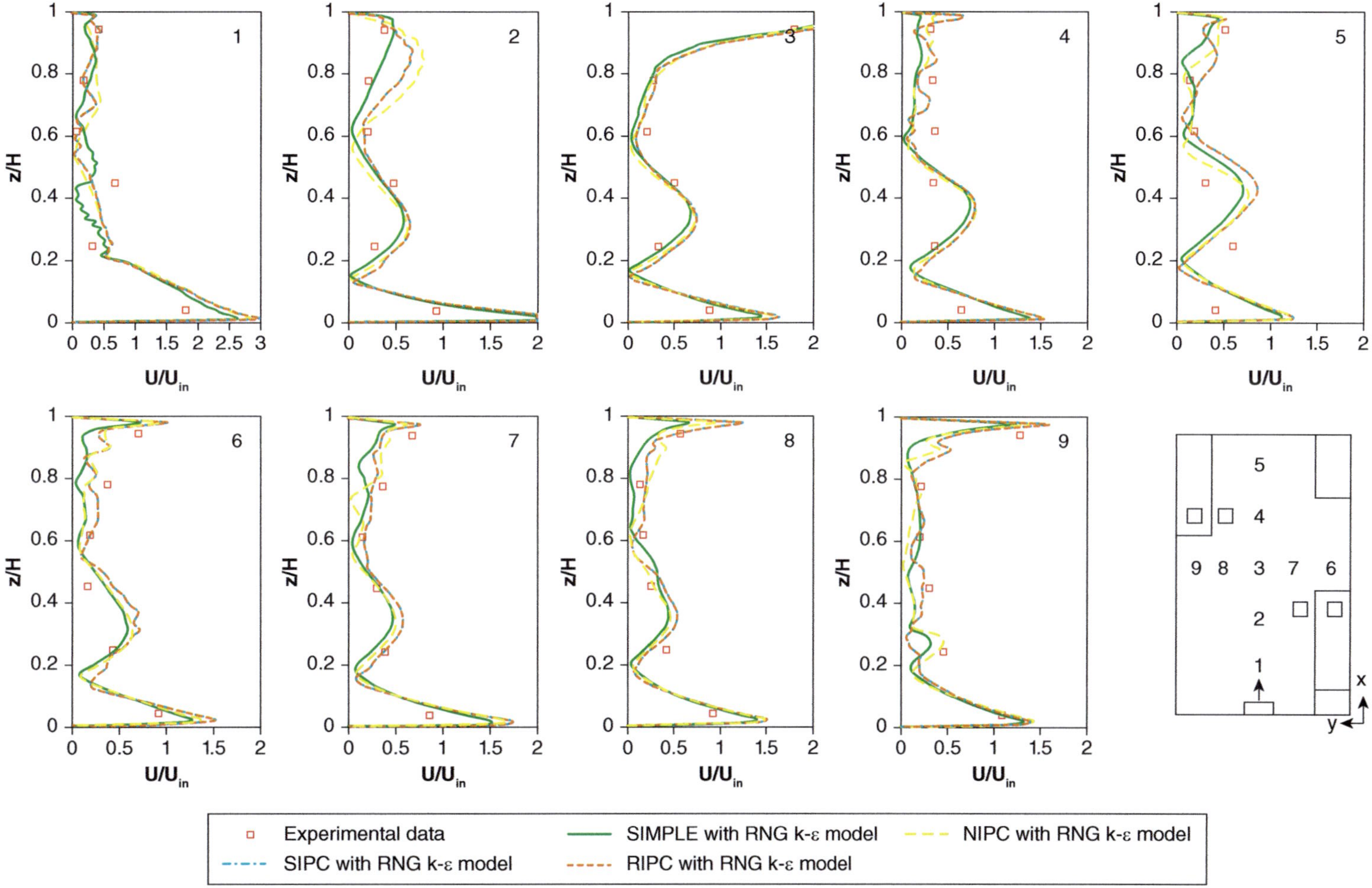

Figure 6.38 Comparison of air velocity profiles predicted by CFD and FFD with the RNG k-ε model, and the experimental data from Yuan et al. (1999)

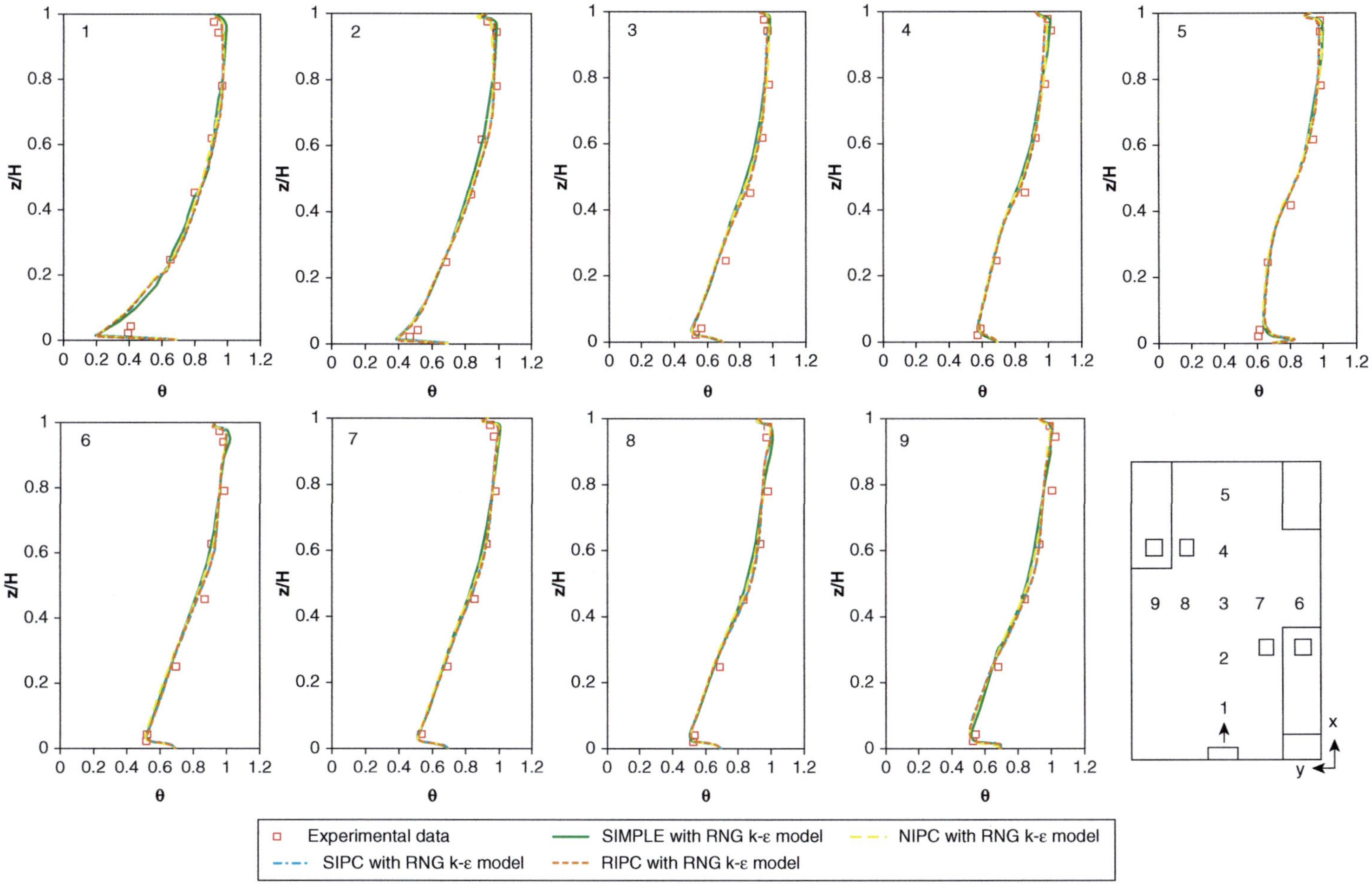

Figure 6.39 Comparison of air temperature profiles predicted by CFD and FFD with the RNG k-ε model, and the experimental data from Yuan et al. (1999)

identify the air supply velocity by setting the measured air velocity as the design objective. Fig. 6.1(a) shows a two-dimensional, non-isothermal case with experimental data from Blay et al. (1992). The measured air velocity on the red dashed line was set as the design objective, and thus the objective function was:

$$O(\mathbf{U}_{in}) = \int_{\Omega} |\mathbf{U} - \mathbf{U}_0|^2 \, d\Omega \tag{6.36}$$

where $\mathbf{U}_0$ is the measured air velocity.

For inverse identification of the air supply velocity, we applied the SIMPLE algorithm and SIPC scheme to solve the Navier-Stokes and adjoint equations, respectively. The SIMPLE algorithm solved the steady-state Navier-Stokes/adjoint equations for 7,000 iterations to ensure convergence. The SIPC scheme solved the transient Navier-Stokes/adjoint equations for 100 s with a time step size of 0.05 s in each design cycle. The RNG k-ε model was adopted to simulate the turbulence. The initialized air supply velocity was $\mathbf{U} = (1.0, 0.2)$ m/s.

Fig. 6.40 compares the air flow fields, temperature fields, and adjoint velocity fields predicted by the SIMPLE algorithm and SIPC scheme in the first design cycle. With the same initialized air supply velocity, the predicted flow fields and temperature fields were almost the same. The two adjoint velocity fields had very similar features, and the field predicted by SIPC had a slightly larger $\mathbf{U}_a$ magnitude. If the adjoint velocity field predicted by the SIMPLE algorithm is reliable, then it is clear that the SIPC scheme performs well in solving the adjoint equations.

6.4.2.3 FFD-based adjoint method vs. CFD-based adjoint method

The previous two sections discussed the validation of the FFD for solving the Navier-Stokes equations and adjoint equations. The present section compares the FFD-based adjoint method and CFD-based adjoint method in two inverse identification cases and one inverse design case.

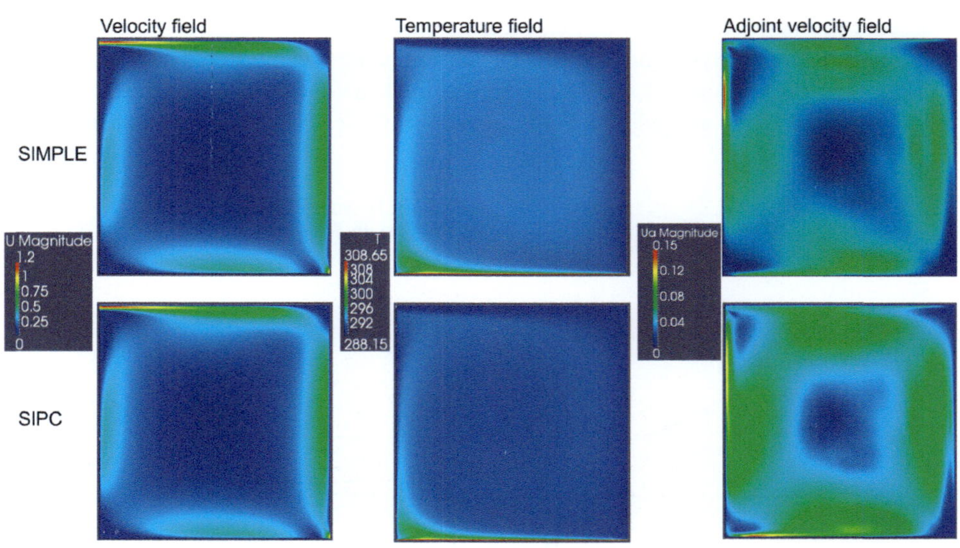

Figure 6.40 Air flow fields, temperature fields, and adjoint velocity fields predicted by the SIMPLE algorithm and SIPC scheme in the first design cycle

6.4.2.3.1 INVERSE IDENTIFICATION OF AIR SUPPLY VELOCITY IN A TWO-DIMENSIONAL CAVITY

This study applied the FFD-based adjoint method and CFD-based adjoint method in the inverse identification of the air supply velocity in Section 6.4.2.2. The FFD-based adjoint method used the SIPC scheme, and the CFD-based adjoint method used the SIMPLE algorithm. One can refer to Section 4.2 for other parameters. The convergence criterion for the inverse identification was $|O_{new} - O_{old}| < 0.001$. Fig. 6.41 shows the objective function versus the design cycle. Both methods led the objective to very small values, and both inverse identifications converged. The convergence speed differed because of the difference in the predicted adjoint fields. We found that in one design cycle, the SIPC scheme was four times faster than the SIMPLE algorithm.

6.4.2.3.2 INVERSE IDENTIFICATION OF AIR SUPPLY CONDITIONS IN A SIMPLIFIED ROOM

To further validate the FFD-based adjoint method for the inverse design process, this study attempted to inversely identify the air supply velocity and temperature in a simplified room as shown in Fig. 6.26. The measured air velocity and temperature at the points indicated by the red circles in the figure were set as the design objective. Therefore, the objective function was:

$$O(\mathbf{U}_{in}, T_{in}) = \frac{1}{|\mathbf{U}_{\mathbf{max}}|^2} \int_\Omega |\mathbf{U} - \mathbf{U}_0|^2 \, d\Omega + \frac{1}{(T_{\max} - T_{\min})^2} \int_\Omega (T - T_0)^2 \, d\Omega \quad (6.37)$$

The FFD-based adjoint method again used the SIPC scheme. This scheme solved the transient Navier-Stokes/adjoint equations for 100 s with time step sizes of 0.05 s, 0.1 s, and 0.2 s. Inverse identification by means of the CFD-based adjoint method was used as a reference. The SIMPLE algorithm solved the steady-state Navier-Stokes/adjoint equations for 5,000 iterations to ensure convergence. The RNG k-ε model was adopted to simulate the turbulence. Since the air supply velocity and temperature in the experiment were $\mathbf{U}_{\mathbf{in}}$ = (1.366, 0, 0) m/s and T_{in} = 295.35 K, respectively, this study initialized them as $\mathbf{U}_{\mathbf{in}}$ = (1.0, 0, 0) m/s and T_{in} = 300.35 K. The convergence criterion for the inverse identification was again $|O_{new} - O_{old}| < 0.001$.

Fig. 6.42 shows the objective function versus the design cycle. All the designs converged and led the objective to very small values. The convergence speed differed slightly between the FFD-based adjoint method and CFD-based adjoint method because of the differences in predicted flow fields and adjoint fields. However, the convergence trajectories for the

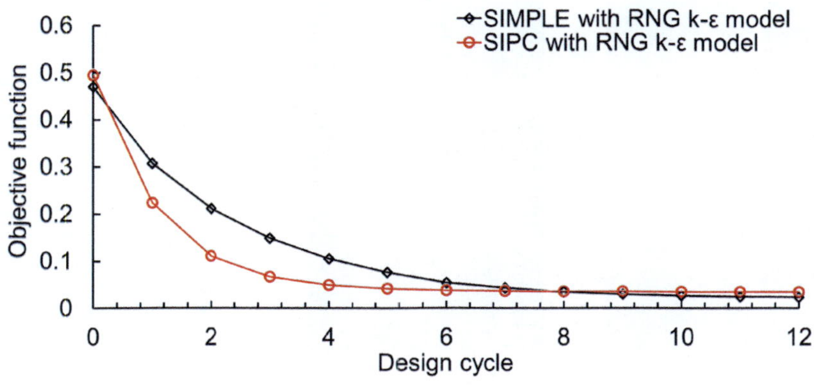

Figure 6.41 Objective function versus design cycle

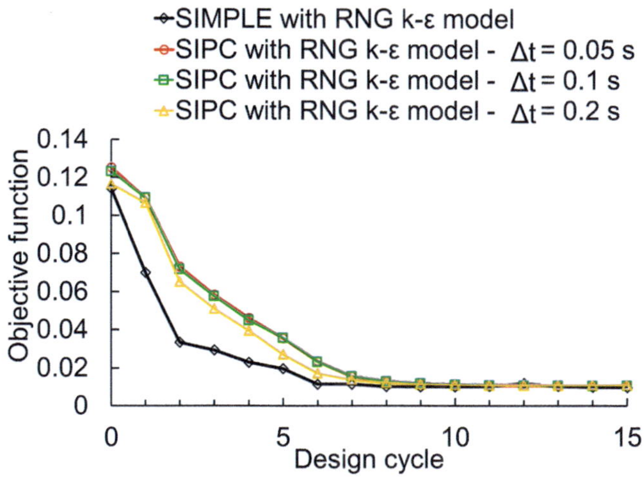

Figure 6.42 Objective function versus design cycle

FFD-based adjoint method were almost the same for different time step sizes. This is because the Navier-Stokes/adjoint equations were solved with a sufficiently long physical flow time in each design cycle. The FFD-based adjoint method with $\Delta t = 0.05$ s was again four times faster than the CFD-based adjoint method. In comparison, the FFD-based adjoint method with $\Delta t = 0.2$ s was almost 16 times faster than the CFD-based adjoint method.

Table 6.4 summarizes the objective function and design variables at the optimal conditions. With the inversely identified optimal air supply conditions, this study conducted forward CFD simulations. We then compared the predicted air velocity and temperature profiles on the design domain with the experimental data, as shown in Fig. 6.43. The predicted air velocity and temperature profiles were similar and agreed well with the experimental data, which confirmed the inverse identification.

6.4.2.3.3 INVERSE DESIGN OF THE THERMAL ENVIRONMENT IN AN AIRCRAFT CABIN

Finally, this investigation conducted an inverse design of the thermal environment in a single-aisle, fully occupied aircraft cabin in order to find the optimal air supply locations, size, and parameters for the diffusers. To decrease the computing load, we performed calculations for only half of one row of the cabin, as illustrated in Fig. 6.3. For all the boundary conditions that did not vary in the design, one can refer to Chapter 3. We used a modified predicted mean vote for air cabins (PMVc) developed by Cui and Zhu (2015) to evaluate the thermal

Table 6.4 Objective function and design variables at the optimal conditions

	Objective function	$\boldsymbol{U}_{in}$ *(m/s)*	T_{in} *(K)*
SIMPLE algorithm	0.0096	(1.26, 0.08, 0)	296.23
SIPC with $\Delta t = 0.05$ s	0.0111	(1.30, 0,13, 0)	296.21
SIPC with $\Delta t = 0.1$ s	0.0111	(1.30, 0.13, 0)	296.20
SIPC with $\Delta t = 0.2$ s	0.0112	(1.30, 0.12, 0)	296.22

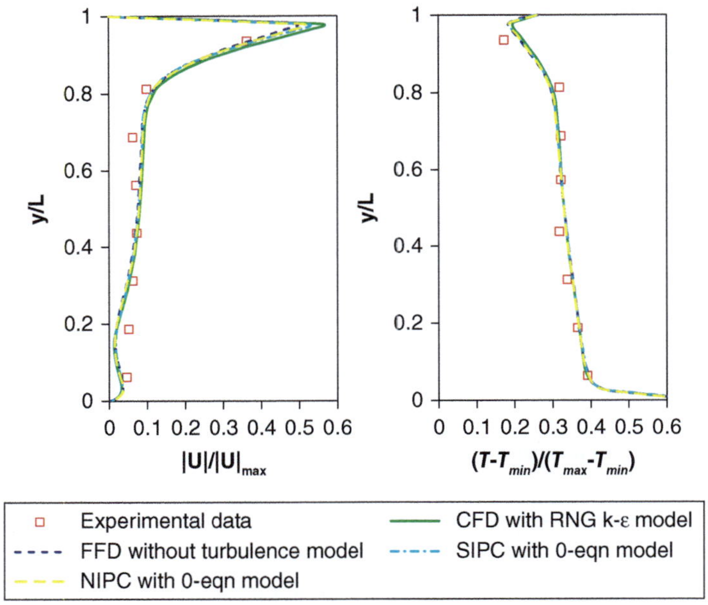

Figure 6.43 Comparison of the predicted air velocity and temperature profiles on the design domain with the experimental data

comfort level. The thermal resistance of the passengers' clothing was assumed to be 0.57 clo for summer (ASHRAE, 2007). The design domain Ω was set as the area 0.1 m away from the passengers. The objective function was again Eq. (6.2).

The FFD-based adjoint method used the SIPC scheme with a time step size of 0.1 s. The SIPC scheme solved the transient Navier-Stokes/adjoint equations for 100 s in each time step. The CFD-based adjoint method with the SIMPLE algorithm solved the steady-state Navier-Stokes/adjoint equations for 5,000 iterations in each design cycle to ensure convergence. The RNG k-ε model was adopted to simulate the turbulence. The initialized air supply velocity and temperature were $\mathbf{U}_{in}$ = (1.5, 0, 0) m/s and T_{in} = 16.1°C, respectively. To avoid longitudinal flow, this study fixed U_z as 0 at the inlet. The convergence criterion for the inverse identification was $|O_{new} - O_{old}| < 0.01$.

Fig, 6.44 shows the changes in objective function and air supply temperature with the design cycle. The two inverse designs both led the objective function to very small values, although the convergence speed differed slightly. The air supply temperature, which was more sensitive to the objective function, varied in the same direction in both designs. Furthermore, the optimal air supply temperatures were almost identical. This result implies that the FFD-based adjoint method is applicable to the inverse design of a complex indoor environment. The computing speed of the FFD-based adjoint method in this case was seven times faster than the CFD-based adjoint method.

6.5 Enhancement of computing speed by semi-Lagrangian PISO algorithm

The previous section discussed our implementation of existing numerical algorithms from the literature for the fast simulation of indoor environments. The present section addresses our

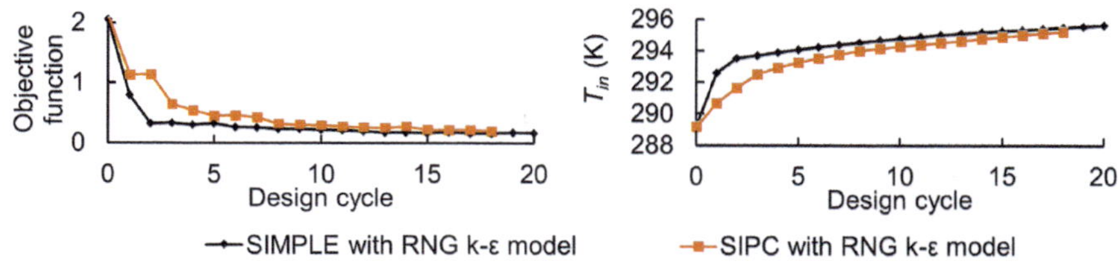

Figure 6.44 Changes in the objective function and air supply temperature with the design cycle

development of a new numerical algorithm that combines the semi-Lagrangian scheme with the PISO algorithm (SLPISO) to increase the computing speed without compromising accuracy.

6.5.1 Principle of semi-Lagrangian scheme in combination with PISO algorithm

A fully implicit algorithm is unconditionally stable and has no Courant–Friedrichs–Lewy restriction (Issa, 1986), and therefore such algorithms are widely used in CFD. However, in solving the momentum equations numerically, the advection term is fundamentally different from others because it contributes significant non-linearity. A semi-Lagrangian scheme (Courant et al., 1952) shows potential for resolving this dilemma. The idea of a semi-Lagrangian scheme originated from the advection of a scalar, but it can also be applied directly to a vector. One can refer to Section 6.4.1.1.4 for the details of this scheme. We propose an algorithm that integrates the semi-Lagrangian advection with the PISO algorithm, as follows.

Step 1 Solution of the advection term for velocity The SLPISO algorithm first solves the advection term by using the semi-Lagrangian scheme to obtain an intermediate velocity U*:

$$\frac{U^* - U^n}{\Delta t} = -(U^* \cdot \nabla)U^* \Rightarrow U^*(X) = U^n(X - \Delta t U^n(X)) \tag{6.38}$$

Step 2 Prediction of velocity The SLPISO algorithm then uses the intermediate velocity U* to solve the diffusion term. The pressure term and source term are added explicitly. A second intermediate velocity U** can be obtained by:

$$\frac{U^{**} - U^*}{\Delta t} = v\nabla^2 U^{**} - \frac{1}{\rho^n}\nabla p^n + \frac{1}{\rho^n}F \tag{6.39}$$

Since this step uses p^n instead of p^{**}, the U** does not satisfy the continuity equation.

Step 3 First pressure correction An approximation of the velocity field U*** together with the corresponding new pressure field p^{***} are sought that satisfy the continuity equation:

$$\nabla U^{***} = 0 \tag{6.40}$$

The momentum equation is then expressed as:

$$\frac{U^{***} - U^*}{\Delta t} = v\nabla^2 U^{**} - \frac{1}{\rho^n}\nabla p^{***} + \frac{1}{\rho^n}F \tag{6.41}$$

Subtracting Eq. (6.39) from Eq. (6.41) yields:

$$\frac{U^{***} - U^{**}}{\Delta t} = -\frac{1}{\rho^n}(\nabla p^{***} - \nabla p^n) \tag{6.42}$$

Taking the divergence of both sides of Eq. (6.42) and coupling with Eq. (6.40) yields:

$$\nabla^2 p^{***} - \nabla^2 p^n = \frac{\rho^n}{\Delta t}\nabla \cdot U^{**} \tag{6.43}$$

The SLPISO algorithm then solves Eq. (6.43) to provide intermediate pressure p^{***}.

Step 4 First velocity correction The updated pressure field p^{***} can be substituted into Eq. (6.41) or Eq. (6.42) to obtain the corrected velocity U^{***}.

Step 5 Second pressure correction The SLPISO algorithm replicates Step 2, using the U^{***} from Step 3 and U^{*} from Step 1 in order to obtain a newer pressure field p^{****}. An updated velocity field can be calculated by:

$$\frac{U^{****} - U^{*}}{\Delta t} = \nu\nabla^2 U^{***} - \frac{1}{\rho^n}\nabla p^{****} + \frac{1}{\rho^n}F \tag{6.44}$$

The U^{****} is expected to satisfy the continuity equation:

$$\nabla U^{****} = 0 \tag{6.45}$$

Subtracting Eq. (6.41) from Eq. (6.44) yields:

$$\frac{U^{****} - U^{***}}{\Delta t} = -\frac{1}{\rho^n}(\nabla p^{****} - \nabla p^{***}) + \nu\nabla^2 U^{***} - \nu\nabla^2 U^{**} \tag{6.46}$$

After taking the divergence of both sides of Eq. (6.46), together with the continuity equations for ∇U^{****} and ∇U^{***}, we obtain the pressure increment equation, which allows us to solve the $p^{****} - p^{***}$ field:

$$\nabla^2 p^{****} - \nabla^2 p^{***} = \rho^n \nabla \cdot (\nu\nabla^2 U^{***} - \nu\nabla^2 U^{**}) \tag{6.47}$$

Step 6 Second correction of velocity The updated pressure field p^{****} can be plugged into Eq. (6.44) or Eq. (6.46) to calculate the velocity U^{****}. Additional correction steps can be used, but, as will be shown later, the accuracy of two correction steps is sufficient for approximating the exact solutions U^{n+1} and p^{n+1}.

The temperature field is solved separately from the velocity field, but the procedure is very similar. Taking into account the coupling of temperature and velocity, the current study used the state equation for ideal gases to update the density of air, as shown in the steps below.

Step 7 Solution of the advection term for temperature The SLPISO algorithm solves the advection term for temperature by the semi-Lagrangian scheme to obtain an intermediate temperature field by:

$$\frac{T^* - T^n}{\Delta t} = -(U^n \cdot \nabla)T^* \Rightarrow T^*(X) = T^n(X - \Delta t U^n(X)) \tag{6.48}$$

Step 8 Correction of temperature Intermediate temperature field T^* is used in the solution of the diffusion term to yield the temperature field T^{n+1}:

$$\frac{T^{n+1} - T^*}{\Delta t} = \kappa \nabla^2 T^{n+1} + S_T \tag{6.49}$$

Step 9 Update of density Finally, the SLPISO algorithm updates the air density with the state equation for ideal gases:

$$\rho^{n+1} = \frac{p^{n+1} M_m}{R_g T^{n+1}} \tag{6.50}$$

Fig. 6.45 summarizes the computational sequence of the complete SLPISO algorithm.

The proposed SLPISO algorithm, without the correction steps (Steps 5 and 6), is similar to FFD with the NIPC-SL scheme, except that the SLPISO algorithm takes into account the pressure field from the previous time step. The FFD algorithm with the NIPC-SL scheme neglects the influence of pressure from the previous time step and assumes that pressure is determined solely by the velocity field under the continuity restriction. The advection term is completely separated from the rest of the momentum equation and is solved by using the semi-Lagrangian scheme, which is faster and more stable than the conventional method of directly solving the advection equation. However, the accuracy of PISO, theoretically and practically, has further advantages over the NIPC-SL scheme in FFD. The integrated algorithm (SLPISO) is expected to improve the accuracy of the NIPC-SL scheme without sacrificing much in terms of computing speed. The semi-Lagrangian scheme is anticipated to greatly reduce the computing cost for the direct solving of the advection term in the original PISO algorithm. The following two sections describe the analysis of the proposed SLPISO algorithm in accuracy and stability.

6.5.2 Accuracy analysis

The methodology presented above provides only an approximation of the exact solutions for velocity, temperature, and pressure, because of the linearization of the differential equations. It is therefore important to estimate the accuracy of the approximation. For incompressible fluids, with a focus on the velocity and pressure, the state function of the SLPISO algorithm is:

$$\frac{U^{n+1} - U^n}{\Delta t} = \nu \nabla^2 U^{n+1} - \frac{1}{\rho^n} \nabla p^{n+1} + \frac{1}{\rho^n} F \tag{6.51}$$

$$\nabla U^{n+1} = 0 \tag{6.52}$$

The corresponding pressure equation can be obtained by taking the divergence of Eq. (6.51) and substituting with Eq. (6.52):

$$\frac{1}{\rho^n} \nabla^2 p^{n+1} = \nabla(\nu \nabla^2 U^{n+1}) + \frac{1}{\Delta t} \nabla U^n + \frac{1}{\rho^n} \nabla F \tag{6.53}$$

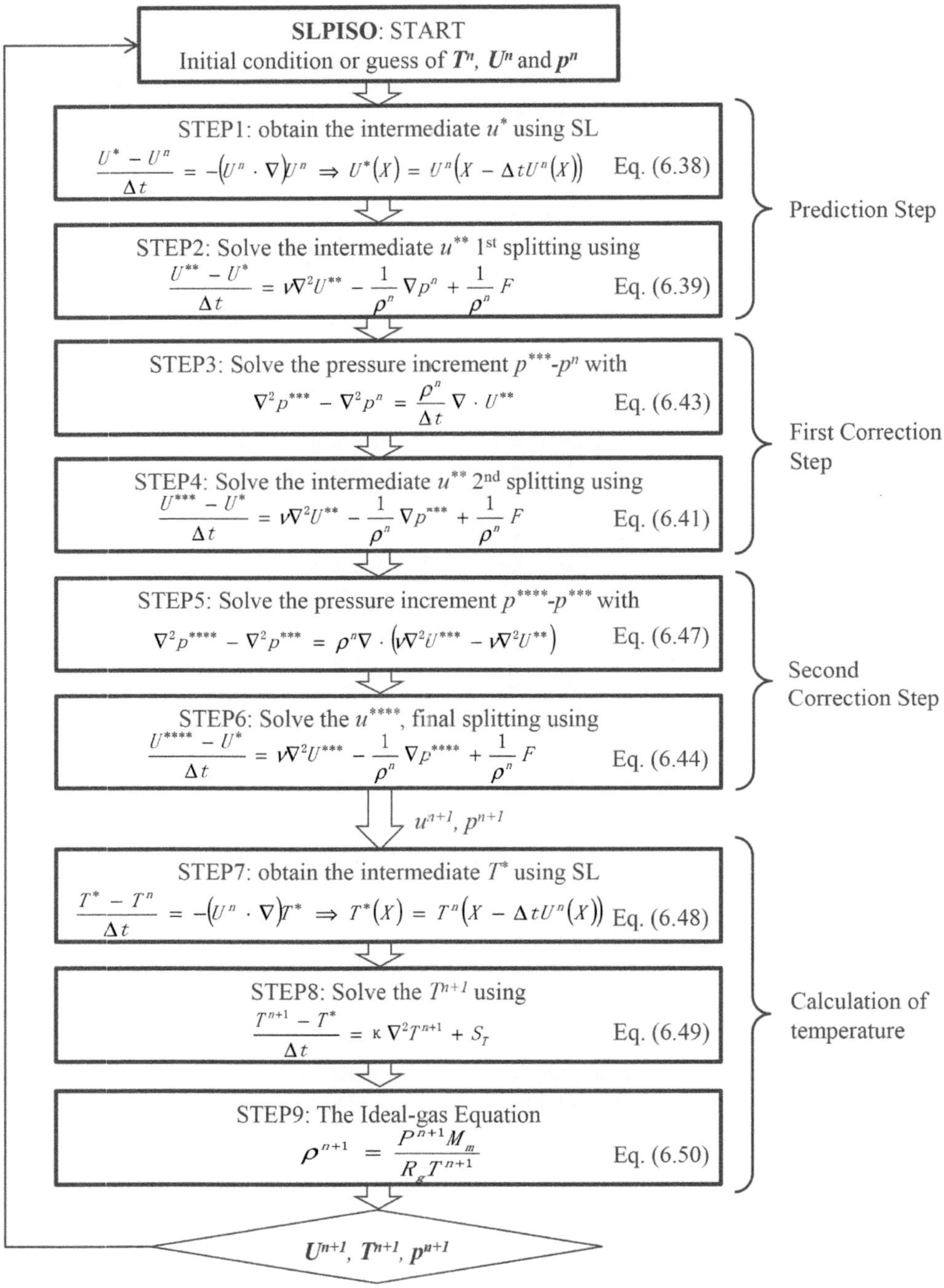

Figure 6.45 Operation sequence of the SLPISO algorithm

Let ε_U and ε_p be the errors in approximating the velocity U and pressure p. We then define:

$$\varepsilon_U^k = U^{n+1} - U^k \tag{6.54}$$

$$\varepsilon_p^l = p^{n+1} - p^l \tag{6.55}$$

where k stands for any one of the superscripts **, ***, and ****; l represents any one of the superscripts n, ***, and ****. We subtract Eq. (6.39) of the prediction stage from Eq. (6.51). Because U^* is obtained from U^n by the semi-Lagrangian scheme, we can assume that $U^n \cong U^*$, and then:

$$\frac{U^{n+1} - U^{**}}{\Delta t} = \nu\nabla^2 (U^{n+1} - U^{**}) - \frac{1}{\rho^n}\nabla(p^{n+1} - p^n) \tag{6.56}$$

Further,

$$\frac{1}{\Delta t}\varepsilon_U^{**} = \nu\nabla^2 \varepsilon_U^{**} - \frac{1}{\rho^n}\nabla\varepsilon_p^n \tag{6.57}$$

where

$$\varepsilon_U^{**} = U^{n+1} - U^{**} \tag{6.58}$$

$$\varepsilon_p^n = p^{n+1} - p^n \tag{6.59}$$

Expanding p by the Taylor series in t yields:

$$\varepsilon_p^n = p^n + O(\Delta t) - p^n = O(\Delta t) \tag{6.60}$$

where $O(\Delta t)$ indicates the order of truncation errors. Similarly, we have:

$$\varepsilon_U^{**} = O\left(\Delta t^2\right) \tag{6.61}$$

Taking the divergence of both sides of Eq. (6.41) that governs the first corrector step, then:

$$\frac{1}{\rho^n}\nabla^2 p^{***} = \nabla(\nu\nabla^2 U^{**}) + \frac{1}{\Delta t}\nabla U^* + \frac{1}{\rho^n}\nabla F \tag{6.62}$$

Subtracting Eq. (6.41) and Eq. (6.62) from Eq. (6.51) and Eq. (6.53) yields:

$$\frac{1}{\Delta t}\varepsilon_U^{***} = \nu\nabla^2 \varepsilon_U^{**} - \frac{1}{\rho^n}\nabla\varepsilon_p^{***} \tag{6.63}$$

$$\frac{1}{\rho^n}\nabla^2 \varepsilon_p^{***} = \nabla\left(\nu\nabla^2 \varepsilon_U^{**}\right) \tag{6.64}$$

Since ε_U^{**} is of the order Δt^2, Eq. (6.64) dictates that ε_p^{***} is also of $O\left(\Delta t^2\right)$, while Eq. (6.63) gives ε_U^{***} as $O\left(\Delta t^3\right)$. Similarly, for the second correction step:

$$\frac{1}{\nabla t}\varepsilon_U^{****} = \nu\nabla^2 \varepsilon_U^{***} - \frac{1}{\rho^n}\nabla\varepsilon_p^{****} \tag{6.65}$$

$$\frac{1}{\rho^n}\nabla^2\varepsilon_p^{****} = \nabla\left(\nu\nabla^2\varepsilon_U^{***}\right) \tag{6.66}$$

Since ε_U^{***} is of the order Δt^3, ε_p^{***} is also of $O\left(\Delta t^3\right)$. Eq. (6.65) provides ε_U^{****} as $O\left(\Delta t^4\right)$.

Thus, the approximations of the exact u^{n+1} and p^{n+1} fields produced by the two-step correction process are on the order of $O\left(\Delta t^4\right)$ and $O\left(\Delta t^3\right)$. The more correction steps, the higher the accuracy. However, Eqs. (6.51) and (6.53) contain discretization errors of $O\left(\Delta t^3\right)$ for even second-order-accurate schemes. Therefore, two correction steps are sufficient here for accuracy.

6.5.3 Stability analysis

As for the algorithm's stability, the numerical scheme used by Eqs. (6.51) and (6.53) is fully implicit, and the system would therefore be unconditionally stable. However, these equations are solved numerically, and the correction steps use explicit numerical schemes. The residual errors and explicit numerical scheme affect the stability of the overall scheme. The present study analyzed the stability of the correction steps in order to determine the best way to obtain reliable results with the SLPISO algorithm. Using Eq. (6.41) as an example, because U^* and U^{**} are known values and U^{***} is unknown, the study replaced U^* and U^{**} with U^n and replaced U^{***} with U^{n+1} as convenient notation. The initial-value stability of the time discretization scheme (neglecting the pressure and source terms) was investigated by discretizing Eq. (6.41) at node *i*:

$$\frac{U_i^{n+1} - U_i^n}{\Delta t} = \nu\frac{U_{i+1}^n - 2U_i^n + U_{i-1}^n}{\Delta x^2} \tag{6.67}$$

We define $\tilde{U}_i^n$ as the calculated result, and then round-off error $\varepsilon_{U,i}^n = U_i^n - \tilde{U}_i^n$. Checking the initial-value stability of the time discretization scheme with the von Neumann analysis method [22] yields:

$$\varepsilon_{U,i}^n = \Psi_i^n e^{Ii\theta} \tag{6.68}$$

where Ψ_i^n is the amplitude of ε_U at node *i* and time step *n*, $I = \sqrt{-1}$, and θ is the phase angle. To ensure that error $\varepsilon_{U,i}^n$ will not be magnified in the next time step, the following should be true:

$$\left|\frac{\Psi_i^{n+1}}{\Psi_i^n}\right| \le 1 \tag{6.69}$$

Then

$$\frac{\psi_i^{n+1} - \psi_i^n}{\Delta t}e^{Ii\theta} = \nu\frac{\psi_i^n}{\Delta x^2}[e^{I(i+1)\theta} - 2e^{Ii\theta} + e^{I(i-1)\theta}] \tag{6.70}$$

$$\psi_i^n\left\{\frac{e^{Ii\theta}}{\Delta t} + \nu\frac{1}{\Delta x^2}\left[e^{I(i+1)\theta} - 2e^{Ii\theta} + e^{I(i-1)\theta}\right]\right\} = \psi_i^{n+1}\frac{e^{Ii\theta}}{\Delta t} \tag{6.71}$$

We define

$$\mu = \frac{\psi_i^{n+1}}{\psi_i^n} = \frac{e^{Ii\theta} + \nu\frac{\Delta t}{\Delta x^2}\left[e^{I(i+1)\theta} - 2e^{Ii\theta} + e^{I(i-1)\theta}\right]}{e^{Ii\theta}} \tag{6.72}$$

$$\mu = \frac{\psi_i^{n+1}}{\psi_i^n} = 1 + v\frac{\Delta t}{\Delta x^2}\left[e^{I\theta} - 2 + e^{-I\theta}\right] \tag{6.73}$$

Then

$$\mu = \frac{\psi_i^{n+1}}{\psi_i^n} = 1 - 2v\frac{\Delta t}{\Delta x^2}(1 - \cos\theta) = 1 - 4v\frac{\Delta t}{\Delta x^2}\sin^2\frac{\theta}{2} \tag{6.74}$$

Hence, for the SLPISO algorithm, we have the following limitations:

$$\left|1 - 4v\frac{\Delta t}{\Delta x^2}\sin^2\frac{\theta}{2}\right| \le 1 \tag{6.75}$$

$$\Delta t \le \frac{1}{v\dfrac{2}{\Delta x^2}} = \frac{\Delta x^2}{2v} \tag{6.76}$$

Eq. (6.76) limits the time step for the SLPISO algorithm to guarantee the stability of the initial value. Applying this analysis to the PISO algorithm yields:

$$\Delta t \le \frac{1}{v\dfrac{2}{\Delta x^2} + \dfrac{U}{\Delta x}} \tag{6.77}$$

The limitation of Δt means that the time steps of the PISO and SLPISO algorithms cannot be too large. The time step constraints will be similar when the velocity is small in comparison with the grid size. However, for simulation of a building environment with a mechanical ventilation system, the inlet flow velocity could be 2 m/s or higher, the mesh size near the inlet would be 2 cm or smaller, so the value of $u/\Delta x$ would be greater than 100. Thus, the time step size of the SLPISO solver can be much larger than that of the PISO solver, which makes SLPISO much faster than PISO when they are used for transient simulation. Considering the influence of the pressure and source terms on stability, the actual time step limit should be smaller than the value determined by Eqs. (6.76) and (6.77).

6.5.4 Validation of the SLPISO algorithm

Implementing the developed SLPISO algorithm in OpenFOAM, this study evaluated its performance in predicting both steady and unsteady airflow. Predictions made by the SIMPLE algorithm, the PISO algorithm, and FFD with the NIPC-SL scheme were used as a reference.

6.5.4.1 Evaluation of steady-state simulation performance

The study first evaluated the SLPISO algorithm with a laminar flow case (Ghia et al., 1982). Fig, 6.46 shows a lid-driven cavity flow under isothermal condition. We compared the performance of four algorithms: SIMPLE, PISO, FFD (NIPC-SL), and SLPISO. The mesh size was 50×50, and the time step size was 0.005 s. Fig. 6.47 shows the predicted U_y at line y = 0.5 m. The PISO and SLPISO results are almost identical. The SIMPLE algorithm predicted similar air velocity, while the FFD results were considerably different. The results for the four algorithms show that the SLPISO predictions were as accurate as those made by the PISO algorithm. These two predictions deviated slightly from the experimental data. FFD with the NIPC-SL scheme had the worst performance. Fig. 6.48 compares the computational costs. The SIMPLE algorithm required the most computational time. SLPISO and FFD had similar speeds but were both slower than the PISO algorithm in this case. This difference in speed will be addressed later.

This study also evaluated the SLPISO algorithm with a two-dimensional mixed convection case (Blay et al., 1992) as shown in Fig. 6.1(a). Since the application of turbulence models would increase the calculation effort, we used the effective kinematic viscosity and heat transfer coefficient, which are equal to about one hundred times the physical values, to

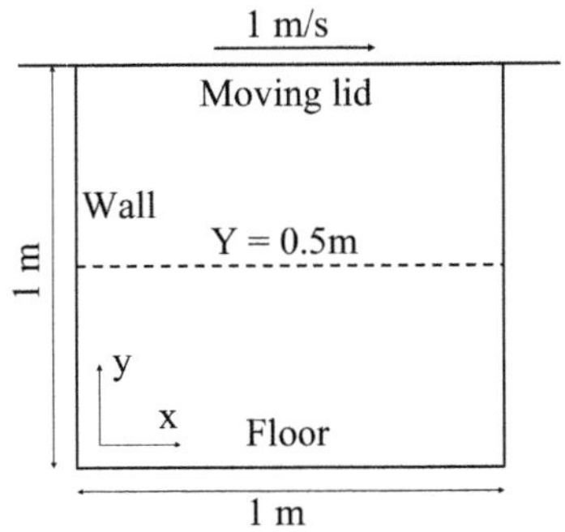

Figure 6.46 Geometry of the lid-driven cavity case

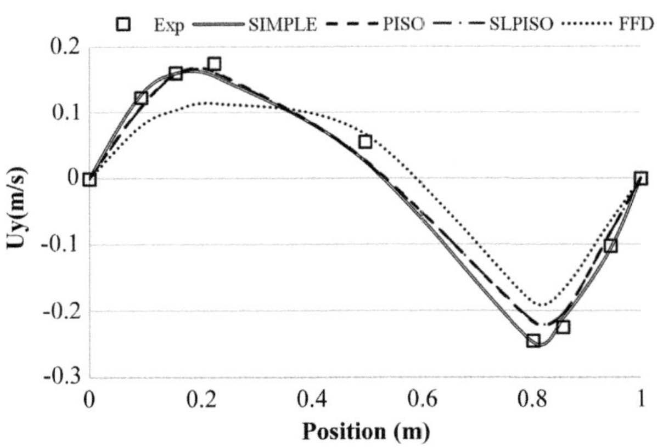

Figure 6.47 Predicted Uy at line y = 0.05 m in the lid-driven cavity case

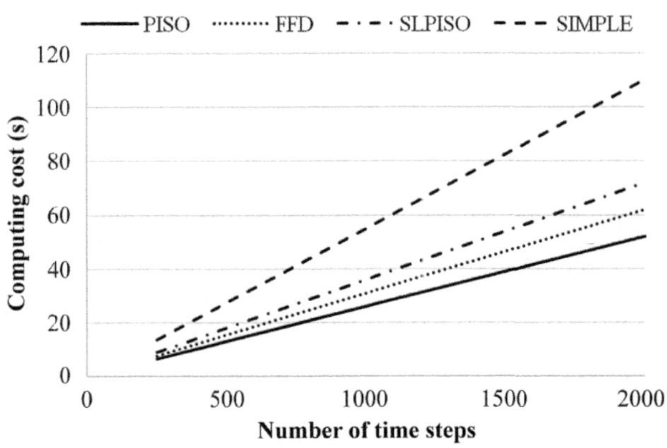

Figure 6.48 Comparison of computational costs

compare the actual calculation speeds of the solvers. The mesh size was 80 × 80, and the time step size was 0.005 s. As shown in Fig. 6.49, the SLPISO algorithm provided similar results to those of the PISO method. The results for FFD with the NIPC-SL scheme exhibit a large disparity in the temperature prediction. Fig. 6.50 shows that SLPISO and FFD had similar computational speed, and both were again slower than the PISO algorithm.

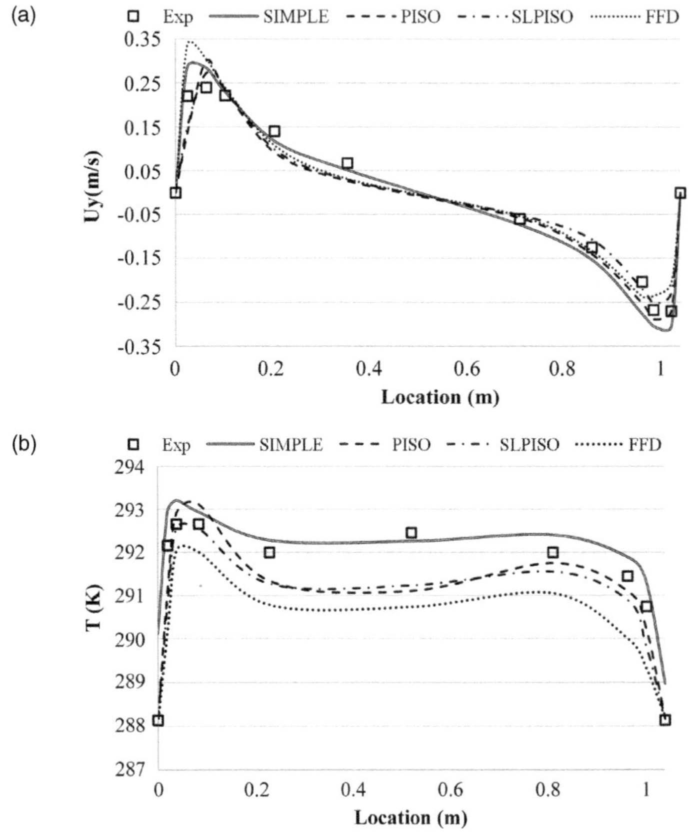

Figure 6.49 Results for the mixed convection case: (a) Comparison of air velocity predictions; (b) Comparison of air temperature predictions

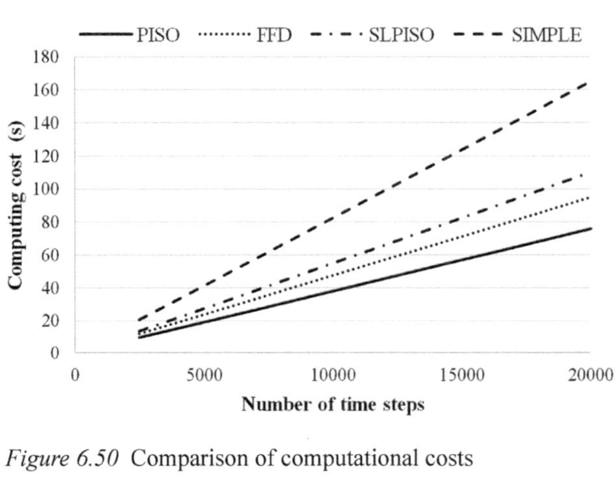

Figure 6.50 Comparison of computational costs

When we increased the grid number from 80 × 80 to 300 × 300, and then to 1000 × 1000, the computational cost for these algorithms changed as shown in Figs. 6.51(a) and (b). As the grid number increased, SLPISO and FFD became faster than PISO. The reason for this was the inherent characteristics of the semi-Lagrangian scheme. As the grid number increased, the computing cost for the traditional solvers, SIMPLE and PISO, grew exponentially, while the growth in cost

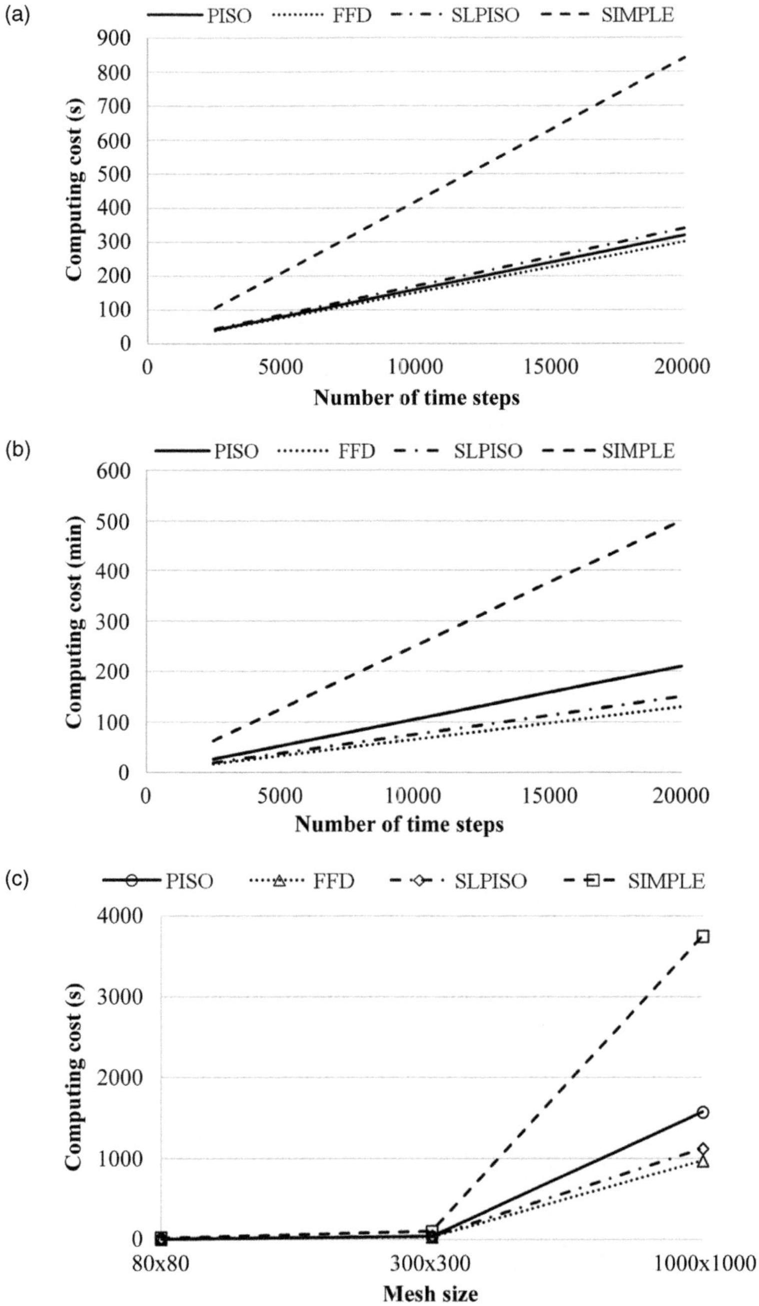

Figure 6.51 Comparison of computational costs for different solvers with different grids: (a) Grid resolution of 300 × 300; (b) Grid resolution of 1000 × 1000; (c) Number of time steps = 2500

for the semi-Lagrangian scheme was linear, as displayed in Fig. 6.51(c). (Under the influence of the correction steps, the growth in calculation cost for FFD and SLPISO was not exactly linear.)

In the comparison of simulation speed above, the same time step was used for all the solvers. However, the stability analysis indicated that the SLPISO solver can tolerate a larger time step than PISO. This study used the Blay case with mesh size of 1000 × 1000 to determine the actual calculation speeds of different solvers with different time steps. Fig. 6.52 shows the computing times for the largest time steps that the different solvers can handle. For the SIMPLE, PISO, SLPISO, and FFD algorithms, the largest time steps at which stable and acceptable results could be obtained were 0.02 s, 0.005 s, 0.08 s, and 0.1 s, respectively. The gray columns in Fig. 6.52 represent the relative computing cost with a time step size of 0.005 s for all the solvers, using the SIMPLE algorithm as the benchmark. The black columns represent the relative computing cost using each solver's own largest time step. While the predicted results for velocity and temperature were similar to those shown in Figs. 6.49(a) and (b), the modeling speeds of SLPISO and FFD were significantly increased with larger time steps.

6.5.4.2 Transient simulation accuracy

To evaluate the transient simulation accuracy of the SLPISO and FFD algorithms, this study simulated a transient flow in the 2D mixing convection case shown in Fig. 6.1(a). Because no experimental transient-flow results existed for this case, the results of the SIMPLE algorithm were used as a reference. These "experimental" data were taken from the SIMPLE prediction at the center point of the test chamber in five-second increments. We used the effective kinematic viscosity and heat transfer coefficient, which are equal to one hundred times the physical values, to account for the impact of turbulence. The mesh size was 80 × 80. The time step varied from 0.005 s to 0.08 s. As the step size increased, the transient results and steady-state results of the SLPISO and FFD algorithms deviated, but SLPISO performed better overall than FFD (Fig. 6.53). FFD must use smaller time steps to obtain similar results to those of the SLPISO algorithm. The increasing deviation of the SLPISO results can be attributed to the false diffusion of the time term. The equation of the prediction step can be simplified as

$$\frac{U^{n+1}-U^{n}}{\Delta t}=\nu\nabla^{2}U-\frac{1}{\rho^{n}}\nabla p+\frac{1}{\rho^{n}}F \tag{6.78}$$

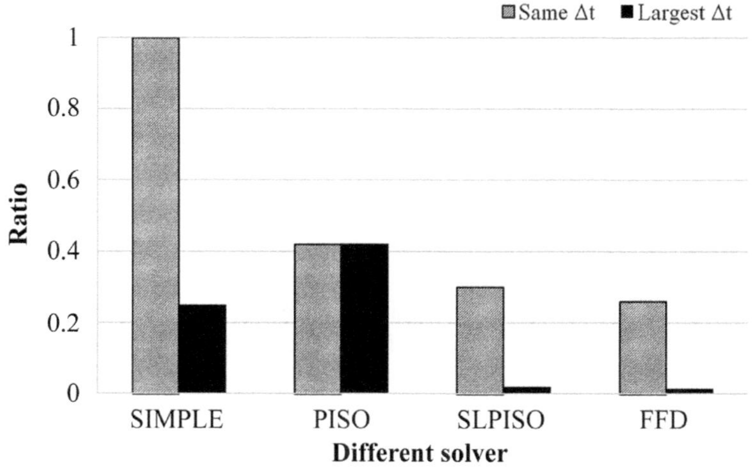

Figure 6.52 Comparison of computational cost with different time steps

Expanding U^n by the Taylor series in time step $n+1$ yields

$$U^n = U^{n+1} - \left.\frac{\partial U}{\partial t}\right|_{n+1} \Delta t + \left.\frac{\partial^2 U}{\partial t^2}\right|_{n+1} \frac{\Delta t^2}{2} + \cdots \tag{6.79}$$

Eq. (6.79) can then be transformed into

$$\frac{U^{n+1} - U^n}{\Delta t} = \left.\frac{\partial U}{\partial t}\right|_{n+1} - \left.\frac{\partial^2 U}{\partial t^2}\right|_{n+1} \frac{\Delta t}{2} + O(\Delta t^2) \tag{6.80}$$

Subtracting Eq. (6.80) from Eq. (6.78) yields

$$\left.\frac{\partial U}{\partial t}\right|_{n+1} - \left.\frac{\partial^2 U}{\partial t^2}\right|_{n+1} \frac{\Delta t}{2} + O(\Delta t^2) = \nu\nabla^2 U - \frac{1}{\rho^n}\nabla p + \frac{1}{\rho^n} F \tag{6.81}$$

$$\left.\frac{\partial U}{\partial t}\right|_{n+1} = \nu\nabla^2 U - \frac{1}{\rho^n}\nabla p + \frac{1}{\rho^n} F + \left.\frac{\partial^2 U}{\partial t^2}\right|_{n+1} \frac{\Delta t}{2} - O(\Delta t^2) \tag{6.82}$$

Since

$$\frac{\partial^2 U}{\partial t^2} = \frac{\partial}{\partial t}\left(\frac{\partial U}{\partial t}\right) \cong \frac{\partial}{\partial t}\left(-U\frac{\partial U}{\partial x}\right) \cong -U\frac{\partial}{\partial x}\left(-U\frac{\partial U}{\partial x}\right) = U^2\frac{\partial^2 U}{\partial x^2} = U^2\nabla^2 U \tag{6.83}$$

Eq. 6.82 becomes approximately

$$\left.\frac{\partial U}{\partial t}\right|_{n+1} = \left(\nu + \frac{U^2\Delta t}{2}\right)\nabla^2 U - \frac{1}{\rho^n}\nabla p + \frac{1}{\rho^n} F - O(\Delta t^2) \tag{6.84}$$

Comparing Eq. (6.86) with the original equation, we see that the discretization of the time term leads to an additional false diffusion $\frac{U^2\Delta t}{2}\nabla^2 U$ that is related to the time step size. An increase in the time step size enhances the false diffusion, which explains why the transient simulation result in this case was less responsive than the reference curve to the transient velocity. If the effective kinematic viscosity is lowered, compensating for the use of a larger time step, then the SLPISO results with a time step of 0.08 s can be similar to the results for the 0.005 s time step. This has been verified in numerical trials.

6.6 Discussion

To solve the adjoint equations by FFD, this study did not conduct systematic validations as we had done in solving the Navier-Stokes equations. There were two reasons for this. First, because the adjoint equations are partial differential equations produced mathematically, there were no analytical solutions or experimental data for use in validation. Second, it was difficult to obtain converged numerical solutions for the adjoint equations. Applying the SIMPLE algorithm required very small relaxation factors. Since we did not know the exact solutions for the adjoint equations, we only compared the solutions obtained by the SIMPLE algorithm and FFD models in a simple inverse identification case, which should be sufficient in the current stage.

This study found that FFD simulations were 20 times faster than CFD simulations in predicting the transient indoor airflow. Therefore, the FFD models could be extended to the fast

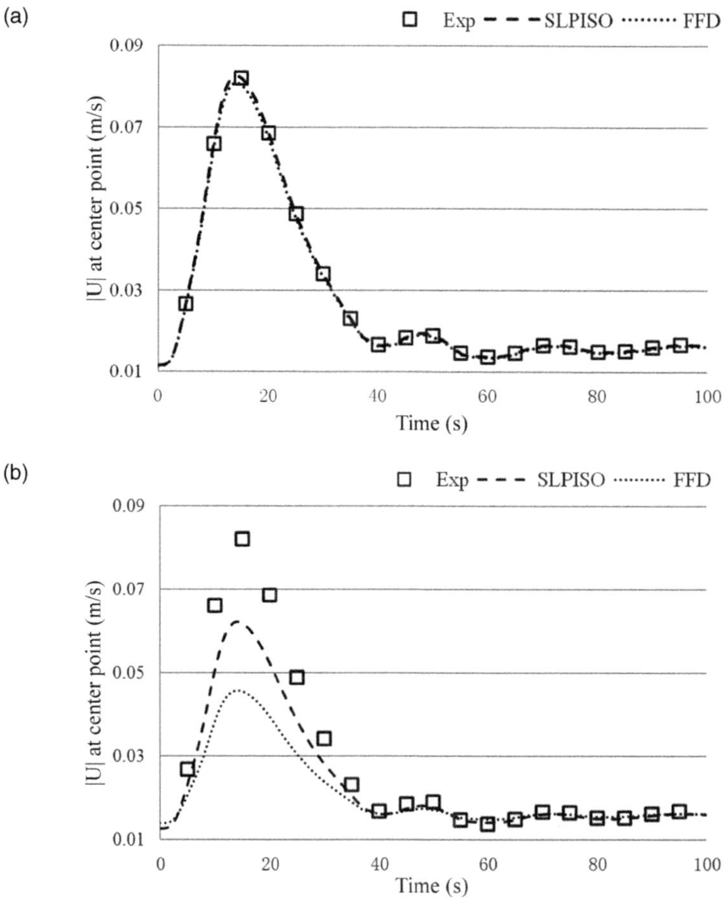

Figure 6.53 Comparison of accuracy with different time steps for transient simulation with the SLPISO and FFD algorithms: (a) Time step of 0.005 s; (b) Time step of 0.05 s.

simulation of gaseous and particulate contaminant transport in indoor or outdoor environments (Chen and Zhao, 2011). For predicting steady-state indoor airflow, the acceleration, which depends on the time step size in FFD and the number of iterations in the SIMPLE algorithm, would not be so significant. However, in the steady-state airflow prediction by the SIMPLE algorithm, it was difficult to obtain a converged solution in the two complex indoor environments, an office with displacement ventilation and an aircraft cabin, unless small relaxation factors were used. It was then necessary to increase the number of iterations because of the reduced convergence speed. For the FFD simulations, there was no need to add relaxation in solving the Navier-Stokes equations, and the solutions were found to be convergent. Therefore, the FFD models are also recommended for solving the steady-state indoor airflow.

Because of the inherent characteristics of the semi-Lagrangian scheme, the SLPISO algorithm may not provide a significant reduction in computing cost over the conventional SIMPLE algorithm used in CFD when the grid number is relatively small. As the grid number increases, the SLPISO algorithm can provide much faster predictions. In most engineering problems, grid-independent solutions require grid numbers greater than one million, and the SLPISO algorithm shows great potential for fast simulation. This potential is further enhanced by the algorithm's ability to provide good results with a larger time step size.

6.7 Conclusions

This study compared the performance of the CFD-based adjoint method, genetic algorithm method, and trial-and-error method for indoor environmental design. The adjoint method converged the fastest. While the GA method had the highest accuracy, it required 20 times as much computing time as the adjoint method. The trial-and-error process required a similar computing effort to that of the GA method but was the least accurate. External integration of the CFD-based GA method and adjoint method can increase the convergence speed, but the timing for introduction of the adjoint method needs further investigation. Internal integration of the two methods is prominent in multi-variable problems.

The integration of GA and ANN for the inverse design was presented in this study. Compared to the GA method, the computational cost of the integrated method decreased when the ANN instead of CFD was used to obtain the values of design objectives, but the design accuracy decreased due to the existence of the prediction error of ANN. When the self-updating method was adopted by ANN, the design accuracy can be maintained but requires more CFD simulations. Compared to the ANN method, the efficiency of the design process was significantly improved when GA was used to search the solutions. The convergence and computational cost of the integrated method was affected by the prediction accuracy of ANN and searching process of GA. The fuzzy control of the computing probability and evolution space can be used to improve the performance of this method

This chapter implemented four FFD models, NIPC-SL, NIPC, SIPC, and RIPC, in OpenFOAM and validated them for predicting steady-state and transient flow in indoor environments. We then used the FFD models to solve the adjoint equations, and validated the FFD-based adjoint method for inverse identification problems and inverse designs in indoor environments. The NIPC-SL scheme with a turbulence model made the flow too diffusive and had the worst performance. The NIPC, SIPC, and RIPC schemes behaved in a similar manner and performed well in predicting both the steady-state and transient flow in an indoor environment. When the same turbulence model was used in these schemes, the prediction accuracy was almost the same as that with the SIMPLE algorithm. The FFD models were more than 20 times faster than the SIMPLE algorithm in predicting the transient flow. A larger time step size affected the accuracy of transient FFD predictions at the beginning of the simulation. For the steady-state flow predictions, the effect of time step size was minimal. The FFD models performed well in solving the adjoint equations. The FFD-based adjoint method was able to identify the optimal air supply conditions in the indoor environment. The FFD-based adjoint method was 4-16 times faster than the CFD-based adjoint method in each design cycle.

The SLPISO algorithm that we have developed is slightly slower than the NIPC-SL scheme but more accurate, especially for transient cases. Larger time steps can be used in the SLPISO algorithm than in the PISO algorithm or NIPC-SL scheme without compromising the accuracy of steady-state results.

Notes

1 This chapter is partially reproduced from the paper entitled A simulation-based inverse design of preset aircraft cabin environment, *Building and Environment*, 82: 20–26. The reprint is with permission from Elsevier.

2 This chapter is partially reproduced from the paper entitled: New semi-Lagrangian-based PISO method for fast and accurate indoor environment modeling, *Building and Environment*, 105, 236–244. The reprint is with permission from Elsevier.

References

ASHRAE. Standard 161-2007, Air quality within commercial aircraft. Atlanta, GA: ASHRAE Inc. 2007.

Blay D, Mergui S, Niculae C. Confined turbulent mixed convection in the presence of a horizontal buoyant wall jet. *Fundamentals of Mixed Convection, ASME HTD*. 1992;213:65–72.

Boithias F, El Mankibi M, Michel P. Genetic algorithms based optimization of artificial neural network architecture for buildings' indoor discomfort and energy consumption prediction. *Building Simulation*. 2012;5(2):95–106. Tsinghua Press.

Caldas LG, Norford LK. Genetic algorithms for optimization of building envelopes and the design and control of HVAC systems. *Journal of Solar Energy Engineering*. 2003;125(3):343–51.

Chen C, Zhao B. Review of relationship between indoor and outdoor particles: I/O ratio, infiltration factor and penetration factor. *Atmospheric Environment*. 2011;45(2):275–288.

Chen Q, Xu W. A zero-equation turbulence model for indoor airflow simulation. *Energy and Buildings*. 1998;28(2), pp.137–144.

Chorin AJ. A numerical method for solving incompressible viscous flow problems. *Journal of Computational Physics*. 1967;2(1):12–26.

Chorin AJ. Numerical solution of the Navier-Stokes equations. *Mathematics of Computation*. 1968;22(104):745–762.

Chow TT, Zhang GQ, Lin Z, Song CL. Global optimization of absorption chiller system by genetic algorithm and neural network. *Energy and Buildings*. 2002;34(1):103–9.

Courant R, Isaacson E, Rees M. On the solution of nonlinear hyperbolic differential equations by finite differences. *Communications on Pure and Applied Mathematics*. 1952;5(3):243–55.

Cui W, Zhu Y. Systematic study on passengers' thermal comfort under low-air-pressure environment in commercial aircraft cabin. *Annual Meeting of the Center for Cabin Air Reformative Environment, Chongqing, China*. 2015.

Dounis AI, Caraiscos C. Advanced control systems engineering for energy and comfort management in a building environment—A review. *Renewable and Sustainable Energy Reviews*. 2009;13(6):1246–1261.

Ferreira PM, Ruano AE, Silva S, Conceicao EZ. Neural networks based predictive control for thermal comfort and energy savings in public buildings. *Energy and Buildings*. 2012;55:238–51.

Ferziger JH, Perić M. *Computational methods for fluid dynamics*. Berlin and New York: Springer Science & Business Media. 2002.

Ghia UK, Ghia KN, Shin CT. High-Re solutions for incompressible flow using the Navier-Stokes equations and a multigrid method. *Journal of Computational Physics*. 1982;48(3):387–411.

Giraldo FX. The Lagrange–Galerkin spectral element method on unstructured quadrilateral grids. *Journal of Computational Physics*. 1998;147(1),:114–146.

Goda K. A multistep technique with implicit difference schemes for calculating two- or three-dimensional cavity flows. *Journal of Computational Physics*. 1979;30(1):76–95.

Gossard D, Lartigue B, Thellier F. Multi-objective optimization of a building envelope for thermal performance using genetic algorithms and artificial neural network. *Energy and Buildings*. 2013;67:253–60.

Guermond JL, Minev P, Shen J. Error analysis of pressure-correction schemes for the time-dependent Stokes equations with open boundary conditions. *SIAM Journal on Numerical Analysis*. 2005;43(1):239–58.

Issa RI. Solution of the implicitly discretised fluid flow equations by operator-splitting. *Journal of Computational Physics*. 1986;62(1):40–65.

Jameson A. Optimum aerodynamic design using CFD and control theory. *AIAA paper*. 1995;1729:124–131.

Jin M, Zuo W, Chen Q. Improvements of fast fluid dynamics for simulating air flow in buildings. *Numerical Heat Transfer, Part B: Fundamentals*. 2012;62(6):419–438.

Lin CH, Dunn KH, Horstman RH, Topmiller JL, Ahlers MF, Bennett JS, Sedgwick LM, Wirogo S. Numerical Simulation of Airflow and Airborne Pathogen Transport in Aircraft Cabins—Part I: Numerical Simulation of the Flow Field. *ASHRAE Transactions*. 2005;111(1).

Liu W, Chen Q. Optimal air distribution design in enclosed spaces using an adjoint method. *Inverse Problems in Science and Engineering*. 2015;23(5):760–79.

Liu W, Jin M, Chen C, You R, Chen Q. Implementation of a fast fluid dynamics model in OpenFOAM for simulating indoor airflow. *Numerical Heat Transfer, Part A: Applications*. 2016;69(7):748–62.

Magnier L, Haghighat F. Multiobjective optimization of building design using TRNSYS simulations, genetic algorithm, and artificial neural network. *Building and Environment*. 2010;45(3):739–46.

Nassif N. Modeling and optimization of HVAC systems using artificial neural network and genetic algorithm. *Building Simulation*. 2014;7(3):237–245. Berlin Heidelberg: Springer.

Nguyen AT, Reiter S, Rigo P. A review on simulation-based optimization methods applied to building performance analysis. *Applied Energy*. 2014;113:1043–1058.

Nielsen PV, Restivo A, Whitelaw JH. The velocity characteristics of ventilated rooms. *Journal of Fluids Engineering*. 1978;100(3):291–298.

OpenFOAM. The Open Source CFD Toolbox, http://www.opencfd.co.uk/openfoam.html. 2007.

Othmer C, de Villiers E, Weller HG. Implementation of a continuous adjoint for topology optimization of ducted flows. *In 18th AIAA Computational Fluid Dynamics Conference*. June 2007.

Patankar SV, Spalding DB. A calculation procedure for heat, mass and momentum transfer in three-dimensional parabolic flows. *International Journal of Heat and Mass Transfer*. 1972;15(10): 1787–1806.

Song Z, Murray BT, Sammakia B, Lu S. Multi-objective optimization of temperature distributions using artificial neural networks. *Thermal and Thermomechanical Phenomena in Electronic Systems (ITherm), 2012 13th IEEE Intersociety Conference on 2012 May 30* (pp. 1209–1218). IEEE.

Stam J. Stable fluids. *In Proceedings of the 26th annual conference on computer graphics and interactive techniques*. 1999:121–128. ACM Press/Addison-Wesley Publishing Co.

Staniforth A, Côté J. Semi-Lagrangian integration schemes for atmospheric models-a review. *Monthly Weather Review*. 1991;119(9):2206–2223.

Temam R. Sur l'approximation de la solution des équations de Navier-Stokes par la méthode des pas fractionnaires (II). *Archive for Rational Mechanics and Analysis*. 1969;33(5):377–385.

Timmermans LJP, Minev PD, Van De Vosse FN. An approximate projection scheme for incompressible flow using spectral elements. *International Journal for Numerical Methods in Fluids*. 1996;22(7):673–688.

Wang M, Chen Q. Assessment of various turbulence models for transitional flows in an enclosed environment (RP-1271). *HVAC&R Research*. 2009;15(6):1099–1119.

Xiu D, Karniadakis GE. A semi-Lagrangian high-order method for Navier–Stokes equations. *Journal of computational physics*. 2001;172(2):658–684.

Xue Y, Liu W, Zhai ZJ. New semi-Lagrangian-based PISO method for fast and accurate indoor environment modeling. *Building and Environment*. 2016;105:236–244.

Yakhot VS, Orszag SA, Thangam S, Gatski TB, Speziale CG. Development of turbulence models for shear flows by a double expansion technique. *Physics of Fluids A: Fluid Dynamics (1989–1993)*. 1992;4(7):1510–1520.

Yu W, Li B, Jia H, Zhang M, Wang D. Application of multi-objective genetic algorithm to optimize energy efficiency and thermal comfort in building design. *Energy and Buildings*. 2015;88:135–143.

Yuan X, Chen Q, Glicksman LR, Hu Y, Yang X. Measurements and computations of room airflow with displacement ventilation. *ASHRAE Transactions*. 1999;105:340.

Zhang T, Chen Q. Novel air distribution systems for commercial aircraft cabins. *Building and Environment*. 2007;42(4):1675–1684.

Zhang T, You X. Inverse design of aircraft cabin environment by coupling artificial neural network and genetic algorithm. *HVAC&R Research*. 2014a;20(7):836–843.

Zhang T, You X. A simulation-based inverse design of preset aircraft cabin environment. *Building and Environment*. 2014b;82:20–26.

Zhang T, You X. The use of genetic algorithm and self-updating artificial neural network for the inverse design of cabin environment. *Indoor and Built Environment*. 2015:1420326X15609772.

Zhang Z, Cheng H, Zhang S. An optimization model for the identification of temperature in intelligent building. *Interdisciplinary Advances in Information Technology Research*. 2013:116.

Zhang Z, Zhang W, Zhai Z, Chen Q. Evaluation of various turbulence models in predicting airflow and turbulence in enclosed environments by CFD: Part 2—Comparison with experimental data from literature. *HVAC&R Research*. 2007;13(6):871–886.

Zhou L, Haghighat F. Optimization of ventilation system design and operation in office environment, Part I: Methodology. *Building and Environment*. 2009a;44(4):651–656.

Zhou L, Haghighat F. Optimization of ventilation systems in office environment, Part II: Results and discussions. *Building and Environment*. 2009b;44(4):657–665.

Zitzler E, Thiele L, Bader J. On set-based multiobjective optimization. *IEEE Transactions on Evolutionary Computation*. 2010;14(1):58–79.

Zuo W, Chen Q. Real-time or faster-than-real-time simulation of airflow in buildings. *Indoor Air*. 2009;19(1):33–44.

Zuo W, Hu J, Chen Q. Improvements in FFD modeling by using different numerical schemes. *Numerical Heat Transfer, Part B: Fundamentals*. 2010;58(1):1–6.

Index